Supersymmetric Methods in Quantum, Statistical and Solid State Physics

Enlarged and revised edition

Supersymmetric Methods in Quantum, Statistical and Solid State Physics

Enlarged and revised edition

Georg Junker

*European Organisation for Astronomical Research in the Southern Hemisphere,
Garching, Germany*
Erlangen-Nürnberg University, Erlangen, Germany

IOP Publishing, Bristol, UK

1st edition: © Springer, 1996

© IOP Publishing Ltd 2019

ISBN 978-0-7503-2026-9 (ebook)
ISBN 978-0-7503-2024-5 (print)
ISBN 978-0-7503-2025-2 (mobi)

DOI 10.1088/2053-2563/aae6d5

Version: 20190101

IOP Expanding Physics
ISSN 2053-2563 (online)
ISSN 2054-7315 (print)

British Library Cataloguing-in-Publication Data: A catalogue record for this book is available from the British Library.

Published by IOP Publishing, wholly owned by The Institute of Physics, London

IOP Publishing, Temple Circus, Temple Way, Bristol, BS1 6HG, UK

US Office: IOP Publishing, Inc., 190 North Independence Mall West, Suite 601, Philadelphia, PA 19106, USA

To Akira Inomata

To Akira Inomata

Contents

Preface

This is a revised and enlarged edition of the monograph previously published by Springer in 1996 under the slightly shorter title *Supersymmetric Methods in Quantum and Statistical Physics*. Despite the tremendous efforts undertaken to observe the supersymmetric partners of the currently known elementary particles, such super-symmetric particles have not yet been observed. Many physicists currently even believe that the end of supersymmetry has arrived—to be more precise, what is meant is the end of space–time supersymmetry.

However, supersymmetry (SUSY) has gone far beyond the idea of an extended space–time symmetry. During the last four decades the algebraic structure of SUSY has been utilised in various models of physical systems. Here SUSY is *not* an extended space–time symmetry. It is a relation between different states of a given system for the same energy. Hence, SUSY in this respect is an algebraic tool rather than a symmetry relating bosons and fermions. Since the previous version of this book was published, SUSY as a mathematical method has been developed further and its range of applications has been extended significantly. This was the major motivation to work on a revised and enlarged edition. Some chapters have also been rearranged or have been split into two to make the material more comprehensible. In addition, each chapter, except the introduction and conclusion, ends with a problems section which may be found useful, in particular to newcomers to this subject.

The first chapter starts with a brief introduction and a historical review of symmetries in general and space–time supersymmetry in particular.

This is then followed by chapter 2, which contains the basic definitions and properties of supersymmetric quantum mechanics. Here the focus is on so-called $N = 2$ SUSY quantum systems, which exhibit all the typical properties of such systems. This chapter is mandatory to understand the supersymmetric methods applied to various systems discussed in the following chapters.

The third chapter is of particular interest to students with some basic knowledge in quantum mechanics. Here we extensively discuss the properties of the Witten model which is the prototype of a supersymmetric one-dimensional quantum system. All the typical properties of an $N = 2$ SUSY system are explicitly discussed.

Chapter 4 is dedicated to applications in one-dimensional quantum systems. Here we discuss the supersymmetrisation of arbitrary one-dimensional systems, the so-called shape-invariance of potentials leading to exact solutions, and the relation to the Darboux method. The latter is used to construct a family of SUSY partners of the harmonic oscillator and various quantum mechanical properties, such as generalised raising and lowering operators and coherent states of this family, are discussed in detail. An application to classical field models in $(1 + 1)$ dimensions is also considered.

Supersymmetric classical mechanics is discussed in chapter 5. Supersymmetric classical mechanics is a special case of so-called pseudo-classical models which, in addition to the Cartesian degrees of freedom, also has Grassmann-valued degrees of freedom. It is shown that for the simplest one-dimensional model an explicit solution of the classical equations of motion can be obtained. In particular, a classical quantity called the fermionic phase is introduced, which also becomes relevant in the quasi-classical approximation. The canonical as well as the path-integral quantisation of this model is performed, resulting in the Witten model of chapter 3.

Chapter 6 discusses the quasi-classical approximation of Witten's model using a path-integral evaluation of the quasi-classical paths discussed in the previous chapter. It turns out that the presence of the fermionic phase results in quasi-classical quantisation conditions which respect the SUSY structure as opposed to the usual WKB approximation. This quasi-classical approximation is found to yield the exact energy eigenvalues for all shape-invariant potentials.

Applications to statistical physics are studied in chapter 7. The supersymmetric structure of the Fokker–Planck and that of the Langevin equation are shown to be closely connected to the quantum mechanical and the pseudo-classical Witten model, respectively. Implications of supersymmetry are discussed and supersymmetric approximation methods for decay rates are also presented. Using the Darboux method of chapter 4 we construct new Fokker–Planck potentials whose decay rates and decay modes are known in closed form. As explicit examples a family of conditionally exactly solvable drift potentials is constructed, being the SUSY partner of the harmonic oscillator. Continuous Markov processes are also discussed and the shape-invariance property from chapter 4 is utilised to obtain relations between the transition probabilities for the corresponding stochastic processes.

Chapter 8 is exclusively dedicated to supersymmetric Pauli systems. Indeed, the Pauli Hamiltonian characterising an electron in an external magnetic field is the prototype of a three-dimensional supersymmetric quantum system. Here we discuss the paramagnetism of a two- and of a three-dimensional non-interacting electron gas. We also consider the Pauli Hamiltonian for an electron with spin $\frac{1}{2}$ in an external spherical symmetric scalar potential and find that, within the subspace of a fixed total angular momentum, this Hamiltonian exhibits a SUSY structure relating states with an orbital angular momentum differing by one unit of $\hbar$. The case of spherically symmetric tensor potentials is also discussed.

The last chapter is dedicated to Dirac systems and their SUSY structure. Here the definition of a supersymmetric Hamiltonian differs from that of non-relativistic quantum mechanics as the Hamiltonian itself in essence constitutes the supercharge. However, it is closely related to the non-relativistic supersymmetric Pauli Hamiltonian. The Dirac Hamiltonian is of particular interest for the characterisation of the electronic structure of solids. Here we limit our discussion to that of graphene, i.e. a two-dimensional Dirac system, in an orthogonal magnetic field. We also discuss the phenomenon of band inversion in semiconductors which is an example of a three-dimensional Dirac Hamiltonian with a scalar field exhibiting a SUSY structure. We also discuss Dirac systems with spherically symmetric scalar and tensor potentials and show that the one-dimensional supersymmetric Dirac system in essence is the relativistic version of Witten's model.

Acknowledgements

This revised version has benefitted from the contribution of various people. First, let me thank all the students who have taken my courses on this topic over the last two decades. Their feedback had been very valuable and they are in essence responsible for having added the problems sections to each chapter. A special thanks goes to the team of the ESO library at Garching—Uta Grothkopf and her colleagues, Silvia Meakins and Dominic Bordelon, did an incredible job of providing me with all the material I was looking for. I am very grateful to Simone Warzel and Mikhail Plyushchay for their valuable comments and suggestions. I am also grateful to John Navas and Daniel Heatley from IOP Publishing for their support during this project.

It is my pleasure to dedicate this book to Akira Inomata. I met Akira Inomata as a student in the autumn of 1984 when we started our first shared research project. Since than we have had a continuous collaboration on various topics in physics. Discussions with him are always enjoyable and regularly provide me with new insights. The last 35 years of collaboration with him has been very valuable for me and, among other achievements, helped form my view of supersymmetric quantum mechanics. Needless to say that his comments have also improved the current publication substantially. I am looking forward to many more years of joint collaboration.

Author biography

Priv.-Doz. Dr Georg Junker

Georg Junker started his academic education as a physicist in 1981 at the University of Würzburg. His studies during the academic year 1984/1985 at the State University of New York at Albany resulted in a Master of Science degree. In 1986 he received a Diploma and in 1989 a PhD in physics from Würzburg University. From 1989 to 1999 he held an assistant position at the University of Erlangen-Nürnberg where he also received his Habilitation and became a Privatdozent. After a few years outside science he accepted a position as programme controller at the European Southern Observatory in 2005.

Georg teaches graduate courses in mathematical physics at his home university in Erlangen and serves as a member of the Advisory Panel for *Journal of Physics A: Mathematical and Theoretical*. In his research activities he focusses on all kinds of symmetry aspects of physical systems. The symmetries of such systems, being of geometric or dynamical origin, usually help in understanding their physical properties and may even lead to explicit results. He has made major contributions within Feynman's path-integral approach to quantum mechanics, and has also successfully applied group theoretical methods to problems in quantum chaos and statistical physics. Originally he concentrated on Lie symmetries and their implications, however, in the early 1990s he also started to look into the supersymmetric aspects of physical systems.

Symbols

Symbol	Description (first occurrence)
$[\cdot,\cdot]$	commutator (chapter 1)
$\{\cdot,\cdot\}$	anticommutator (chapter 1)
$\{\cdot,\cdot\}_{DB}$	Dirac bracket (section 5.5.1)
$\mathbf{1}$	unit matrix, unit operator (chapter 1)
$\vec{\nabla}$	nabla or gradient vector operator (section 2.1.1)
$\approx$	weak equality (section 5.5.1)
$\simeq$	approximation (section 6.1.1)
$\sim$	asymptotic behaviour (section 3.5)
$\lvert 0 \rangle$	vacuum, ground state (chapter 1)
$\lvert \uparrow \rangle, \lvert \downarrow \rangle$	positive, negative Witten-parity states (section 2.2)
$a, a^\dagger$	bosonic annihilation operator, creation operator (section 2.2)
a	$a(x) = \arcsin\left(\Phi(x)/\sqrt{2E}\right)$ (section 5.4)
a	$a(x) = \arcsin\left(\Phi(x)/\sqrt{E}\right)$ (section 6.1.2)
$A, A^\dagger$	generalised annihilation, creation operators (section 2.2.1)
$\vec{A}$	vector potential (section 2.1.1)
$b, b^\dagger$	fermionic annihilation, creation operators (section 2.2)
$B, B^\dagger$	generalised annihilation, creation operators (section 4.4.1)
$\vec{B}, B$	magnetic field, magnetic field strength (section 2.1.1)
c	speed of light (section 2.1.1)
C	normalisation constant (section 3.3.1)
$\mathcal{C}(\mathcal{Q}, x_0)$	space of continuous paths in $\mathcal{Q}$ starting at x_0 (section 7.5)
$\mathbb{C}$	complex numbers (section 2.1.1)
D	diffusion constant (section 7.1)
e	electric charge, elementary charge (section 2.1.1)
dim	dimension of a space (section 2.2.4)
deg	degree of a Grassmann number (section 5.5.1)
det	determinant of a matrix or operator (section 5.5.2)
E, E_n	energy, energy eigenvalues (section 2.2.2)
E_0	ground-state energy (section 2.1)
$E[\phi]$	energy functional (section 4.5)
$E_{qc-SUSY}$	quasi-classical SUSY approx. to energy eigenvalue (section 6.4.2)
$E_{WKB\pm}$	WKB approximation to energy eigenvalue (section 6.4.2)
$\mathcal{E}$	Grassmann-valued classical energy (section 5.3)
F	soul of Grassmann-valued energy (section 5.3)
F	mapping of a set of parameters (section 4.2)
F	drift coefficient (section 7.1)
F	magnetic flux (section 8.1)
F_r	family of drift coefficients (section 7.5)
$_1F_1$	confluent hypergeometric function (section 4.4)
$_pF_q$	generalised hypergeometric function (section 4.4.2)
$\mathcal{F}$	fermion-number operator (section 2.2.1)
g	Landé or g-factor (section 2.1.1)
$g, g^\pm$	iterated resolvent (section 9.1.3)
G	resolvent of the Dirac operator (section 9.1.3)
G_{rs}^{pq}	Meijer's G-function (section 4.2.2)
$\hbar$	Planck's constant divided by 2π (section 2.1.1)

H,	Hamilton operator, SUSY Hamiltonian (section 2.1)
$H_\pm$	SUSY partner Hamiltonians (section 2.1.2)
H_{loop}	loop correction (section 3.1)
$H_{\text{D}}, H_{\text{P}}$	Dirac Hamiltonian, Pauli Hamiltonian (section 2.1.1)
$H_{\text{P}}^{(2)}, H_{\text{P}}^{(3)}$	Pauli Hamiltonian in 2, 3 space dimensions (section 8.1)
H_n	Hermite polynomial of order n (section 4.4)
H_s	hierachy of isospectral Hamiltonians (section 4.2)
H_{SUSY}	SUSY Hamiltonian associated with H_{D} (section 9.1)
H_{tree}	tree Hamiltonian (section 3.1)
H_{T}	total Hamiltonian (section 5.5.1)
H_V	standard Schrödinger Hamiltonian for potential V (section 4.1)
$\mathcal{H}$	abstract Hilbert space (section 2.1)
$\mathcal{H}^\pm$	subspaces of positive/negative Witten parity (section 2.2.1)
i	imaginary unit, $\text{i}^2 = -1$ (chapter 1)
ind	Fredholm index of an operator (section 2.2.4)
inf	infimum (section 2.1)
j_ℓ	spherical Bessel function (section 8.3.2)
k_{F}	Fermi momentum (section 8.2.2)
k_L, k_R	wave number on left/right side of potential barrier (section 3.5)
ker	kernel of an operator (section 2.2.1)
$K_t^\pm$	heat kernel or Euclidean propagator (section 7.2)
K	relativistic spin–orbit operator (section 9.4)
K_{nr}	non-relativistic spin–orbit operator (section 8.3)
$K(\vec{r}'', \vec{r}'; t)$	Feynman kernel (section 9.1.3)
L_0, L	Lagrangian (section 5.1, section 5.2)
$\mathcal{L}$	Lagrange density (section 5.2)
$L_{\text{qc}}, \tilde{L}_{\text{qc}}^\pm$	quasi-classical Lagrangians (section 5.3)
$L^2(\mathbb{R}^d)$	Hilbert space of square integrable functions on $\mathbb{R}^d$ (section 2.1.1)
$L^2(\mathcal{M}, \text{d}\mu)$	Hilbert space of square integrable functions on domain $\mathcal{M}$ with Lebesgue measure μ (section 3.1)
m	mass (section 2.1.1)
$m_t, m_t^\pm$	transition probability density (section 7.1)
M	magnetisation (section 8.2)
$M_{\mu\nu}$	angular momentum tensor (chapter 1)
$\mathcal{M}$	configuration space of Witten's model (section 3.1)
$\mathcal{M}$	probability measure on path space (section 7.5)
$\mathbf{M}$	generator of a stochastic process (section 7.5)
$\mathbf{M}_s$	family of generators of stochastic processes (section 7.5)
n	non-negative integer or index of energy eigenvalues (section 3.7)
$n_\pm$	number of zero modes of $H_\pm$ (section 2.2.4)
N	number of self-adjoint supercharges (section 2.1)
$\mathcal{N}_\pm$	number of electrons with spin up/down (section 8.2)
$\mathbb{N}, \mathbb{N}_0$	natural numbers excluding/including zero (section 2.2)
$\vec{p}, p$	momentum variable or operator (section 2.1.1)
p_E	magnitude of classical momentum for given energy E (section 6.1.1)
p_E^{qc}	magnitude of quasi-classical momentum for given energy E (section 6.1.2)
P	stationary distribution (section 7.2)
P	linear momentum operator (section 7.5)
$P^\pm$	projectors onto $\mathcal{H}^\pm$ (section 2.2.1)
$P_\zeta(\vec{r}'', \vec{r}'; t)$	promotor (section 9.1.3)

Q	position operator (section 7.5)
$\mathcal{Q}$	configuration space of a stochastic process (section 7.5)
Q_i	self-adjoint supercharges (section 2.1)
$\tilde{Q}_i, Q, Q^\dagger$	complex supercharge (section 2.2)
q	soul of Grassmann-valued classical path (section 5.3)
$q_\mathrm{L}, q_\mathrm{R}$	classical left/right turning point (section 6.1.1)
$\vec{r}, r$	position variable or operator (section 2.1.1)
R	residual part of shape-invariance condition (section 4.2)
$R^\pm(E)$	reflection amplitude (section 3.5)
$\mathbb{R}, \mathbb{R}^d$	real numbers, d-dimensional Euclidean space (section 2.1.1)
Res	residuum (section 6.3)
$S, S^\pm$	action functional (section 6.1.1, section 6.1.2)
S_tree	action for quasi-classical paths (section 6.1.2)
sgn	sign function (section 3.4)
spec	spectrum of an operator (section 2.1)
t	time (section 2.1)
$T^\pm(E)$	transition amplitude (section 3.5)
$T_E,$	period of classical motion with energy E (section 6.1.1)
$T_E^\mathrm{qc},$	period of quasi-classical motion with energy E (section 6.3)
U	superpotential (section 3.3.1)
$U_\pm$	drift potential (section 7.2)
$\vec{v}$	velocity operator (section 8.1)
$\vec{v}$	arbitrary vector operator (section 8.3)
$V_\pm$	partner potentials in Witten's model (section 3.1)
W	Witten parity (section 2.2.1)
W	SUSY potential (section 4.5)
W, W_tree	Hamilton's characteristic function (section 6.1.1, section 6.1.2)
x	position variable or operator (section 2.1.2)
x	Grassmann-valued classical path (section 5.3)
x_qc	quasi-classical solution or path (section 5.3)
$x_\mathrm{L}, x_\mathrm{R}$	quasi-classical left/right turning point (section 5.4)
$\mathbb{Z}$	integers (section 5.5.2)
$(z)_n$	Pochhammer symbol (section 4.4.2)
$|\alpha\rangle$	canonical coherent state (section 4.4.2)
$\vec{\alpha}, \beta$	Dirac matrices (section 9.1)
β	inverse temperature (section 8.2.1)
Γ	Euler's gamma function (section 3.6.2)
δ	Dirac's delta function (section 3.6.1)
δ_{ij}	Kronecker's delta symbol (section 2.1)
Δ	Witten index (section 2.2.4)
$\bar{\Delta}, \hat{\Delta}, \tilde{\Delta}$	regularised indices (section 2.2.4)
ε	factorisation energy (section 4.1)
$\varepsilon, \bar{\varepsilon}$	infinitesimal Grassmann numbers (section 5.2)
ε_{ijk}	Levi-Civita symbol (section 8.4)
ε_F	Fermi energy (section 8.2.1)
$\theta, \bar{\theta}$	Grassmann variable (section 5.2)
Θ	unit-step function (section 2.2.4)
κ	negative eigenvalue of (non-)relativistic spin–orbit operator (section 8.3)
κ_R	penetration length (section 3.7)
λ_1, λ_2	Lagrange multipliers (section 5.5.1)

λ_n, $\lambda_n^{\pm}$	decay rates (section 7.2)
Λ	helicity operator (section 2.1.1)
μ	Morse index (section 6.1.2)
μ_B	Bohr magneton (section 8.1)
$\lvert\mu\rangle$	generalised non-linear coherent state (section 4.4.2)
μ	Maslov index (section 6.1.1)
$\xi^{\pm}$	canonical momenta of $\tilde{L}_{\mathrm{qc}}^{\pm}$ (section 5.3)
ξ, $\xi^{\pm}$	white noise variables (section 7.1, section 7.3)
π, $\bar{\pi}$	classical fermionic momenta (section 5.5.1)
Π	space-parity operator (section 3.4)
σ_1, σ_2, σ_3, $\sigma_{\pm}$, $\vec{\sigma}$	Pauli matrices (chapter 1)
φ, $\varphi_k^{(i)}$	fermionic phase functional (section 5.3)
$\lvert\phi^{\pm}\rangle$	states in $\mathcal{H}^{\pm}$ (section 2.2.2)
$\lvert\phi_E^{\pm}\rangle$	eigenstates of $H_{\pm}$ in $\mathcal{H}^{\pm}$ (section 2.2.2)
ϕ_0	zero-energy eigenfunction of Witten's model (section 3.3.1)
ϕ_s	static solution (section 4.5)
Φ	SUSY potential (section 2.1.2)
$\Phi_{\pm}$	asymptotic values of the SUSY potential (section 3.4)
χ	paramagnetic susceptibility (section 8.2.1)
χ_0	paramagnetic zero-field susceptibility (section 8.2.1)
χ_1, χ_2	second-class constraints (section 5.5.1)
$\chi_{\pm}$	spinors (section 8.3.1)
ψ, $\bar{\psi}$	classical fermionic degrees of freedom (section 5.2)
ψ_0, $\bar{\psi}_0$	generators of a Grassmann algebra (section 5.1)
$\lvert\psi\rangle$	states in $\mathcal{H}$ (section 2.2.1)
$\lvert\psi_0\rangle$, $\lvert\psi_0^j\rangle$	SUSY ground state(s) (section 2.1)
$\lvert\psi^{\pm}\rangle$	positive/negative Witten-parity states (section 2.2.2)
$\lvert\psi_E^{\pm}\rangle$	energy eigenstates with positive/negative Witten parity (section 2.2.2)
ω	frequency (section 2.3)
ω	element of a probability space (section 7.5)
ω_c	cyclotron frequency (section 8.1)
Ω	probability space (section 7.5)

IOP Publishing

Supersymmetric Methods in Quantum, Statistical and Solid State Physics
Enlarged and revised edition
Georg Junker

Chapter 1

Introduction

'Dass der Raum, als Ort für Puncte aufgefasst, nur drei Dimensionen hat, braucht vom mathematischen Standpuncte aus nicht discutirt zu werden; ebenso wenig kann man aber vom mathematischen Standpuncte aus Jemanden hindern, zu behaupten, der Raum habe eigentlich vier, oder unbegränzt viele Dimensionen, wir seien aber nur im Stande, drei wahrzunehmen.' [1]

Felix Klein (1849–1925)

Symmetry plays important roles in physics. There are numerous kinds of symmetries in nature. Some are visible and some are hidden. Some are static and some are dynamic. Some belong to simple individual systems and some may be seen in the collective behaviour of many systems. Often, various symmetries reveal themselves at the same time and complicate the appearance of a physical phenomenon. Mathematically, symmetry is handled by group theory; the symmetry of a physical system is seen as an invariance under a group action. The visual symmetry of a crystal may be described by a discrete group. The symmetry of a system in motion is represented by a continuous group. The mirror-image symmetry of a system can be described by a discrete group of reflections and uniformity of time flow may be seen as a consequence of invariance under a continuous group of time translations.

In 1872, Felix Klein [1], in his inaugural lecture at Erlangen, commonly known as the 'Erlangen program', made an observation that the geometry of space is associated with a mathematical group. According to Klein, the Euclidean space, for instance, is characterised by its transformation groups which consist of rotations, translations, and reflections. In 1918, Emmy Noether [2] put forth another important theorem that if a system is invariant under a continuous group of n parameters and satisfies the equations of motion, there exist n constants of motion. As a result of Noether's theorem, we see that if a system moves freely in D-dimensional Euclidean space then the D-component linear momentum and

doi:10.1088/2053-2563/aae6d5ch1

$(D^2 - D)/2$-component angular momentum must be conserved in association with the translation group of D parameters and the rotation group of $(D^2 - D)/2$ parameters, respectively. The symmetry considerations are useful not only in the study of classical systems but also in understanding quantum phenomena. Soon after the development of quantum mechanics, symmetry methods were found to be powerful in analysing quantum spectra [3–6].

In modern physics we recognise two kinds of symmetry, *external symmetry* and *internal symmetry*. In classical physics, we are concerned with external symmetries. In understanding the detailed structure of an atomic spectrum, it became necessary to introduce the concept of quantised spin. Since it is difficult to ascribe the spin concept associated with a point particle to the classical spinning of an extended body, the spin in non-relativistic quantum mechanics was understood as an internal degree of freedom. As quantum mechanics was applied to nuclear physics and high-energy physics, numerous additional internal degrees of freedom and their associated internal symmetries were introduced. There have been attempts to unify external symmetries and internal symmetries. However, such a grand unification has not been fully achieved so far.

In Minkowskian space–time, where quantum field theory is formulated, the maximal symmetry group is the Poincaré (or inhomogeneous Lorentz) group. It contains the homogeneous Lorentz group as a subgroup which allows for a classification of fundamental particles by their spin [7]. Since the allowed values of spin are integers or half-integers, as long as our physical world is Minkowskian, fundamental particles must be either bosons or fermions. Thus, in the relativistic formulation, the spin is no longer associated with an internal symmetry but with a manifestation of the external space–time symmetry. Internal degrees of freedom such as isospin, baryon number, colour, strangeness, charm, etc, are still associated with internal symmetries.

In 1967 Coleman and Mandula [8] investigated all possible symmetries of the scattering matrix in relativistic quantum field theory. They considered a field theory in more than two space–time dimensions with a finite number of massive one-particle states and a non-vanishing scattering amplitude. Their result is as follows. If one restricts the set of continuous symmetries to those generated by a *Lie algebra*, then the set of all possible generators consists only of the angular momentum Lorentz tensor $M_{\mu\nu}$, the linear momentum Lorentz vector P_μ, and Lorentz scalars. While $M_{\mu\nu}$ and P_μ generate the Poincaré group, additional symmetries must be generated by Lorentz scalars, which are indeed internal symmetries. However, the restriction to Lie symmetries made in the work of Coleman and Mandula has no *a priori* grounds.

In 1968, Miyazawa [9] suggested a possible unification scheme for mesons and baryons based on a *superalgebra*. While a Lie algebra consists only of commutators, a superalgebra is closed with respect to commutators and anticommutators. For details see, for example, the books by Scheunert [10] and Cornwell [11]. Gol'fand and Likhtman [12] were the first to embed the Poincaré algebra into a superalgebra and construct a supersymmetric Lagrangian. Independently a supersymmetric field theory was formulated in 1972/1973 by Volkov and Akulov [13, 14] but was not renormalisable. In 1974, Wess and Zumino [15, 16] succeeded in formulating another simpler and renormalisable supersymmetric field theory. Subsequently, Haag, Łopuszánski, and Sohnius [17] constructed 'all possible generators of supersymmetries of the S-matrix'. They found that within the framework of superalgebras, besides the generators of the Poincaré group and possible Lorentz scalars, there are spinor operators Q_i allowed. In the Weyl representation where the *supercharges Q_i, $i \in \{1, 2\}$*, are given by two components of a left-handed Weyl spinor, the superalgebra reads (without the Poincaré subalgebra and extra Lorentz scalars)

$$[Q_i, M_{\mu\nu}] = \frac{1}{2}(\sigma_{\mu\nu})_i^j \, Q_j, \qquad [Q_i, P_\mu] = 0,$$
$$\{Q_i, Q_j\} = 0, \qquad \{Q_i, Q_j^\dagger\} = 2(\sigma^\mu)_{ij} P_\mu. \tag{1.1}$$

Here $[A, B] := AB - BA$ and $\{A, B\} := AB + BA$ denote the *commutator* and *anticommutator*, respectively, and the Einstein summation convention is used. Furthermore,

$$\sigma^\mu := (\mathbf{1}, \vec{\sigma}) \equiv (\mathbf{1}, \sigma_1, \sigma_2, \sigma_3), \qquad \sigma^{\mu\nu} := \frac{i}{2}[\sigma^\mu, \sigma^\nu], \tag{1.2}$$

with Pauli matrices

$$\sigma_1 := \begin{pmatrix} 0 & 1 \\ 1 & 0 \end{pmatrix}, \qquad \sigma_2 := \begin{pmatrix} 0 & -i \\ i & 0 \end{pmatrix}, \qquad \sigma_3 := \begin{pmatrix} 1 & 0 \\ 0 & -1 \end{pmatrix}. \tag{1.3}$$

From this superalgebra we notice for the energy operator

$$H \equiv P_0 = \frac{1}{4}\sum_{i=1}^{2}\{Q_i, Q_i^\dagger\}, \qquad Q_i^2 = 0. \tag{1.4}$$

From the first commutator of equation (1.1), it is also obvious that the supercharge operators change the eigenvalues of the third component M_{12} of the angular momentum operator by $\frac{1}{2}$. Therefore, each of the supercharge operators converts a bosonic state to a fermionic state and a fermionic state to a bosonic state, and becomes a generator of the so-called *supersymmetry transformations*, see figure 1.1.

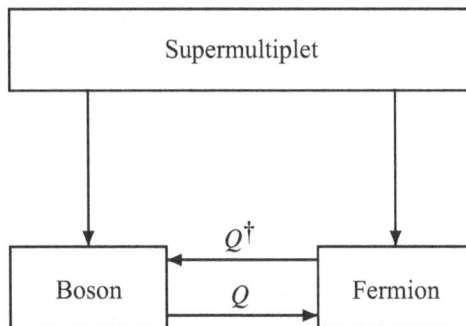

Figure 1.1. The SUSY transformations in relativistic quantum field theories generated by supercharges Q, $Q^\dagger$.

In a nucleus, the proton, and the neutron are constantly transmuted into each other, so that they are not physically distinguishable. It is more appropriate to consider them as two possible states of a single nucleon, forming an iso-spinor. The indistinguishability of the proton and the neutron in a nucleus led Heisenberg [18] to the idea of iso-symmetry. In much the same sense, the idea of supersymmetry assumes that there may be environments where bosons and fermions become indistinguishable. They are to be viewed as members of a single supermultiplet. In order for a physical system to have supersymmetry (SUSY), the ground state (vacuum) $|0\rangle$ of the system should be invariant under any SUSY transformation. This means that if SUSY is a good symmetry

$$Q_i|0\rangle = 0 \quad \text{and} \quad Q_i^\dagger|0\rangle = 0, \qquad \text{for all } i. \tag{1.5}$$

If SUSY is spontaneously broken, then we have

$$Q_i|0\rangle \neq 0 \quad \text{and/or} \quad Q_i^\dagger|0\rangle \neq 0, \qquad \text{for at least one } i. \tag{1.6}$$

As was mentioned earlier, SUSY was originally introduced to physics in search of a possible non-trivial unification of space–time and internal symmetries within four-dimensional relativistic quantum field theory. However, application of the SUSY idea is not limited to high-energy particle physics. SUSY has been successfully applied to other areas of theoretical physics such as nuclear, atomic, solid-state, and statistical physics [19, 20]. Even for mathematical aspects of theoretical physics, it has been found to be a useful concept. As will be shown in later chapters, SUSY has become a powerful tool in various areas of physics. The first SUSY application in statistical physics was done in 1976 by Nicolai [21]. The SUSY in non-relativistic quantum mechanics became popular with Witten's toy model [22] introduce in 1981. The supersymmetric structure of Dirac's formulation of relativistic quantum mechanics, see for example the book by Thaller [23], is currently being successfully used in the understanding of various phenomena in solid-state physics, in particular, related to topological superconductors [24] and carbon-based nano-structures such as graphene [25].

Although the SUSY idea is fascinating, we know that a boson and a fermion are quite clearly separate objects. In our surrounding environments, we can find no phenomenon in which a boson is converted into a fermion. The situation is different within so-called topological superconductors, where space–time SUSY indeed has been found to emerge under certain conditions [24]. Nevertheless, despite tremendous efforts at LHC, so far we have no signals of any elementary SUSY particles [26]. In other words, as far as we see, SUSY is not a good symmetry. SUSY is in fact broken in our physical world. Yet, if we have faith in SUSY, then it is reasonable to assume that SUSY has spontaneously been broken at some point in time and temperature. In 1981, Witten [22] introduced SUSY quantum mechanics, based on the simplest superalgebra, in order to provide a simple non-relativistic model for the spontaneous SUSY breaking mechanism. Witten's formulation of non-relativistic SUSY quantum mechanics has attracted considerable attention in the last few decades and still serves as a useful tool in quantum physics. It has become part of

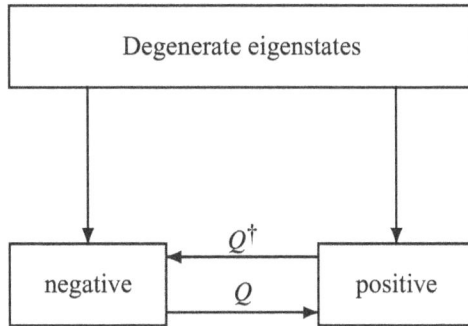

Figure 1.2. The SUSY transformation of SUSY quantum mechanics transforms between states of positive and negative Witten parity.

standard textbooks on quantum mechanics [23, 27–30] and several monographs have been dedicated to the topic of SUSY quantum mechanics and its applications [20, 31–34].

Despite the fact that SUSY quantum mechanics is indeed the (0 + 1)-dimensional limit of SUSY quantum field theory, it is rather independent of the latter. SUSY in SUSY quantum mechanics is not the original supersymmetry relating bosons and fermions. The supercharges of SUSY quantum mechanics do not generate transformations between bosons and fermions. They generate transformations between two orthogonal eigenstates of a given Hamiltonian for the same degenerate eigenvalue (see figure 1.2). These two orthogonal states are eigenstates of the so-called Witten-parity operator (see chapter 2) with eigenvalues +1 and −1. Although in supersymmetric quantum field theory the degeneracy arises from the fact that the supercharge operators change the eigenvalues of the spin by $\frac{1}{2}$, the Witten-parity states in SUSY quantum mechanics may be taken in general as independent from the real spin states. Some of the examples we shall discuss later on are, indeed, models of a particle carrying a physical spin-$\frac{1}{2}$ degree of freedom. In order to avoid a possible confusion, we shall refer to those states that are independent of spin as the Witten-parity states.

References

[1] Klein F 1872 *Vergleichende Betrachtungen über neuere geometrische Forschungen, Programm zum Eintritt in die philosophische Facultät und den Senat der k. Friedrich-Alexanders-Universität zu Erlangen* (Erlangen: Verlag von Andreas Deichert) p 42

[2] Noether E 1918 Invariante Variationsprobleme *Nachr. König. Gesellsch. Wiss. Göttingen* **1918** 235–57

[3] Pauli W 1926 Über das Wasserstoffspektrum vom Standpunkt der neuen Quantenmechanik *Z. Phys.* **36** 336–63

[4] Weyl H 1931 *Gruppentheorie und Quantenmechanik* (Leipzig: Hirzel)

[5] Wigner E P 1931 *Gruppentheorie und ihre Anwendung auf die Quantenmechanik der Atomspektren* (Braunschweig: Vieweg)

[6] van der Waerden B L 1932 *Die Gruppentheorische Methode in der Quantenmechanik* (Berlin: Springer)

[7] Wigner E 1939 On unitary representations of the inhomogeneous Lorentz group *Ann. Math.* **40** 149–204

[8] Coleman S and Mandula J 1967 All possible symmetries of the S matrix *Phys. Rev.* **159** 1251–6

[9] Miyazawa H 1968 Spinor currents and symmetries of baryons and mesons *Phys. Rev.* **170** 1586–90

[10] Scheunert M 1979 *The Theory of Lie Superalgebras Lecture Notes in Mathematics* vol 716 (Berlin: Springer)

[11] Cornwell J F 1989 *Group Theory in Physics: Supersymmetries and Infinite-Dimensional Algebras* vol 3 (London: Academic)

[12] Gol'fand Y A and Likhtam E P 1971 Extension of the algebra of Poincaré group generators and violation of P-invariance *JETP Lett.* **13** 323–6

[13] Volkov D V and Akulov V P 1972 Possible universal neutrino interaction *JETP Lett.* **16** 438–40

[14] Volkov D V and Akulov V P 1973 Is the neutrino a Goldstone particle? *Phys. Lett.* B **46** 109–10

[15] Wess J and Zumino B 1974 Supergauge transformations in four dimensions *Nucl. Phys.* B **70** 39–50

[16] Wess J and Zumino B 1974 Supergauge invariant extension of quantum electrodynamics *Nucl. Phys.* B **78** 1–13

[17] Haag R, Łopuszański J T and Sohnius M 1975 All possible generators of supersymmetries of the S-matrix *Nucl. Phys.* B **88** 257–74

[18] Heisenberg W 1932 Über den Bau der Atomkerne. I *Z. Phys.* **77** 1–11

[19] Kostelecký V A and Campbell D K (ed) 1985 *Supersymmetry in Physics* (Amsterdam: North-Holland)

[20] Junker G 1996 *Supersymmetric Methods in Quantum and Statistical Physics* 1st edn (Berlin: Springer)

[21] Nicolai H 1976 Supersymmetry and spin systems *J. Phys. A: Math. Theor.* **9** 1497–506

[22] Witten E 1981 Dynamical breaking of supersymmetry *Nucl. Phys.* B **188** 513–54

[23] Thaller B 1992 *The Dirac Equation* (Berlin: Springer)

[24] Grover T, Sheng D N and Vishwanath 2014 Emergent space–time supersymmetry at the boundary of a topological phase *Science* **344** 280–3

[25] Sarma S D, Adam S, Hwang E H and Rossi E 2011 Electronic transport in two-dimensional graphene *Rev. Mod. Phys.* **83** 407–70

[26] ATLAS Collaboration 2018 ATLAS extends searches for natural supersymmetry *CERN Courier* **58** 11–2

[27] Cycon H L, Froese R G, Kirsch W and Simon B 1987 *Schrödinger Operators with Application to Quantum Mechanics and Global Geometry* (Berlin: Springer)

[28] Grosse H 1991 Supersymmetric quantum mechanics *Recent Developments in Quantum Mechanics, Mathematical Physics Studies Nr. 12* ed A Boutet de Monvel *et al* (Dordrecht: Kluwer) pp 299–327

[29] de Lange O L and Raab R E 1991 *Operator Methods in Quantum Mechanics* (Oxford: Clarendon)

[30] Schwabl F 2007 *Quantum Mechanics* 4th edn (Berlin: Springer)

[31] Kalka H and Soff G 1997 *Supersymmetrie* (Stuttgart: Teubner)
[32] Cooper F, Khare A and Sukhatme U 2001 *Supersymmetry in Quantum Mechanics* (Singapore: World Scientific)
[33] Bagchi B K 2001 *Supersymmetry in Quantum and Classical Mechanics* (Boca Raton, FL: Chapman and Hall)
[34] Hirshfeld A 2012 *The Supersymmetric Dirac Equation* (London: Imperial College Press)

IOP Publishing

Supersymmetric Methods in Quantum, Statistical and Solid State Physics
Enlarged and revised edition
Georg Junker

Chapter 2

Supersymmetric quantum mechanics

This chapter will introduce a very generic definition of a so-called N-extended supersymmetric quantum system. The $N = 2$ supersymmetric quantum mechanics, as a $(0 + 1)$-dimensional limit of a supersymmetric field theory, was first formulated by Nicolai [1, 2] some 40 years ago in his search for supersymmetry in non-relativistic quantum systems related to models of statistical physics. Supersymmetric quantum mechanics became popular with Witten's model [3, 4] where he studied the dynamical breaking of supersymmetry using a simple quantum mechanical model. Witten's approach has been for a general $N \geqslant 1$ but, for $N \geqslant 2$, has the same features as Nicolai's formulation.

Here we will present the definition due to Witten in order to include the $N = 1$ case, which has some interesting applications in the theory of a three-dimensional electron gas in a magnetic field. However, only for $N \geqslant 2$ will we find the typical properties of a supersymmetric quantum system such as the pairing of energy eigenstates indicated in figure 2.1. In particular, $N \geqslant 2$ allows the introduction of the so-called Witten parity and thus implies a natural grading of the Hilbert space into two subspaces.

2.1 Definition of SUSY quantum mechanics

Let us start with an abstract quantum system defined by a Hamiltonian H acting on some Hilbert space $\mathcal{H}$. Furthermore, we will postulate the existence of N self-adjoint operators $Q_i = Q_i^{\dagger}, i = 1, 2, \ldots , N$, that also act on $\mathcal{H}$. With this setup we spell out the following definition.

Definition 2.1. A quantum system characterised by the set $\{H, Q_1, \ldots , Q_N; \mathcal{H}\}$, is called *supersymmetric* if the following anticommutation relation is valid for all $i, j = 1, 2, \ldots , N$:

doi:10.1088/2053-2563/aae6d5ch2

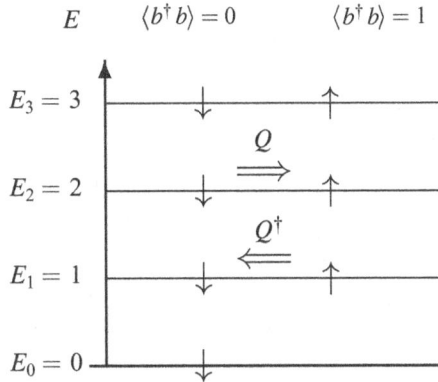

Figure 2.1. The spectrum of the supersymmetric harmonic oscillator and the associated SUSY transformations.

$$\{Q_i, Q_j\} = H\delta_{ij},$$ (2.1)

where δ_{ij} denotes Kronecker's delta symbol. The self-adjoint operators Q_i are called *supercharges* and the Hamiltonian H is called the *SUSY Hamiltonian*. The symmetry described by the *superalgebra* (2.1) is called *N-extended supersymmetry*.

Let us remark that sometimes SUSY quantum mechanics is defined with non-self-adjoint supercharges. For example, for even N, say $N = 2M$, one can define the 'complex' supercharges $\tilde{Q}_k := (Q_{2k-1} + iQ_{2k})/\sqrt{2}$, $k = 1, 2, \ldots, M$. Because of this, an $N = 2M$ extended supersymmetry, in the above sense, is sometimes called an M-extended SUSY. For complex supercharges the superalgebra reads

$$\left\{\tilde{Q}_k, \tilde{Q}_l^\dagger\right\} = H\delta_{kl}, \qquad \tilde{Q}_k^2 = 0 = \left(\tilde{Q}_k^\dagger\right)^2.$$ (2.2)

For a discussion on the various definitions used in the literature and their interconnection see [5]. We will exclusively use the notion of the above definition, that is, counting the real supercharges.

The above superalgebra implies that different supercharges anticommute, that is,

$$\{Q_i, Q_j\} = 0 \quad \text{for} \quad i \neq j.$$ (2.3)

It also puts restrictions on the SUSY Hamiltonian H. First we notice that from the fundamental anticommutation relation (2.1) immediately follows:

$$H = 2Q_1^2 = 2Q_2^2 = \cdots = 2Q_N^2 = \frac{2}{N}\sum_{i=1}^{N}Q_i^2,$$ (2.4)

which may be compared with equation (1.4). The supercharges of a supersymmetric quantum system are *square roots* of the SUSY Hamiltonian. A direct consequence of the relation $H = 2Q_i^2$ is

$$[H, Q_i] = 0 \quad \text{for all} \quad i = 1, 2, \dots, N. \tag{2.5}$$

Therefore, the supercharges are *constants of motion* if they do not depend explicitly on time, that is when $\partial Q_i/\partial t = 0$. The same is valid for the complex supercharges as obviously $[H, \tilde{Q}_k] = 0 = [H, \tilde{Q}_k^\dagger]$.

As a second consequence of equation (2.4) we note that the Hamiltonian does not have negative eigenvalues. In other words, the ground-state energy

$$E_0 := \inf \text{spec } H \tag{2.6}$$

of a supersymmetric quantum system is non-negative:

$$E_0 \geqslant 0. \tag{2.7}$$

The symbol spec H denotes the spectrum (set of all eigenvalues) of the operator H.

In analogy to supersymmetric field theories, see equation (1.5), we introduce the notion of good and broken SUSY.

Definition 2.2. A supersymmetric quantum system $\{H, Q_1, \dots, Q_N; \mathcal{H}\}$ is said to have a *good* (or unbroken) SUSY if its ground-state energy vanishes, that is, $E_0 = 0$. For a strictly positive ground-state energy $E_0 > 0$ SUSY is said to be (spontaneously) *broken*. The analogy to equation (1.5) becomes visible through the following.

Proposition 2.3. For good SUSY all ground states $|\psi_0^j\rangle$ (j enumerates a possible degeneracy of the ground-state energy $E_0 = 0$) are annihilated by all supercharges:

$$Q_i|\psi_0^j\rangle = 0 \quad \text{for all } i \text{ and } j. \tag{2.8}$$

If SUSY is broken ($E_0 > 0$) there exists at least one pair (i,j) for which

$$Q_i|\psi_0^j\rangle \neq 0. \tag{2.9}$$

This is actually a so-called spontaneous breaking of supersymmetry, where the ground states are not invariant under SUSY transformation. From now on we will omit the word spontaneous for simplicity.

Proof. Obviously, $H|\psi_0^j\rangle = E_0|\psi_0^j\rangle$ implies by equation (2.4) $\sum_{i=1}^{N}\||Q_i|\psi_0^j\rangle\|^2 = E_0 N/2$. Hence, $E_0 = 0$ implies equation (2.8) and $E_0 > 0$ implies equation (2.9), respectively.

It should be pointed out that supersymmetry imposes even more and stronger constraints on quantum systems. A rather general construction procedure for quantum systems with a SUSY structure as defined above has been given by de Crombrugghe and Rittenberg [6]. Below we will present two examples of super-symmetric quantum systems for $N = 1$ and 2.

2.1.1 The Pauli Hamiltonian ($N = 1$)

As a first example let us mention a system which is characterised by the Pauli Hamiltonian for a spin-$\frac{1}{2}$ particle in an external magnetic field [7]. The mass of this particle is denoted by m, its charge by e with $e < 0$ for an electron, and the external magnetic field by $\vec{B}(\vec{r}) := \vec{\nabla} \times \vec{A}(\vec{r})$. Here the real-valued function $\vec{A}: \mathbb{R}^3 \to \mathbb{R}^3$ denotes the vector potential. Furthermore, c and $\hbar$ denote the speed of light and Planck's constant (divided by 2π), respectively. Because the presence of spin the Hilbert space is given as the tensor product $\mathcal{H} := L^2(\mathbb{R}^3) \otimes \mathbb{C}^2$. Here $L^2(\mathbb{R}^3)$ is the infinite-dimensional Hilbert space of Lebesgue square-integrable complex-valued functions on the three-dimensional Euclidean space $\mathbb{R}^3$ and $\mathbb{C}^2$ is the two-dimensional Hilbert space equipped with the standard scalar product.

Let us define this supersymmetric system by introducing [6] the self-adjoint supercharge ($N = 1$):

$$Q_1 := \frac{1}{\sqrt{4m}} \left(\vec{p} - \frac{e}{c} \vec{A}(\vec{r}) \right) \cdot \vec{\sigma}, \qquad (2.10)$$

where $\vec{r}$ and $\vec{p}$ denote the usual position and momentum operators acting on $L^2(\mathbb{R}^3)$ and $\vec{\sigma}$ is the three-dimensional vector with components consisting of the Pauli matrices (1.3) acting on $\mathbb{C}^2$. Due to the SUSY requirement (2.4) the Hamiltonian is necessarily given by $H_P := 2Q_1^2$. An explicit calculation leads to

$$H_P = \frac{1}{2m} \left(\vec{p} - \frac{e}{c} \vec{A} \right)^2 - \frac{e\hbar}{2mc} \vec{B} \cdot \vec{\sigma}, \qquad (2.11)$$

and coincides with the well-known Pauli Hamiltonian [7] for a spin-$\frac{1}{2}$ particle with Landé g-factor $g = 2$. It is an amusing and interesting observation [6] that supersymmetry suggests $g = 2$. Note that the Pauli Hamiltonian is given by $H_P = \frac{1}{2m}(\vec{p} - \frac{e}{c}\vec{A})^2 - \vec{M} \cdot \vec{B}$, where the magnetic moment $\vec{M} := g\mu_B \vec{S}/\hbar$ with Bohr's magneton $\mu_B = e\hbar/2mc$ and spin operator $\vec{S} = (\hbar/2)\vec{\sigma}$. Hence SUSY requires a g-factor $g = 2$. Without supersymmetry this usually follows from the relativistic covariant Dirac Hamiltonian only. In fact, relativity and supersymmetry are closely related [8]. To be more explicit, the supercharge (2.10) is related to the Dirac Hamiltonian (in the standard Pauli–Dirac representation) for the same point mass in the same magnetic field $\vec{B}$:

$$H_D := \begin{pmatrix} mc^2 & (c\vec{p} - e\vec{A}) \cdot \vec{\sigma} \\ (c\vec{p} - e\vec{A}) \cdot \vec{\sigma} & -mc^2 \end{pmatrix} = \begin{pmatrix} mc^2 & 2\sqrt{mc^2}\, Q_1 \\ 2\sqrt{mc^2}\, Q_1 & -mc^2 \end{pmatrix}. \qquad (2.12)$$

See chapter 9 for more details on supersymmetric Dirac Hamiltonians. Let us also mention that the last expression for H_D immediately leads to a relation between the Pauli and Dirac Hamiltonians:

$$H_D^2 = m^2 c^4 \begin{pmatrix} 1 + 2H_P/mc^2 & 0 \\ 0 & 1 + 2H_P/mc^2 \end{pmatrix}. \tag{2.13}$$

Obviously, the supercharge (2.10) commutes with H_P as well as with H_D and, therefore, for $\partial \vec{A}/\partial t = 0$ it is a constant of motion. This in fact was already known [9] before the SUSY idea was introduced.

Note that for the general $N = 1$ case one can define on the orthogonal complement of the kernel of H, that is, on $\mathcal{H} \backslash \ker H$ a generalised helicity operator

$$\Lambda := \frac{Q_1}{|Q_1|} = \operatorname{sgn} Q_1. \tag{2.14}$$

Remember, the kernel of H is the space spanned by the zero-energy eigenstates of H, $\ker H := \operatorname{span}\{|\psi_0^j\rangle | j = 1, 2, \ldots\}$. Obviously, Λ commutes with H and $\Lambda^2 = 1$. In other words, H and Λ have a common set of eigenstates on $\mathcal{H} \backslash \ker H$ with $H|\psi_E^\pm\rangle = E|\psi_E^\pm\rangle$ and $\Lambda|\psi_E^\pm\rangle = \pm|\psi_E^\pm\rangle$. However, there is no guarantee that for each eigenvalue E there exists such a pair of helicity eigenstates. In fact, the supercharge does not transform states with positive helicity into states with negative helicity and therefore does not generate a SUSY transformation as indicated by figure 1.2. However, there may be situations, that is, a special form of the magnetic field, where it is possible to establish an $N \geqslant 2$ SUSY. The $N = 2$ case of a unidirectional magnetic field is discussed in chapter 8. For cases with $N = 3$ and $N = 4$ see [10, 11].

For a further discussion on the supersymmetric structure in the Pauli Hamiltonian where Λ is indeed representing the helicity see chapter 8. See exercise 2.1 at the end of this chapter for a rather trivial $N = 1$ example, which is the free particle on the real line.

2.1.2 Witten's SUSY quantum mechanics ($N = 2$)

The most popular model of an $N = 2$ SUSY quantum system has been introduced by Witten [3, 4]. This model describes a Cartesian degree of freedom which carries an additional internal spin-$\frac{1}{2}$-like degree of freedom. Hence, the Hilbert space is given by $\mathcal{H} := L^2(\mathbb{R}) \otimes \mathbb{C}^2$. The two supercharges have been defined by Witten as follows:

$$Q_1 := \frac{1}{\sqrt{2}} \left(\frac{p}{\sqrt{2m}} \otimes \sigma_1 + \Phi(x) \otimes \sigma_2 \right), \qquad Q_2 := \frac{1}{\sqrt{2}} \left(\frac{p}{\sqrt{2m}} \otimes \sigma_2 - \Phi(x) \otimes \sigma_1 \right), \tag{2.15}$$

where Φ is a real-valued function, $\Phi: \mathbb{R} \to \mathbb{R}$, which, for convenience, is assumed to be continuously differentiable. Note that Φ is customarily called the *SUSY potential*. This should not be confused with the *superpotential* usually introduced in supersymmetric quantum field theories. In fact, the SUSY potential Φ is basically the derivative of the superpotential.

The SUSY Hamiltonian is necessarily given by $H := 2Q_1^2 = 2Q_2^2$ and has the explicit form

$$H = \left(\frac{p^2}{2m} + \Phi^2(x) \right) \otimes 1 + \frac{\hbar}{\sqrt{2m}} \Phi'(x) \otimes \sigma_3, \qquad (2.16)$$

where the prime denotes differentiation with respect to the argument, that is, $\Phi' := \frac{d\Phi}{dx}$. In the eigenbasis of σ_3 the Hamiltonian becomes diagonal in the $\mathbb{C}^2$-space,

$$H = \begin{pmatrix} H_+ & 0 \\ 0 & H_- \end{pmatrix}, \qquad (2.17)$$

where

$$H_\pm := \frac{p^2}{2m} + \Phi^2(x) \pm \frac{\hbar}{\sqrt{2m}} \Phi'(x) \qquad (2.18)$$

are standard Schrödinger operators acting on states in $L^2(\mathbb{R})$.

This *Witten model* is the simplest one which shows all typical features of supersymmetric quantum mechanics. We will discuss this model extensively in chapter 3. The discussion in the sections below is on general $N = 2$ SUSY systems and is not limited to the one-dimensional case of Witten's model.

2.2 Properties of $N = 2$ SUSY quantum mechanics

So far the definition 2.1 does not necessarily give rise to degenerated energy eigenvalues of the Hamiltonian H, which in turn allow for a construction of supercharges generating a SUSY transformation (cf. figure 1.2). In this section we are going to show that for $N = 2$, and therefore also for $N > 2$, one can define an additional operator, which we will call Witten-parity operator. Analysing the properties of this operator we will find that only $N \geqslant 2$ in general implies a degeneracy of the eigenvalues of H and thus allows for the construction of the corresponding SUSY transformation which transforms the associated energy eigenstates into each other.

The $N = 2$ SUSY quantum mechanics consists of two supercharges Q_1, Q_2, and a Hamiltonian H which obey the following relations:

$$Q_1 Q_2 = -Q_2 Q_1, \qquad H = 2Q_1^2 = 2Q_2^2 = Q_1^2 + Q_2^2. \qquad (2.19)$$

Let us introduce the complex supercharges

$$Q := \frac{1}{\sqrt{2}}(Q_1 + iQ_2), \qquad Q^\dagger = \frac{1}{\sqrt{2}}(Q_1 - iQ_2). \qquad (2.20)$$

These operators together with the Hamiltonian H close the superalgebra

$$\boxed{Q^2 = 0 = (Q^\dagger)^2, \qquad \{Q, Q^\dagger\} = H}, \qquad (2.21)$$

which implies

$$[H, Q] = 0 = [H, Q^\dagger]. \tag{2.22}$$

It should be noted that the algebra (2.21) is sometimes used as an alternative definition of SUSY quantum mechanics. Obviously, for $N > 1$ it is identical to the others. This is because this case is sufficiently general to establish all typical properties of SUSY quantum mechanics. Nevertheless, from a purely algebraic point of view there is no reason to exclude $N = 1$ which, in addition to the Pauli Hamiltonian discussed above, also serves as the most simple supersymmetric classical model in a (1,1) superspace [12].

Example: the supersymmetric harmonic oscillator

Consider a quantum system on $\mathcal{H} = L^2(\mathbb{R}) \otimes \mathbb{C}^2$ characterising a spin-$\frac{1}{2}$ quantum particle in one dimension. Let us define a 'bosonic' and a 'fermionic' annihilation operator by

$$a := \frac{1}{\sqrt{2}}(\partial_x + x), \quad b := \sigma_- = \begin{pmatrix} 0 & 0 \\ 1 & 0 \end{pmatrix}. \tag{2.23}$$

Obviously we have

$$[a, a^\dagger] = \mathbf{1}, \qquad \{b, b^\dagger\} = \mathbf{1}, \qquad b^2 = 0 = (b^\dagger)^2, \tag{2.24}$$

which allows us to define a complex supercharge (from now on we omit the tilde for complex supercharges as it will be obvious that Q is not self-adjoint),

$$Q := a \otimes b^\dagger = \begin{pmatrix} 0 & a \\ 0 & 0 \end{pmatrix} \tag{2.25}$$

and the Hamiltonian is given by

$$H := \{Q, Q^\dagger\} = a^\dagger a + b^\dagger b. \tag{2.26}$$

The eigenstates of H are given by

$$|n, \downarrow\rangle = |n\rangle \otimes \begin{pmatrix} 0 \\ 1 \end{pmatrix}, \qquad |n, \uparrow\rangle = |n\rangle \otimes \begin{pmatrix} 1 \\ 0 \end{pmatrix}, \qquad n \in \mathbb{N}_0, \tag{2.27}$$

where

$$a|n\rangle = \sqrt{n}\,|n - 1\rangle, \qquad a^\dagger|n\rangle = \sqrt{n + 1}\,|n + 1\rangle \tag{2.28}$$

and

$$b|\uparrow\rangle = |\downarrow\rangle, \qquad b|\downarrow\rangle = 0, \qquad b^\dagger|\downarrow\rangle = |\uparrow\rangle, \qquad b^\dagger|\uparrow\rangle = 0. \tag{2.29}$$

It is evident that $a^\dagger a|n\rangle = n|n\rangle$, $b^\dagger b|\uparrow\rangle = |\uparrow\rangle$ and $b^\dagger b|\downarrow\rangle = 0$. Hence the eigenvalues of H are given by $E_n = n, n \in \mathbb{N}_0$. See also figure 2.1 where the spectrum of H is shown including the twofold degeneracy of the positive-energy eigenvalues. Here SUSY is

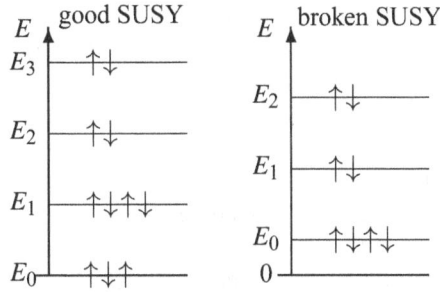

Figure 2.2. Typical spectra for good ($E_0 = 0$) and broken ($E_0 > 0$) SUSY. Strictly positive-energy eigenvalues occur in positive–negative-parity pairs (↑ ↓). Note that this chapter is not limited to one-dimensional systems. For systems in one dimension a typical spectrum is similar to that of the supersymmetric harmonic oscillator shown in figure 2.1.

unbroken as $E_0 = 0$. Furthermore, the supercharges act as following on the eigenstates of H,

$$Q|n, \downarrow\rangle = \sqrt{n}|n - 1, \uparrow\rangle, \qquad Q|n, \uparrow\rangle = 0,$$
$$Q^\dagger|n, \uparrow\rangle = \sqrt{n + 1}|n + 1, \downarrow\rangle, \qquad Q^\dagger|n, \downarrow\rangle = 0. \tag{2.30}$$

Q transforms a spin-down into a spin-up state with same energy eigenvalue and vice versa for $Q^\dagger$. This is indeed a SUSY transformation as anticipated in figure 2.1.

Note that despite the fact that this example is in one dimension the discussion through this chapter is not limited to one-dimensional systems.

2.2.1 The Witten parity

Let us now, in addition to the operators forming the superalgebra (2.21), postulate the existence of a self-adjoint operator W, which obeys the relations

$$\boxed{[W, H] = 0, \qquad \{W, Q\} = 0 = \{W, Q^\dagger\}, \qquad W^2 = \mathbf{1}.} \tag{2.31}$$

That is, W should commute with the Hamiltonian, anticommute with the complex supercharges and define a unitary involution on $\mathcal{H}$.

Definition 2.4. A self-adjoint operator W which obeys relations (2.31) is called *Witten parity* or the *Witten operator*. The quantum system $\{H, Q, Q^\dagger, W; \mathcal{H}\}$ will be called a *supersymmetric quantum system with Witten parity*.

It should be noted that for $N = 2$ such an operator may always be found by setting

$$W := \frac{2}{H}QQ^\dagger - \mathbf{1} = \frac{1}{iH}[Q_1, Q_2] = \frac{[Q, Q^\dagger]}{\{Q, Q^\dagger\}}, \tag{2.32}$$

which is only well defined on the orthogonal complement of ker H. That is, on the subspace spanned by the eigenvectors of H with strictly positive-energy eigenvalues. However, the above explicit form (2.32) of W often has a natural extension to ker H. For example, in Witten's model, where Q_1, Q_2, and H are given by equations (2.15) and (2.16), respectively, we find $[Q_1, Q_2] = \mathrm{i}H\sigma_3$. Hence, according to equation (2.32) the Witten operator is given by σ_3 and is well defined on the full Hilbert space $\mathcal{H}$. If there does not exist such a natural extension one may define W on ker H, for example, by postulating that all zero modes of H are positive or negative states. See definition 2.6 below. It should be remarked that such a problem may only occur for unbroken SUSY.

Being a unitary involution the Witten operator takes only the eigenvalues ± 1 justifying the name parity. Instead of W, Witten [4] originally considered the operator $(-1)^{\mathcal{F}} = -W$ where $\mathcal{F}$ denotes the so-called 'fermion-number operator' [13]. Actually, it is even possible to introduce (on the orthogonal complement of ker H) a 'fermionic' annihilation operator

$$b := Q^{\dagger}/\sqrt{H} \qquad (2.33)$$

obeying the relations

$$\{b, b^{\dagger}\} = \mathbf{1}, \qquad b^2 = 0 = (b^{\dagger})^2. \qquad (2.34)$$

Hence, the 'fermion-number' operator

$$\mathcal{F} := b^{\dagger}b = QQ^{\dagger}/H = \mathcal{F}^{\dagger} = \mathcal{F}^2 \qquad (2.35)$$

obeys the algebra

$$[\mathcal{F}, H] = 0, \qquad [\mathcal{F}, Q] = Q, \qquad [\mathcal{F}, Q^{\dagger}] = -Q^{\dagger}, \qquad (2.36)$$

and is related to the Witten parity by

$$W = 2\mathcal{F} - \mathbf{1} = (-\mathbf{1})^{\mathcal{F}+1}. \qquad (2.37)$$

Note that the relation (2.35) implies that $\mathcal{F}$ has only the two eigenvalues 0 and 1. The eigenspace of W with eigenvalue $+1$, i.e. $\mathcal{F} = 1$, is sometimes called 'fermionic' subspace, whereas the eigenspace of W corresponding to the eigenvalue -1 or $\mathcal{F} = 0$ is called 'bosonic' subspace. This terminology might be confusing, as SUSY quantum mechanics does not deal with the real boson–fermion symmetry, we call them subspaces of positive and negative (Witten) parity, respectively.

Definition 2.5. Let $P^{\pm} := \frac{1}{2}(1 \pm W)$ be the orthogonal projection of $\mathcal{H}$ onto the eigenspace of the Witten operator with eigenvalue ± 1. The subspace

$$\mathcal{H}^{\pm} := P^{\pm}\mathcal{H} = \{|\psi\rangle \in \mathcal{H}\colon W|\psi\rangle = \pm|\psi\rangle\} \qquad (2.38)$$

is called the space of *positive* $(\mathcal{H}^+)$ and *negative* $(\mathcal{H}^-)$ Witten parity, respectively.

It is natural to decompose the Hilbert space into these eigenspaces of W,

$$\mathcal{H} = \mathcal{H}^+ \oplus \mathcal{H}^-, \tag{2.39}$$

and represent each linear operator acting on $\mathcal{H}$ by 2×2 matrices. Because of this grading of $\mathcal{H}$ the Witten operator is sometimes called *grading operator*. The Witten parity W and the projectors $P^\pm$ in this representation read

$$W = \begin{pmatrix} 1 & 0 \\ 0 & -1 \end{pmatrix}, \qquad P^+ = \begin{pmatrix} 1 & 0 \\ 0 & 0 \end{pmatrix}, \qquad P^- = \begin{pmatrix} 0 & 0 \\ 0 & 1 \end{pmatrix}. \tag{2.40}$$

Let us now consider the matrix representation for the supercharges. For this we make the general ansatz

$$Q = \begin{pmatrix} \alpha & A \\ B & \beta \end{pmatrix}. \tag{2.41}$$

The condition $\{W, Q\} = 0$ immediately leads us to the conclusion that $\alpha = 0 = \beta$. Because of $Q^2 = 0$ we also conclude that $AB = 0 = BA$ and hence we arrive at

$$H = \{Q, Q^\dagger\} = \begin{pmatrix} AA^\dagger + BB^\dagger & 0 \\ 0 & A^\dagger A + BB^\dagger \end{pmatrix}. \tag{2.42}$$

In other words the most general $N = 2$ SUSY system can be decomposed into two subsystems,

$$H_A := \begin{pmatrix} AA^\dagger & 0 \\ 0 & A^\dagger A \end{pmatrix}, \qquad Q_A := \begin{pmatrix} 0 & A \\ 0 & 0 \end{pmatrix} \quad Q_A^\dagger = \begin{pmatrix} 0 & 0 \\ A^\dagger & 0 \end{pmatrix} \tag{2.43}$$

and

$$H_B := \begin{pmatrix} B^\dagger B & 0 \\ 0 & BB^\dagger \end{pmatrix}, \qquad Q_B := \begin{pmatrix} 0 & 0 \\ B & 0 \end{pmatrix} \quad Q_B^\dagger = \begin{pmatrix} 0 & B^\dagger \\ 0 & 0 \end{pmatrix}. \tag{2.44}$$

Note that all A-operators commute with all B-operators. Hence the SUSY quantum system $\{H, Q, Q^\dagger, W; \mathcal{H}\}$ can be decomposed into two independent subsystems,

$$\{H_A, Q_A, Q_A^\dagger, W_A; \mathcal{H}\} \qquad \text{and} \qquad \{H_B, Q_B, Q_B^\dagger, W_B; \mathcal{H}\}, \tag{2.45}$$

where $H = H_A + H_B$, $Q = Q_A + Q_B$, $W_A = W$, and $W_B = -W$. Obviously, system A is transformed into system B by the replacement $A \rightarrow B^\dagger$. Therefore, without loss of generality we can restrict our further discussion to one of these subsystems, say A, and omit the subindex A from now on.

Hence, without loss of generality, the matrix representation of the supercharges is given by

$$Q = \begin{pmatrix} 0 & A \\ 0 & 0 \end{pmatrix}, \qquad Q^\dagger = \begin{pmatrix} 0 & 0 \\ A^\dagger & 0 \end{pmatrix}, \tag{2.46}$$

which implies

$$Q_1 = \frac{1}{\sqrt{2}} \begin{pmatrix} 0 & A \\ A^\dagger & 0 \end{pmatrix}, \qquad Q_2 = \frac{i}{\sqrt{2}} \begin{pmatrix} 0 & -A \\ A^\dagger & 0 \end{pmatrix}. \tag{2.47}$$

Here $A: \mathcal{H}^- \to \mathcal{H}^+$ denotes a generalised 'annihilation' operator and $A^\dagger: \mathcal{H}^+ \to \mathcal{H}^-$ is its adjoint, which may be understood as a generalised 'creation' operator [14]. Let us also note that the real supercharges obey the commutation relations

$$[W, Q_i] = 2i\varepsilon_{ij}Q_j, \tag{2.48}$$

where a sum over j is implied and ε_{ij} is the two-dimensional antisymmetric Levi-Civita symbol with $\varepsilon_{12} = 1$.

The SUSY Hamiltonian becomes diagonal in the representation (2.40),

$$H = \begin{pmatrix} AA^\dagger & 0 \\ 0 & A^\dagger A \end{pmatrix}. \tag{2.49}$$

Hence, for an $N = 2$ supersymmetric quantum system the total SUSY Hamiltonian H consists of two so-called *SUSY partner Hamiltonians*,

$$H_+ := AA^\dagger \geqslant 0, \qquad H_- := A^\dagger A \geqslant 0. \tag{2.50}$$

Let us mention that an arbitrary operator O acting on $\mathcal{H}$ can be decomposed into its diagonal (even) part O_e and its off-diagonal (odd) part O_o. That is, $O = O_e + O_o$ with

$$[W, O_e] = 0, \qquad \{W, O_o\} = 0. \tag{2.51}$$

In particular, the SUSY Hamiltonian H is an even operator, whereas the supercharges Q and $Q^\dagger$ are odd operators.

Finally, we note that for $N = 1$ there does not necessarily exist a Witten operator. For this reason, some authors [14–17] include the existence of such a Witten operator in their definition of supersymmetric quantum systems. Indeed, with a given self-adjoint Q_1 and a Witten operator W satisfying $\{W, Q_1\} = 0$ one can define a second self-adjoint supercharge $Q_2 := iWQ_1$, which leads to an $N = 2$ SUSY.

2.2.2 SUSY transformation

According to the definition 2.5 we introduce the notion of positive and negative Witten-parity states.

Definition 2.6. Eigenstates of W are called *positive* and *negative (Witten-) parity states*, respectively. They are denoted by $|\psi^\pm\rangle$:

$$W|\psi^\pm\rangle = \pm|\psi^\pm\rangle. \tag{2.52}$$

For simplicity we will also call them positive and negative states.

In the matrix notation of equation (2.40) they are represented in the form

$$|\psi^+\rangle = \begin{pmatrix} |\phi^+\rangle \\ 0 \end{pmatrix}, \qquad |\psi^-\rangle = \begin{pmatrix} 0 \\ |\phi^-\rangle \end{pmatrix}, \tag{2.53}$$

where $|\phi^\pm\rangle \in \mathcal{H}^\pm$.

It is obvious that the supercharges (2.46) generate a SUSY transformation in the sense that they map negative states into positive states and vice versa:

$$Q\mathcal{H}^- \subset \mathcal{H}^+, \qquad Q^\dagger \mathcal{H}^+ \subset \mathcal{H}^-. \tag{2.54}$$

Furthermore, $Q^\dagger \mathcal{H}^- = 0 = Q\mathcal{H}^+$. This SUSY transformation can be made explicit for eigenstates of H. Note that $[W, H] = 0$ and therefore, the Hamiltonian and the Witten operator have simultaneous eigenstates.

Proposition 2.7. To each positive (negative) eigenstate $|\psi_E^+\rangle$ ($|\psi_E^-\rangle$) of the Hamiltonian H with eigenvalue $E > 0$ there exists a negative (positive) eigenstate of H with the same eigenvalue. These eigenstates are related by the *SUSY transformation*

$$\boxed{|\psi_E^-\rangle = \frac{1}{\sqrt{E}} \, Q^\dagger |\psi_E^+\rangle, \qquad |\psi_E^+\rangle = \frac{1}{\sqrt{E}} \, Q|\psi_E^-\rangle.} \tag{2.55}$$

Proof. Let, for example, $|\psi_E^-\rangle$ be a negative eigenstate of H, that is, $H|\psi_E^-\rangle = E|\psi_E^-\rangle$. Then, because of $[H, Q] = 0$, we have $HQ|\psi_E^-\rangle = QH|\psi_E^-\rangle = EQ|\psi_E^-\rangle \in \mathcal{H}^+$. Hence, the normalised vector $(1/\sqrt{E})Q|\psi_E^-\rangle$ is a positive eigenstate of H for the same eigenvalue $E > 0$.

Note that the normalisation constant may have an arbitrary phase factor which we have ignored in the above relation (2.55). This is acceptable as long as we consider bound states. In the case of scattering states this phase can no longer be ignored.

Corollary 2.8. The spectra of the two SUSY partner Hamiltonians H_+ and H_- are identical away from zero:

$$\mathrm{spec}\,(H_+)\backslash\{0\} = \mathrm{spec}\,(H_-)\backslash\{0\}. \tag{2.56}$$

If $|\phi_E^\pm\rangle \in \mathcal{H}^\pm$ denotes an eigenstate of $H_\pm$ with eigenvalue $E > 0$, that is, $H_\pm|\phi_E^\pm\rangle = E|\phi_E^\pm\rangle$, then the corresponding SUSY transformation (2.55) reads

$$\boxed{|\phi_E^-\rangle = \frac{1}{\sqrt{E}} \, A^\dagger |\phi_E^+\rangle, \qquad |\phi_E^+\rangle = \frac{1}{\sqrt{E}} \, A|\phi_E^-\rangle.} \tag{2.57}$$

In other words, the strictly positive eigenvalues of the SUSY partner Hamiltonians $H_\pm$ coincide. Operators with the property (2.56) are said to be *essential iso-spectral*

[18, 19]. Hence, the Hamiltonian of $N = 2$ SUSY quantum mechanics consists of a pair of essential iso-spectral SUSY partner Hamiltonians. For more properties on essential iso-spectral operators we refer the reader to [17, 18]. We note that the above relations (2.55) and (2.57) are also valid for continuous eigenstates. The normalisation factor $1/\sqrt{E}$ is only defined up to an unknown phase.

2.2.3 Ground-state properties for good SUSY

Until now we have considered eigenstates of the SUSY Hamiltonian H with strictly positive energies $E > 0$. For good SUSY, by definition, there exists (at least) one state in $\mathcal{H}$ with vanishing energy eigenvalue. Let us denote such a state by $|\psi_0\rangle$, that is, $H|\psi_0\rangle = 0$. Necessarily, $|\psi_0\rangle$ is annihilated by the supercharges Q and $Q^\dagger$:

$$Q|\psi_0\rangle = 0, \qquad Q^\dagger|\psi_0\rangle = 0. \tag{2.58}$$

For a negative ground state $|\psi_0^-\rangle = \begin{pmatrix} 0 \\ |\phi_0^-\rangle \end{pmatrix}$ this implies

$$A|\phi_0^-\rangle = 0, \tag{2.59}$$

whereas, a positive ground state $|\psi_0^+\rangle = \begin{pmatrix} |\phi_0^+\rangle \\ 0 \end{pmatrix}$ requires

$$A^\dagger|\phi_0^+\rangle = 0. \tag{2.60}$$

In general, the ground state $|\psi_0\rangle$ could be of positive or negative Witten parity. If the ground-state energy $E_0 = 0$ is degenerate even both types of states may occur. It should be noted, that in general, these states are not paired like those for strictly positive-energy eigenvalues. For typical spectra of a SUSY Hamiltonian in the case of good and broken SUSY see figure 2.2.

2.2.4 The Witten index

In the above we have seen that the ground-state properties of a SUSY quantum system may be of particular interest. In order to decide whether there are such *zero modes*, that is, states with zero energy, Witten [4] introduced the following quantity.

Definition 2.9. Let us denote by $n_\pm$ the number of zero modes of $H_\pm$ in the subspace $\mathcal{H}^\pm$. For finite n_+ and n_- the quantity

$$\boxed{\Delta := n_- - n_+} \tag{2.61}$$

is called the *Witten index*.

It is obvious that for $n_+ + n_- > 0$, by definition, SUSY is good. Hence, whenever the Witten index is a non-zero integer SUSY is good. However, if $\Delta = 0$ it is not clear whether SUSY is broken ($n_+ = 0 = n_-$) or not ($n_+ = n_- \neq 0$).

It should be remarked that the Witten index is related to the so-called Fredholm index [20] of the annihilation operator A characterising the supercharge Q:

$$\text{ind } A := \dim \ker A - \dim \ker A^\dagger = \dim \ker A^\dagger A - \dim \ker AA^\dagger, \qquad (2.62)$$

where $\dim \ker A$ denotes the dimension of the space spanned by the linearly independent zero modes of the operator A (the kernel of A). Clearly, equation (2.62) only makes sense if both $\dim \ker A$ and $\dim \ker A^\dagger$ are finite. Operators having this property are called Fredholm operators. The most remarkable property of the Fredholm index is its 'topological invariance' [20]. The Witten index obviously is related to the Fredholm index by

$$\Delta = \text{ind } A = \dim \ker H_- - \dim \ker H_+. \qquad (2.63)$$

For systems where definition 2.9 does not apply it has been suggested [4, 21–23] to consider so-called regularised indices:

$$\bar{\Delta}(\beta) := \text{Tr}\left(-W e^{-\beta H}\right) = \text{Tr}_-(e^{-\beta A^\dagger A}) - \text{Tr}_+(e^{-\beta AA^\dagger}),$$

$$\hat{\Delta}(z) := \text{Tr}\left(-W \frac{z}{H - z}\right) = \text{Tr}_-\left(\frac{z}{A^\dagger A - z}\right) - \text{Tr}_+\left(\frac{z}{AA^\dagger - z}\right), \qquad (2.64)$$

$$\tilde{\Delta}(\varepsilon) := \text{Tr}\left(-W \Theta(\varepsilon - H)\right) = \text{Tr}_-(\Theta(\varepsilon - A^\dagger A)) - \text{Tr}_+(\Theta(\varepsilon - AA^\dagger)),$$

where $\beta > 0$, $z < 0$ and $\varepsilon > 0$, respectively, $\text{Tr}_\pm(\cdot)$ stands for the trace in the subspaces $\mathcal{H}^\pm$, and Θ denotes the unit-step function, $\Theta(x) = 0$ for $x < 0$ and $\Theta(x) = 1$ for $x > 0$. The above indices are called the *heat kernel regularised, resolvent kernel regularised,* and *IDOS kernel regularised* index, respectively, for obvious reasons. Here IDOS stands for integrated density of states. The heat kernel regularised index has been introduced by Atiyah, Bott, and Patodi [24] for an alternative proof of the Atiyah–Singer index theorem. It has also been used by Atiyah, Padoti, and Singer [25] in their study of the spectral asymmetry of certain elliptic self-adjoint operators. For a derivation of the Atiyah–Singer index theorem based on supersymmetric quantum mechanics see [26–29].

For A being a Fredholm operator these regularised indices are independent of their argument and are identical to the Fredholm and Witten index:

$$\Delta = \bar{\Delta}(\beta) = \hat{\Delta}(z) = \tilde{\Delta}(\varepsilon) = \text{ind } A. \qquad (2.65)$$

As already mentioned, for A not being a Fredholm operator the above definition for the Witten index cannot be used. In this case there are several alternative definitions available in the literature,

$$\Delta := \lim_{\beta \to \infty} \bar{\Delta}(\beta), \qquad \Delta := \lim_{z \uparrow 0} \hat{\Delta}(z), \qquad \Delta := \lim_{\varepsilon \downarrow 0} \tilde{\Delta}(\varepsilon), \qquad (2.66)$$

whenever the quantity on the right-hand side is well defined. The heat kernel and resolvent kernel regularised indices are well studied. For more details see, for example, [16–17, 21, 30–36]. The IDOS kernel regularised index has been shown to be related to the magnetisation of a non-interacting electron gas in an arbitrary magnetic field [23]. See also section 8.2.

At this stage let us also briefly note a close relationship between the partition functions $Z_{\pm}(\beta) := \operatorname{Tr} e^{-\beta H_{\pm}}$ and the associated internal energies $U_{\pm}(\beta) := -\partial_{\beta} \ln Z_{\pm}(\beta)$ of the two partner Hamiltonians for an $N = 2$ SUSY system. Obviously, for A being Fredholm we have the relation

$$Z_-(\beta) = \Delta + Z_+(\beta), \tag{2.67}$$

but in addition it is easy to also prove the following relation between the internal energies:

$$U_-(\beta)Z_-(\beta) = U_+(\beta)Z_+(\beta). \tag{2.68}$$

2.2.5 SUSY and gauge transformations

As a last point we will mention that the $N = 2$ superalgebra (2.19), respectively (2.21), is invariant under the 'rotation'

$$\begin{pmatrix} Q'_1 \\ Q'_2 \end{pmatrix} := \begin{pmatrix} \cos\alpha & \sin\alpha \\ -\sin\alpha & \cos\alpha \end{pmatrix} \begin{pmatrix} Q_1 \\ Q_2 \end{pmatrix}, \qquad \alpha \in [0, 2\pi). \tag{2.69}$$

This invariance becomes explicit by noting that

$$Q'_1 Q'_2 = Q_1 Q_2 = -Q_2 Q_1 = -Q'_2 Q'_1, \qquad Q'^2_1 = H/2 = Q'^2_2. \tag{2.70}$$

Similarly, for the complex supercharges one obtains

$$Q' := e^{-i\alpha}Q, \qquad Q'^{\dagger} = e^{i\alpha}Q^{\dagger}, \tag{2.71}$$

which obviously obey the algebra (2.21).

It is interesting to note that this invariance of the superalgebra can be related to global gauge transformations, that is, changing the states in $\mathcal{H}^{\pm}$ by a constant phase:

$$|\psi'^{\pm}\rangle := e^{-i\beta_{\pm}}|\psi^{\pm}\rangle. \tag{2.72}$$

Note that for $\beta_+ \neq \beta_-$ the above global gauge transformations are different in the two subspaces $\mathcal{H}^+$ and $\mathcal{H}^-$, respectively. The SUSY relations (2.55) remain invariant if we perform the rotation (2.69) with a rotation angle given by $\alpha = \beta_+ - \beta_-$. Hence, the parameter α accounts for different global gauges in the two subspaces $\mathcal{H}^{\pm}$. The parameters $\beta_{\pm}$ may be used to fix the phases in the SUSY transformation relations which we had set to zero in equation (2.55).

The above invariance of equation (2.69) representing a $U(1)$ automorphism of the SUSY algebra is often referred to as *R symmetry*. The presence of such an *R* symmetry has been shown to be an essential ingredient for the dynamical supersymmetry breaking in field theories [37, 38].

2.3 Problems

Problem 2.1. The free particle on the real line

The free motion of a point mass m on the real line is characterised by the Hamiltonian $H = p^2/2m$ with momentum operator p acting on the Hilbert space $\mathcal{H} = L^2(\mathbb{R})$.

(a) Show that $Q_1 := p/\sqrt{4m}$ is a supercharge and $\{H, Q_1; \mathcal{H}\}$ forms a supersymmetric quantum system with unbroken SUSY.

(b) Let $|k\rangle$ denote the eigenstates of p, that is, $p|k\rangle = \hbar k|k\rangle$ with $k \in \mathbb{R}$ being the so-called wave number. Consider the generalised helicity operator $\Lambda := \mathrm{sgn}\, Q_1 = Q_1/|Q_1|$ and show that Λ induces a grading on $\mathcal{H}$. Why does Λ not represent a Witten operator?

Problem 2.2. SUSY transformations for $N \geqslant 2$

Consider a SUSY quantum system $\{H, Q_1, Q_2; \mathcal{H}\}$ and let $|q_1\rangle$ be an eigenstate of the self-adjoint supercharge Q_1, that is, $Q_1|q_1\rangle = q_1|q_1\rangle$. Obviously $|q_1\rangle$ is also an eigenstate of H with eigenvalue $E = 2q_1^2$. Show that for $q_1 \neq 0$ the state $Q_2|q_1\rangle$ is a simultaneous eigenstate of H and Q_1 with eigenvalues $E = 2q_1^2$ and $-q_1$, respectively.

Problem 2.3. The supersymmetric harmonic oscillator

Consider the special case $\Phi(x) = \sqrt{m/2}\,\omega x$, $\omega > 0$, of Witten's SUSY quantum mechanics discussed in section 2.1.2.

(a) Show that this system corresponds to the example discussed in section 2.2.

(b) Discuss conditions (2.58) and explicitly construct the zero-energy ground state.

(c) Show that the SUSY transformations (2.57) reduce to those discussed in equation (2.30) with $\mathcal{H}^+ = \mathcal{H}^- = L^2(\mathbb{R})$.

Problem 2.4. The Witten operator W

(a) Prove explicitly that W as defined in equation (2.32) obeys the conditions (2.31).

(b) Show that the three definitions of W given in equation (2.32) are equivalent.

(c) Verify that the operators $P^\pm$ as defined in definition 2.5 obey the relations

$$(P^\pm)^2 = P^\pm, \qquad P^\mp P^\pm = 0, \qquad P^+ + P^- = 1.$$

Hint: $QQ^\dagger Q = Q(H - QQ^\dagger) = QH = HQ$ and similar $Q^\dagger QQ^\dagger = Q^\dagger H = HQ^\dagger$.

Note: For the proof of (c) only the property $W^2 = 1$ is required. Hence, the generalised helicity Λ introduced in section 2.1.1 for $N = 1$ SUSY, which obeys $\Lambda^2 = 1$, indeed induces a grading but no SUSY transformations.

Problem 2.5. Construction of supercharges from Witten operators

Consider a SUSY system consisting of one self-adjoint supercharge Q_1 and $N - 1$ Witten operators, that is, $\{H, Q_1, W_2, W_3, \ldots W_N; \mathcal{H}\}$ obeying the algebra

$$H = 2Q_1^2, \qquad \{Q_1, W_i\} = 0, \qquad \{W_i, W_j\} = 2\delta_{ij}, \qquad i, j \in \{2, 3, \ldots, N\}.$$

Show that the supercharges defined by $Q_j := iW_jQ_1$ are self-adjoint and obey the superalgebra (2.1) of an N-extended supersymmetric system $\{H, Q_1, \ldots, Q_N; \mathcal{H}\}$.

Note: This approach was used in [10] to construct $N = 3$ and $N = 4$ realisations of the three-dimensional Pauli and Dirac Hamiltonians.

Problem 2.6. Decomposition of the most general representation of supercharges

Show for the operators as defined in equation (2.43) and in equation (2.44), with $AB = 0 = BA$, that all commutators between A- and B-operators vanish. That is, $[H_A, H_B] = 0$, $[H_A, Q_B] = 0 = [H_A, Q_B^\dagger]$, $[H_B, Q_A] = 0 = [H_B, Q_A^\dagger]$, $[Q_A, Q_B] = 0 = [Q_A^\dagger, Q_B^\dagger]$, and $[Q_A, Q_B^\dagger] = 0 = [Q_A^\dagger, Q_B]$.

Problem 2.7. Generalised fermion operators and matrix representation of $\mathcal{F}$

(a) Show that on $\mathcal{H}\backslash \ker H$ the following relations are valid:

$$b := Q^\dagger H^{-1/2} = H^{-1/2}Q^\dagger, \qquad b^\dagger = QH^{-1/2} = H^{-1/2}Q,$$

that is, Q and $H^{-1/2}$ commute. Prove $\{b, b^\dagger\} = \mathbf{1}$.

(b) Show that b and $b^\dagger$ represent normalised SUSY transformations, that is,

$$b|\psi_E^+\rangle = |\psi_E^-\rangle, \qquad b^\dagger|\psi_E^-\rangle = |\psi_E^+\rangle.$$

(c) Show for the Fermion-number operator that

$$\mathcal{F} := b^\dagger b = QQ^\dagger H^{-1} = H^{-1}QQ^\dagger = \begin{pmatrix} 1 & 0 \\ 0 & 0 \end{pmatrix}.$$

Hint: An arbitrary state $|\psi\rangle \in \mathcal{H}\backslash \ker H$ may be decomposed into eigenstates of H_+ and H_- as follows:

$$|\psi\rangle = \sum_{E>0}\left(\alpha_E^+|\psi_E^+\rangle + \alpha_E^-|\psi_E^-\rangle\right).$$

Problem 2.8. On the internal energy relation of SUSY partners

Prove the relation (2.68).

References

[1] Nicolai H 1976 Supersymmetry and spin systems *J. Phys. A: Math. Gen.* **9** 1497–506
[2] Nicolai H 1977 Extension of supersymmetric spin systems *J. Phys. A: Math. Gen.* **10** 2143–51

[3] Witten E 1981 Dynamical breaking of supersymmetry *Nucl. Phys.* B **188** 513–54

[4] Witten E 1982 Constraints on supersymmetry breaking *Nucl. Phys.* B **202** 253–316

[5] Combescure M, Gieres F and Kibler M 2004 Are $N = 1$ and $N = 2$ supersymmetric quantum mechanics equivalent? *J. Phys. A.: Math. Gen.* **37** 10385–96

[6] de Crombrugghe M and Rittenberg V 1983 Supersymmetric quantum mechanics *Ann. Phys.* **151** 99–126

[7] Pauli W 1927 Zur Quantenmechanik des magnetischen Elektrons *Z. Phys.* **43** 601–23

[8] Casalbuoni R 1976 Relativity and supersymmetries *Phys. Lett.* B **62** 49–50

[9] Feynman R P 1961 *Quantum Electrodynamics* (New York: Benjamin) p 50, problems 1 and 2

[10] Tkachuk V M and Vakarchuk S I 1996 The $N = 4$ supersymmetry of electron in magnetic field *J. Phys. Stud.* **1** 39–41 http://physics.lnu.edu.ua/jps/

[11] Tkachuk V M and Vakarchuk S I 1999 Broken supersymmetry for the electron in the magnetic field of straight current *J. Phys. Stud.* **3** 291–4 http://physics.lnu.edu.ua/jps/

[12] Freund P G O 1986 *Introduction to Supersymmetry. Cambridge Monographs on Mathematical Physics* (Cambridge: Cambridge University Press)

[13] Salomonson P and van Holten J W 1982 Fermionic coordinates and supersymmetry in quantum mechanics *Nucl. Phys.* B **196** 509–31

[14] Cycon H L, Froese R G, Kirsch W and Simon B 1987 *Schrödinger Operators with Application to Quantum Mechanics and Global Geometry* (Berlin: Springer)

[15] Grosse H and Pittner L 1987 Supersymmetric quantum mechanics defined as sesquilinear forms *J. Phys. A: Math. Gen.* **20** 4265–84

[16] Jaffe A, Lesniewski A and Lewenstein M 1987 Ground state structure in supersymmetric quantum mechanics *Ann. Phys.* **178** 313–29

[17] Thaller B 1992 *The Dirac Equation* (Berlin: Springer)

[18] Deift P A 1987 Applications of a commutation formula *Duke Math. J.* **45** 267–310

[19] Schmincke U 1978 On Schrödinger factorization method for Sturm–Liouville operators *Proc. R. Soc. Edin.* A **80** 67–84

[20] Kato T 1984 *Perturbation Theory for Linear Operators* 2nd edn (Berlin: Springer)

[21] Callias C 1978 Axial anomalies and index theorems on open spaces *Commun. Math. Phys.* **62** 213–34

[22] Weinberg E J 1979 Parameter counting for multimonopole solutions *Phys. Rev.* D **20** 936–44

[23] Junker G 1995 Recent developments in supersymmetric quantum mechanics *Turk. J. Phys.* **19** 230–48 https://arxiv.org/abs/cond-mat/9403088

[24] Atiyah M, Bott R and Patodi V K 1973 On the heat equation and the index theorem *Invent. Math.* **19** 279–330

[25] Atiyah M F, Padoti V K and Singer I M 1975 Spectral asymmetry and Riemannian geometry *Math. Proc. Camb. Philos. Soc.* **77** 43–69

[26] Alvarez-Gaumé L 1983 Supersymmetry and the Atiyah–Singer index theorem *Commun. Math. Phys.* **90** 161–73

[27] Friedan D and Windey P 1984 Supersymmetric derivation of the Atiyah–Singer index and the chiral anomaly *Nucl. Phys.* B **235** 395–16

[28] Mañes J and Zumino B 1986 WKB method SUSY quantum mechanics and the index theorem *Nucl. Phys.* B **270** 651–86

[29] Jarvis P D 1989 Supersymmetric quantum mechanics and the index theorem *Proc. Center for Mathematical Analysis* vol 22 ed M N Barber and M K Murrays (Canberra: Australian National University) pp 50–81

[30] Cecotti S and Girardello L 1983 Stochastic and parastochastic aspects of supersymmetric functional measures: a new non-perturbative approach to supersymmetry *Ann. Phys.* **145** 81–99

[31] Hirayama M 1983 Supersymmetric quantum mechanics and index theorem *Prog. Theor. Phys.* **70** 1444–53

[32] Niemi A J and Wijewardhana L C R 1984 Fractionization of the Witten index *Phys. Lett.* B **138** 389–92

[33] Akhoury R and Comtet A 1984 Anomalous behavior of the Witten index—exactly soluble models *Nucl. Phys.* B **246** 253–78

[34] Boyanovsky D and Blankenbecler R 1984 Fractional indices in supersymmetric theories *Phys. Rev.* D **30** 1821–4

[35] Bollé D, Gesztesy F, Grosse H, Schweiger W and Simon B 1987 Witten index, axial anomaly, and Krein's spectral shift function in supersymmetric quantum mechanics *J. Math. Phys.* **28** 1512–25

[36] Nakahara M 1990 *Geometry, Topology and Physics* (Bristol: IOP Publishing)

[37] Nelson A E and Seiberg N 1994 R symmetry breaking versus supersymmetry breaking *Nucl. Phys.* B **416** 46–62

[38] Abe H, Kobayashi T and Omura Y 2007 R-symmetry supersymmetry, breaking and metastable vacua in global and local supersymmetric theories *J. High Energy Phys.* **11** 044

IOP Publishing

Supersymmetric Methods in Quantum, Statistical and Solid
State Physics
Enlarged and revised edition
Georg Junker

Chapter 3

The Witten model

In the following chapter we will study a particular supersymmetric quantum model
introduced by Witten in 1981 [1]. Originally, this model was designed to serve as a
simple example for studying the dynamical SUSY-breaking mechanism in quantum
field theories. However, soon after its introduction this model found applications in
many other areas of theoretical physics. Some of these applications will be discussed
in this book.

3.1 Witten's model and its modification

The Witten model, which is an $N = 2$ extended supersymmetric quantum system, is
characterised by the two supercharges

$$Q_1 := \frac{1}{\sqrt{2}}\left(-\frac{p}{\sqrt{2m}} \otimes \sigma_2 + \Phi(x) \otimes \sigma_1\right), \quad Q_2 := \frac{1}{\sqrt{2}}\left(\frac{p}{\sqrt{2m}} \otimes \sigma_1 + \Phi(x) \otimes \sigma_2\right). \quad (3.1)$$

Note that this definition of the supercharges slightly differs from Witten's original
one (2.15) by replacing $Q_1 \to Q_2$ and $Q_2 \to -Q_1$. Actually, this replacement corresponds to a rotation (2.69) with angle $\alpha = -\pi/2$. As a second modification to Witten's
setup we also restrict the configuration space of the point mass m to a one-
dimensional subspace $\mathcal{M} \subseteq \mathbb{R}$. Here $\mathcal{M}$ either denotes the Euclidean line $\mathbb{R}$
(Witten's original model), the half line $\mathbb{R}^+ := [0, \infty)$, or a finite interval $[a, b] \subset \mathbb{R}$.
Hence, the Hilbert space is given by $\mathcal{H} = L^2(\mathcal{M}) \otimes \mathbb{C}^2$. Let us note that for the latter
two cases the supercharges (3.1) are in general symmetric but not self-adjoint. In
addition one has to impose boundary condition for the wave functions at $x = 0$ and
$x = a, b$, respectively [2, 3]. If not stated otherwise we will consider Witten's original
case, that is, $\mathcal{M} = \mathbb{R}$. However, the results derived below are also valid for the other
cases if we do not explicitly state the contrary. As before, we assume that the SUSY
potential $\Phi: \mathcal{M} \to \mathbb{R}$ is a continuous differentiable real-valued function on $\mathcal{M}$.

doi:10.1088/2053-2563/aae6d5ch3

In analogy to section 2.2 we may introduce the complex supercharge

$$Q = \begin{pmatrix} 0 & A \\ 0 & 0 \end{pmatrix}, \quad Q^\dagger = \begin{pmatrix} 0 & 0 \\ A^\dagger & 0 \end{pmatrix}, \tag{3.2}$$

where now the annihilation and creation operators are explicitly given by

$$A := \frac{ip}{\sqrt{2m}} + \Phi(x), \quad A^\dagger = -\frac{ip}{\sqrt{2m}} + \Phi(x). \tag{3.3}$$

The SUSY Hamiltonian reads

$$H = \begin{pmatrix} H_+ & 0 \\ 0 & H_- \end{pmatrix} = \begin{pmatrix} AA^\dagger & 0 \\ 0 & A^\dagger A \end{pmatrix}, \tag{3.4}$$

with *SUSY partner Hamiltonians*

$$H_\pm := \frac{p^2}{2m} + V_\pm(x) \tag{3.5}$$

and *partner potentials*

$$V_\pm(x) := \Phi^2(x) \pm \frac{\hbar}{\sqrt{2m}} \, \Phi'(x). \tag{3.6}$$

Let us also mention that the commutator and anticommutator of the creation and annihilation operators (3.3) can be related to particular parts of the Hamiltonians $H_\pm = H_{\text{tree}} \pm H_{\text{loop}}$:

$$H_{\text{tree}} := \frac{p^2}{2m} + \Phi^2(x) = \frac{1}{2}\{A, A^\dagger\}, \quad H_{\text{loop}} := \frac{\hbar}{\sqrt{2m}}\Phi'(x) = \frac{1}{2}[A, A^\dagger]. \tag{3.7}$$

The interpretation of the two parts H_{tree} and H_{loop} as 'tree Hamiltonian' and 'loop correction' (due to Grassmannian degrees of freedom) will be explained in section 5.5.

3.2 Witten parity and SUSY transformation

It might be instructive to verify some of the general properties of $N = 2$ SUSY systems discussed in section 2.2 for the Witten model. First, let us consider the Witten parity defined in equation (2.32). A simple computation shows that $[Q_1, Q_2] = iH\sigma_3$ and, hence, the Witten parity is indeed given by $W = \sigma_3$. Following the general treatment, we can grade the Hilbert space into eigenspaces of W. Obviously, the positive and negative Witten-parity subspaces are given by $\mathcal{H}^\pm = L^2(\mathcal{M})$. Here we note that for the case of $\mathcal{M} = \mathbb{R}^+$ and $\mathcal{M} = [a, b]$ the two subspaces $\mathcal{H}^\pm$ may have different boundary conditions. In many cases the positive and negative subspaces are identical, but there exist exceptions. Examples of such exceptions are given in equation (3.60) and problem 3.6 below.

Second, for completeness, we give the explicit form of the SUSY transformation (2.55), respectively, (2.57). If $|\phi_E^{\pm}\rangle$ denotes eigenstates of $H_{\pm}$ to the same eigenvalue $E > 0$, that is, $H_{\pm}|\phi_E^{\pm}\rangle = E|\phi_E^{\pm}\rangle$, the SUSY transformation explicitly reads

$$
\begin{aligned}
A|\phi_E^-\rangle &= \left(\frac{ip}{\sqrt{2m}} + \Phi(x)\right)|\phi_E^-\rangle = \sqrt{E}\,|\phi_E^+\rangle, \\
A^{\dagger}|\phi_E^+\rangle &= \left(-\frac{ip}{\sqrt{2m}} + \Phi(x)\right)|\phi_E^+\rangle = \sqrt{E}\,|\phi_E^-\rangle.
\end{aligned}
\tag{3.8}
$$

In the coordinate representation $\phi_E^{\pm}(x) := \langle x|\phi_E^{\pm}\rangle$ these relations take the form

$$
\left(\pm\frac{\hbar}{\sqrt{2m}}\frac{\partial}{\partial x} + \Phi(x)\right)\phi_E^{\mp}(x) = \sqrt{E}\,\phi_E^{\pm}(x).
\tag{3.9}
$$

Note that the factor $\sqrt{E}$ is only unique up to an arbitrary phase, which for convenience we have omitted here. In case such a phase becomes relevant we will explicitly mention that.

3.3 The SUSY potential and zero-energy states

3.3.1 Ground state for good SUSY

One of the important features of the Witten model is that it allows for a rather explicit discussion of its ground-state properties. That is, one can determine from the shape of the SUSY potential whether SUSY is broken or not. In the good-SUSY case it is even possible to give the ground-state wave function explicitly.

Let us assume that SUSY is unbroken, i.e. $E_0 = 0$, and the corresponding zero-energy eigenstate belongs to $\mathcal{H}^-$. That is, the ground state has negative Witten parity. This state necessarily satisfies equation (2.59), which in the present model reads

$$
\left(\frac{\hbar}{\sqrt{2m}}\frac{\partial}{\partial x} + \Phi(x)\right)\phi_0^-(x) = 0.
\tag{3.10}
$$

This first-order differential equation is easily integrated to

$$
\phi_0^-(x) = C\exp\left\{-\frac{\sqrt{2m}}{\hbar}\int_{x_0}^{x} dz\,\Phi(z)\right\},
\tag{3.11}
$$

where $x_0 \in \mathcal{M}$ is an arbitrary but fixed constant and $C = \phi_0^-(x_0)$ stands for a normalisation constant. Introducing the so-called *superpotential*

$$
U(x) := \frac{\sqrt{2m}}{\hbar}\int_{x_0}^{x} dz\,\Phi(z)
\tag{3.12}
$$

the ground-state wave function reads

$$\phi_0^-(x) = C \exp\{-U(x)\}. \tag{3.13}$$

Similarly, the assumption that the ground state belongs to the positive subspace $\mathcal{H}^+$ leads to

$$\phi_0^+(x) = \phi_0^+(x_0)\exp\{+U(x)\} \propto \frac{1}{\phi_0^-(x)}. \tag{3.14}$$

If SUSY is good, that is, there exists a zero-energy eigenstate for the SUSY Hamiltonian H, this state belongs either to $\mathcal{H}^-$ or $\mathcal{H}^+$. This is obvious, because if equation (3.13) is square integrable then equation (3.14) cannot be square integrable and vice versa. In any case, the eigenvalue $E_0 = 0$, if it exists, is not degenerate.

We also note that in the case of good SUSY there is a close connection between the SUSY potential, the superpotential, and the ground-state wave function [4] expressed by the relations

$$U(x) = \pm\ln\frac{\phi_0^\pm(x)}{\phi_0^\pm(x_0)}, \quad \Phi(x) = \pm\frac{\hbar}{\sqrt{2m}}\frac{\left(\phi_0^\pm\right)'(x)}{\phi_0^\pm(x)}, \tag{3.15}$$

where the upper sign is valid if the ground state belongs to $\mathcal{H}^+$ and the lower one if it belongs to $\mathcal{H}^-$, respectively.

3.3.2 An additional symmetry

As the reader may have realised, the two SUSY partner Hamiltonians $H_\pm$ differ only by an overall sign in the SUSY potential. In other words, changing the sign of Φ simply replaces H_+ by H_- and vice versa. Hence, the Witten model is invariant under the simultaneous replacements

$$\Phi \to \tilde{\Phi} := -\Phi, \qquad H_\pm \to \tilde{H}_\pm := H_\mp, \qquad H \to \tilde{H} := \begin{pmatrix} \tilde{H}_+ & 0 \\ 0 & \tilde{H}_- \end{pmatrix}. \tag{3.16}$$

The above transformation implies

$$Q_1 \to \tilde{Q}_1 := \frac{1}{\sqrt{2}}\left(-\frac{p}{\sqrt{2m}} \otimes \sigma_2 - \Phi(x) \otimes \sigma_1\right)$$
$$Q_2 \to \tilde{Q}_2 := \frac{1}{\sqrt{2}}\left(\frac{p}{\sqrt{2m}} \otimes \sigma_1 - \Phi(x) \otimes \sigma_2\right) \tag{3.17}$$

and hence the superalgebra remains invariant, that is,

$$\{\tilde{Q}_i, \tilde{Q}_j\} = \tilde{H}\delta_{ij}. \tag{3.18}$$

However, we note that the transformation (3.17) of the supercharges cannot be generated via a rotation (2.69). Therefore, this is an additional symmetry which we have already encountered in the general discussion of section 2.2.1. To be more explicit let us denote the SUSY annihilation operator with the inverted SUSY potential by B, that is, $B = ip/\sqrt{2m} - \Phi(x)$ then we obviously have

$$\tilde{H} = \begin{pmatrix} H_- & 0 \\ 0 & H_+ \end{pmatrix} = \begin{pmatrix} B^\dagger B & 0 \\ 0 & BB^\dagger \end{pmatrix}. \tag{3.19}$$

This is actually the alternative realisation of an $N = 2$ SUSY as discussed in equation (2.44).

Within the Witten model this invariance can be used to fix the overall sign of the SUSY potential by some convention. The standard convention is to choose the overall sign of Φ such that for good SUSY the unique ground state belongs to $\mathcal{H}^-$. That is, the ground state is a negative state (bosonic vacuum). If SUSY is broken the overall sign for Φ may be chosen arbitrarily. As a consequence of this convention we have the following spectral properties of the partner Hamiltonians:

$$\begin{aligned} \mathrm{spec}(H_-)\backslash\{0\} &= \mathrm{spec}(H_+) \quad \text{for good SUSY,} \\ \mathrm{spec}(H_-) &= \mathrm{spec}(H_+) \quad \text{for broken SUSY.} \end{aligned} \tag{3.20}$$

For broken SUSY, both Hamiltonians H_- and H_+ have identical spectra whereas for good SUSY H_- and H_+ are only essential iso-spectral. If not explicitly mentioned otherwise we will work within the convention (3.20) from now on.

3.3.3 Asymptotic behaviour of the SUSY potential

All properties of the Witten model are exclusively determined by the SUSY potential Φ, respectively its integral, the superpotential U. In this section we will show how the asymptotic behaviour of Φ and U, respectively, provide an answer to the question whether SUSY is good or broken.

The requirement for a good SUSY is the existence of a normalisable zero-energy wave function $\phi_0^-(x)$ (because of our convention). Thus we have to impose the following condition on the superpotential U:

$$\int_{\mathcal{M}} \mathrm{d}x \, \exp\{-2U(x)\} < \infty. \tag{3.21}$$

In other words, the superpotential has to diverge fast enough for $x \to \pm\infty$,

$$U(x) \to +\infty \quad \text{for} \quad x \to \pm\infty. \tag{3.22}$$

Note that we will discuss only the case $\mathcal{M} = \mathbb{R}$. The discussion for the other cases $\mathcal{M} = \mathbb{R}^+$ and $\mathcal{M} = [a, b]$ are similar. There the limit $x \to -\infty$ is to be replaced by $x \to 0$ and $x \to a$, respectively. The limit $x \to \infty$ is the same for the case $\mathcal{M} = \mathbb{R}^+$ but is to be replaced by $x \to b$ in the case $\mathcal{M} = [a, b]$.

Let us explore what could be meant by fast enough. For this, we make the ansatz

$$U(x) \sim a_\pm |x|^{\alpha_\pm} \quad \text{for} \quad x \to \pm\infty. \tag{3.23}$$

Clearly, for $a_\pm > 0$ and $\alpha_\pm > 0$ the integral (3.21) exists. Hence, SUSY is good for any superpotential which diverges algebraically to $+\infty$ as $x \to \pm\infty$. The behaviour (3.23) implies for the SUSY potential the asymptotic form

$$\Phi(x) = \frac{\hbar}{\sqrt{2m}} U'(x) \sim \frac{\hbar a_\pm \alpha_\pm}{\sqrt{2m}} \frac{|x|^{\alpha_\pm}}{x} \quad \text{for} \quad x \to \pm\infty. \tag{3.24}$$

In particular, a sufficient condition for good SUSY is $\Phi(x) \to \pm\infty$ as $x \to \pm\infty$ (i.e. $\alpha_\pm > 1$). See figure 3.1 for the example $U(x) = x^4/4$. In contrast, a SUSY potential which for $x \to \pm\infty$ behaves, for example, like $\Phi(x) \to +\infty$ necessarily leads to a broken SUSY. See figure 3.2 for the example $U(x) = x^3/3$.

However, even a logarithmic divergence of U at infinity can lead to a good SUSY. Making the ansatz

$$U(x) \sim b_\pm \ln|x| \quad \text{for} \quad x \to \pm\infty, \tag{3.25}$$

the square integrability of $\phi_0^-(x)$ requires $b_\pm > \frac{1}{2}$. For the SUSY potential this leads to

$$\Phi(x) \sim \frac{\hbar b_\pm}{\sqrt{2m}} \frac{1}{x} \quad \text{for} \quad x \to \pm\infty, \tag{3.26}$$

which may be considered as a limiting case of equation (3.24) where $\alpha_\pm \to 0$, $a_\pm \to \infty$ such that $b_\pm = \alpha_\pm a_\pm > \frac{1}{2}$.

Let us mention that the conditions $a_\pm > 0$ and $b_\pm > \frac{1}{2}$ are a direct consequence of our convention. If we want to have a positive-parity zero-energy state these conditions change to $a_\pm < 0$ and $b_\pm < -\frac{1}{2}$. In figure 3.3 we display the regimes of the parameters $a_\pm$ and $\alpha_\pm$ which give rise to broken and good SUSY, respectively.

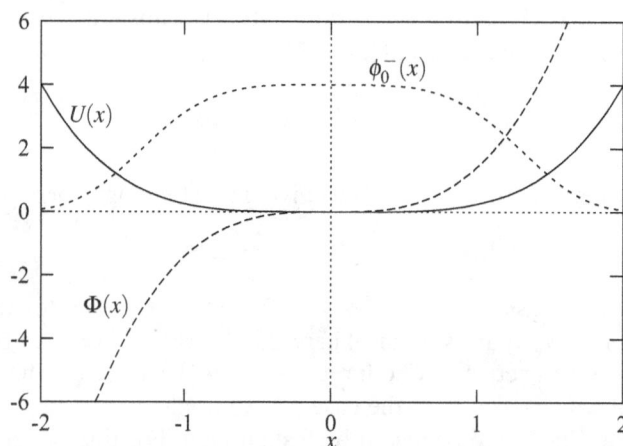

Figure 3.1. A typical example for good SUSY. Shown are the superpotential $U(x) = x^4/4$, the SUSY potential $\Phi(x)$, and the corresponding normalisable ground-state wave function $\phi_0^-(x) = Ce^{-U(x)}$ in units $\hbar = m = C/4 = 1$.

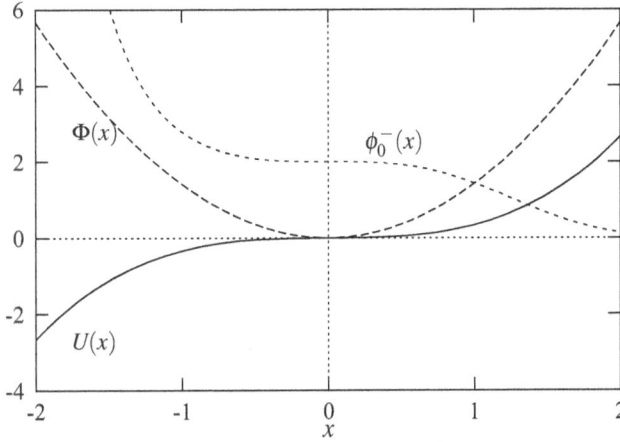

Figure 3.2. An example for broken SUSY. Again the superpotential $U(x) = x^3/3$, the SUSY potential $\Phi(x)$, and the corresponding non-normalisable function $\phi_0^-(x) = Ce^{-U(x)}$ are shown ($\hbar = m = C/2 = 1$).

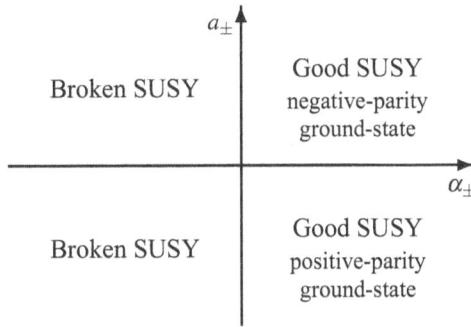

Figure 3.3. Whether SUSY is broken or not is determined by the asymptotic behaviour of the superpotential.

3.4 Broken versus good SUSY

In the above discussion we have demonstrated that the breaking of SUSY is completely determined by the asymptotic behaviour of the SUSY potential Φ. In fact, more generally, the existence of a zero-energy state is obtainable from the quantities

$$\Phi_{\pm} := \lim_{x \to \pm\infty} \Phi(x) \in [-\infty, +\infty]. \tag{3.27}$$

Defining the sign function

$$\mathrm{sgn}(z) := \begin{cases} +1 & \text{for} \quad 0 < z \leqslant +\infty \\ -1 & \text{for} \quad 0 > z \geqslant -\infty \end{cases} \tag{3.28}$$

it is obvious that for $\Phi_{\pm} \neq 0$, this corresponds to $\alpha_{\pm} \geqslant 1$ if Φ has an asymptotic behaviour as in equation (3.24), we have the equivalence relations

$$\text{sgn}(\Phi_+) = \text{sgn}(\Phi_-) \iff \text{SUSY is broken,}$$
$$\text{sgn}(\Phi_+) = -\text{sgn}(\Phi_-) \iff \text{SUSY is good.} \tag{3.29}$$

The asymptotic values $\Phi_{\pm}$ also allow for an explicit representation of the Witten index:

$$\Delta = \frac{1}{2}[\text{sgn}(\Phi_+) - \text{sgn}(\Phi_-)]. \tag{3.30}$$

This result shows that the Witten index does not depend on the details of the SUSY potential Φ. As long as the asymptotic behaviour of Φ does not change the sign, that is, $\Phi_{\pm}$ do not change sign, the Witten index is invariant under deformations of the SUSY potential. This is an example of the 'topological invariance' of Δ mentioned in section 2.2.4.

It should be noted that in the case where one or both of the values $\Phi_{\pm}$ vanish the above criterion (3.29) does not apply. Here more information about the asymptotic behaviour of Φ is necessary. See, for example, the discussion of equation (3.26).

There exist also other criteria for good and broken SUSY, which of course are equivalent to the above. For example, it has already been noted by Witten [1] that for a continuous differentiable SUSY potential with $|\Phi_{\pm}| > 0$ SUSY will be broken if Φ has an even number of zeros (counted with their multiplicity). On the other hand, SUSY will be good if Φ has an odd number of zeros. In particular, if Φ is a polynomial of degree p, this number determines whether SUSY is broken (p even) or not (p odd) [5].

Another interesting criterion for broken and good SUSY can be found if the SUSY potential Φ has a definite space parity [6]. That is, it is an eigenfunction of the space-parity operator Π,

$$\Pi\Phi(x) := \Phi(-x) = \pm\Phi(x), \tag{3.31}$$

which should not be confused with the Witten-parity operator. Obviously, the operator Π commutes with the tree Hamiltonian H_{tree} defined in equation (3.7)

$$[\Pi, H_{\text{tree}}] = 0. \tag{3.32}$$

Now, if Φ has even parity, which implies a broken SUSY, then the loop correction H_{loop}, which is proportional to the derivative of Φ, has odd parity and, hence, anticommutes with Π,

$$[\Pi, \Phi] = 0 \implies \{\Pi, H_{\text{loop}}\} = 0. \tag{3.33}$$

As a consequence, both Hamiltonians $H_{\pm} = H_{\text{tree}} \pm H_{\text{loop}}$ do not have a well-defined parity. For an even Φ parity as well as SUSY are broken.

If, however, Φ has odd parity then

$$\{\Pi, \Phi\} = 0 \implies [\Pi, H_{\text{loop}}] = 0 \implies [\Pi, H_{\pm}] = 0 \tag{3.34}$$

and parity as well as SUSY are good symmetries.

The asymptotic values $\Phi_\pm$ allow also for a discussion of the excited energy eigenvalues of H. For example, if both $|\Phi_+|$ and $|\Phi_-|$ are equal to $+\infty$ then the two SUSY partner Hamiltonians H_+ and H_- have a purely discrete spectrum. In contrast, if one or both of $|\Phi_\pm|$ are finite (including zero) then the SUSY partner Hamiltonians have, in addition to a possible discrete spectrum, identical continuous spectra starting at $\varepsilon_c := \min(\Phi_+^2, \Phi_-^2)$. In particular, for $\varepsilon_c = 0$ both Hamiltonians $H_\pm$ have a continuous spectrum given by the half line $\mathbb{R}^+$. In addition, if SUSY is good, one of them also has a bound state with eigenvalue $E_0 = \varepsilon_c = 0$.

3.5 SUSY transformations for continuum states

Let us assume that the asymptotic limits $\Phi_\pm$ of the SUSY potential, as defined in equation (3.27), are finite. Then for large $x \to \pm\infty$ the partner potentials also have a finite limit as

$$\lim_{x \to -\infty} V_\pm(x) = \Phi_-^2 \quad \text{and} \quad \lim_{x \to \infty} V_\pm(x) = \Phi_+^2. \qquad (3.35)$$

Note that for physical SUSY potentials having finite asymptotic values it is reasonable to assume that $\lim_{x \to \pm\infty} \Phi'(x) = 0$. Obviously eigenstates $\phi_E^\pm$ of the SUSY partner Hamiltonians for energy values $E > \max\{\Phi_-^2, \Phi_+^2\}$ are unbounded and thus allow for reflections and transmissions. Hence a particle with energy E approaching the potentials from the far left is described by a right-moving plan wave $\exp\{ik_L x\}$ with wave number

$$k_L(E) := \frac{1}{\hbar}\sqrt{2m(E - \Phi_-^2)} \qquad (3.36)$$

and will be reflected with an amplitude $R(E)$ and a phase $\exp\{-ik_L x\}$ representing a left-moving plane wave. The transmission results in a right-moving plane wave $\exp\{ik_R x\}$ with amplitude $T(E)$. Here the wave number is given by

$$k_R(E) := \frac{1}{\hbar}\sqrt{2m(E - \Phi_+^2)}. \qquad (3.37)$$

See figure 3.4 where the example $\Phi(x) = \tanh x + 1/2$ is shown indicating the incoming, reflected, and transmitted plane wave.

In summary we have

$$\phi_E^\pm(x) \sim \exp\{ik_L(E)x\} + R^\pm(E)\exp\{-ik_L(E)x\} \quad \text{for} \quad x \to -\infty \qquad (3.38)$$

$$\phi_E^\pm(x) \sim T^\pm(E)\exp\{ik_R(E)x\} \quad \text{for} \quad x \to \infty. \qquad (3.39)$$

Utilising the SUSY transformation (3.9) we obtain the asymptotic relation

$$C\left(\frac{\hbar}{\sqrt{2m}}\frac{\partial}{\partial_x} + \Phi_\pm\right)\Phi_E^-(x) = \Phi_E^+(x) \quad \text{for} \quad x \to \pm\infty. \qquad (3.40)$$

Note that now we will also consider phase factors for normalisation constant C, which in equation (3.9) was fixed to $C = 1/\sqrt{E}$. Equating the coefficients of the left-

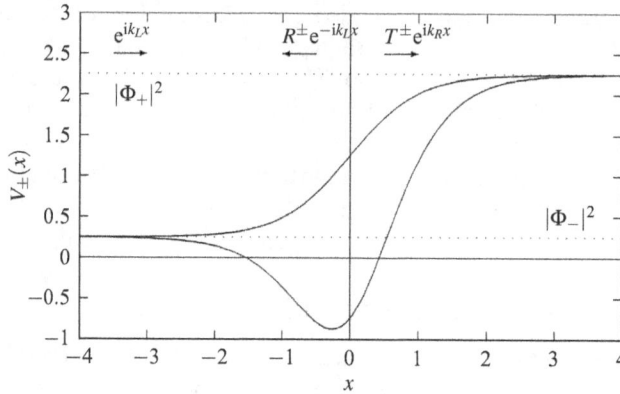

Figure 3.4. An example of reflection and transmission for SUSY partner potentials. Here $\Phi(x) = \tanh(x) + 1/2$ and $2m = \hbar = 1$. Obviously $V_+(x)$ does not have any bound states, whereas $V_-(x)$ has exactly one bound state at $E = 0$ as SUSY is unbroken.

and right-moving wave and eliminating the unknown constant C then leads to a relation between the reflection and transmission amplitudes of the partner potentials,

$$R^+(E) = \frac{\Phi_- - i\sqrt{E - \Phi_-^2}}{\Phi_- + i\sqrt{E - \Phi_-^2}} R^-(E), \qquad (3.41)$$

$$T^+(E) = \frac{\Phi_+ + i\sqrt{E - \Phi_+^2}}{\Phi_- + i\sqrt{E - \Phi_-^2}} T^-(E), \qquad (3.42)$$

From this it obviously follows that $|R^+(E)|^2 = |R^-(E)|^2$ and $|T^+(E)|^2 = |T^-(E)|^2$. In addition, in the case that one of the partner potentials is reflectionless, say $R^+(E) = 0$, then it immediately follows that the partner is also reflexionless, that is, $R^-(E) = 0$. In the next section we will show that the Rosen–Morse potential with a particular potential strength is indeed the SUSY partner of the free particle on the real line and thus has a vanishing reflection coefficient.

Finally we remark that in problem 3.2 below the reader is invited to discuss the case $\Phi_-^2 < E < \Phi_+^2$. That is, the incoming particle from the left has an energy being below the potential barrier and thus no transmission takes place.

3.6 Examples

3.6.1 Systems on the Euclidean line

As a first class of examples for the Witten model we consider the SUSY quantum mechanics of a Cartesian degree of freedom on the Euclidean line $\mathcal{M} = \mathbb{R}$. The corresponding Hilbert subspaces $\mathcal{H}^\pm := L^2(\mathbb{R})$ are identical.

Example 3.1. Supersymmetric anharmonic oscillator
This system is characterised by the SUSY potential

$$\Phi(x) := \frac{a\hbar}{\sqrt{2m}} \, \text{sgn}(x)|x|^{\alpha-1}, \quad \alpha > 1, \quad a > 0, \quad (3.43)$$

which is an odd function of its argument. Hence, by the reasoning of the previous section we expect SUSY to be good. It may easily be verified that the partner potentials are given by

$$V_\pm(x) = \frac{\hbar^2}{2m}[a^2|x|^{2\alpha-2} \pm a(\alpha - 1)|x|^{\alpha-2}] = V_\pm(-x) \quad (3.44)$$

and the ground-state wave function reads

$$\phi_0^-(x) = C \exp\left\{-\frac{a}{\alpha}|x|^\alpha\right\}, \quad (3.45)$$

with normalisation constant $|C|^2 = (\alpha/2)(2a/\alpha)^{1/\alpha}/\Gamma(1/\alpha)$. In figure 3.5 we plot for various values of the parameter α the potential V_- and V_+. Note that $\alpha = 2$ is the special case of the supersymmetric harmonic oscillator and that for large α the potential V_- acquires narrow and deep dips near $x = \pm 1$. In fact, in the limit $\alpha \to \infty$ these dips become singular and lead to stationary wave functions with vanishing slope at $x = \pm 1$. In other words, they simulate Neumann boundary conditions, $\phi'(\pm 1) = 0$. For the partner potential V_+ the limit $\alpha \to \infty$ produces Dirichlet conditions at $x = \pm 1$, that is, $\phi(\pm 1) = 0$. Indeed the supersymmetric square well, as discussed in example 3 below section 3.6.3, exhibits a SUSY structure when taking Neumann and Dirichlet boundary conditions at $x = \pm 1$, respectively.

Example 3.2. Anharmonic oscillator with broken SUSY
Dropping the sign function in equation (3.43) we obtain the SUSY potential

$$\Phi(x) := \frac{a\hbar}{\sqrt{2m}} \, |x|^{\alpha-1}, \quad a > 0, \quad \alpha > 1, \quad (3.46)$$

which now is an even function giving rise to a broken SUSY. The corresponding partner potentials read

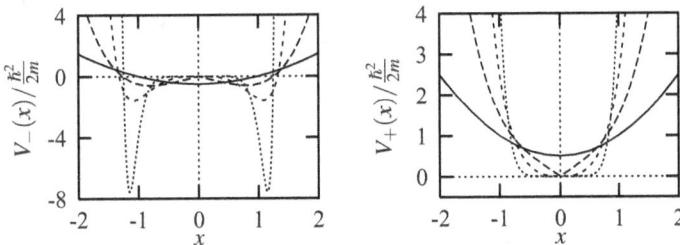

Figure 3.5. The two partner potentials V_- and V_+ for the good-SUSY potential (3.43) for $a = 1$. Displayed are the potentials for $\alpha = 2$ (——), $\alpha = 3$ (– – –), $\alpha = 5$ (- - - -) and $\alpha = 10$ (······).

$$V_\pm(x) = \frac{\hbar^2}{2m}[a^2|x|^{2\alpha-2} \pm a(\alpha - 1)\mathrm{sgn}(x)|x|^{\alpha-2}]. \qquad (3.47)$$

Note that $V_\pm(-x) = V_\mp(x)$ and, hence, it is not surprising that the two partner Hamiltonians have identical spectra as expected from broken SUSY. We have plotted V_- in figure 3.6 for the same parameters as in the previous example.

Note that in addition to SUSY, parity is also broken. Here in the limit $\alpha \to \infty$ the potential V_- simulates at $x = -1$ a Dirichlet condition and at $x = +1$ a Neumann condition (and vice versa for V_+).

Example 3.3. The free particle and its SUSY partner
As a third example we consider the SUSY potential

$$\Phi(x) := \frac{\hbar}{\sqrt{2m}} \tanh(x), \qquad (3.48)$$

which gives rise to the two partner potentials

$$V_+(x) = \frac{\hbar^2}{2m}, \quad V_-(x) = \frac{\hbar^2}{2m}\left[1 - \frac{2}{\cosh^2 x}\right]. \qquad (3.49)$$

Again SUSY is good as $\Phi_+ = -\Phi_-$, cf. equation (3.29). The potential V_+ is constant and hence characterises a free particle. There are no bound states for H_+ and therefore, because of good SUSY, H_- has a single bound state with zero energy. The corresponding wave function reads

$$\phi_0^-(x) = \frac{\sqrt{2}}{\cosh(x)}. \qquad (3.50)$$

The potential V_- is an example of the so-called symmetric Rosen–Morse potentials. The Schrödinger equation for this type of potentials is studied, for example, in connection with soliton solutions of the Korteweg–de Vries equation [7] or relaxation processes in Fermi liquids [8]. Following our discussion in the previous section it is obvious that the potential V_- is reflectionless due to the reflectionlessness

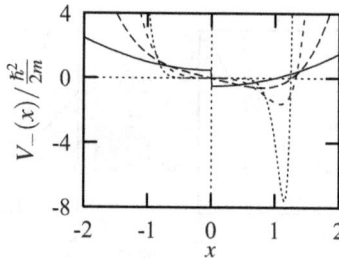

Figure 3.6. The same as figure 3.5 but for the broken-SUSY potential (3.46). Here only V_- is plotted as $V_+(x) = V_-(-x)$.

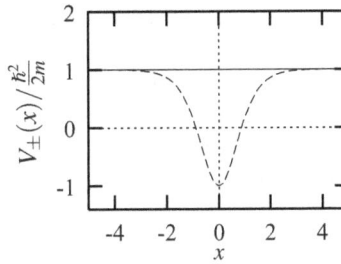

Figure 3.7. The free particle on the real line (——) and its SUSY partner (– – –).

of its SUSY partner V_+, the constant potential [9]. In figure 3.7 we have plotted both potentials.

Example 3.4. Attractive and repulsive δ-potential
As a last example on the Euclidean line we mention the attractive and repulsive δ-potential. For this we consider the SUSY potential of the form

$$\Phi(x) := \frac{a\hbar}{\sqrt{2m}}[\Theta(x) - \Theta(-x)], \quad a > 0. \tag{3.51}$$

Note that this SUSY potential is not continuously differentiable, an assumption we have made so far for all SUSY potentials. Nevertheless, interpreting the derivative of the unit-step function in a distributional sense, that is, $\Theta'(x) = \delta(x)$, we arrive at

$$V_\pm(x) = \frac{\hbar^2}{2m}[a^2 \pm 2a\delta(x)], \quad \phi_0^-(x) = \sqrt{a}\,\exp\{-a|x|\}. \tag{3.52}$$

The attractive and repulsive δ-potential are SUSY partners. Note that SUSY is good as the bound state associated with V_- has zero energy. This model can also be considered as the limiting case $\alpha = 1$ of our first example.

3.6.2 Systems on the half line

The second class of examples which we are going to mention, has as configuration space the positive half line $\mathcal{M} = \mathbb{R}^+$, that is, $x \geqslant 0$. Here the partner Hamiltonians $H^\pm$ are formal differential operators and require a careful specification of their domains $\mathcal{H}^\pm$, which in fact depend on the explicit form of the potentials $V_\pm$. In the following we will consider the Hilbert space with Dirichlet boundary condition at $x = 0$,

$$\mathcal{H}^\pm := \{\phi \in L^2(\mathbb{R}^+): \phi(0) = 0\}, \tag{3.53}$$

and only those potentials $V_\pm$ for which $H_\pm$ will be self-adjoint on $\mathcal{H}^\pm$. We note that on $\mathcal{H}^\pm$ the momentum operator p is symmetric but not self-adjoint [2]. Hence, the supercharges Q_1 and Q_2 are also symmetric but in general not self-adjoint. However, $p^\dagger p$ is still self-adjoint on $\mathcal{H}^\pm$ and therefore, also $H_\pm$ may be well-defined partner Hamiltonians if the SUSY potential Φ is chosen appropriately. For example, a singular SUSY potential which behaves like $\Phi(x) \sim -\eta/x$ near $x = 0$ requires a careful discussion [10–12] showing that not all $\eta \in \mathbb{R}$ are admissible if the Dirichlet condition in equation (3.53) would have been omitted. If such care is not taken, then one may find negative energy eigenstates or unpaired positive-energy eigenvalues for $H_\pm$ [13, 14], which are a direct consequence of the non-self-adjointness of the supercharges. Below, for each example, we simply list the SUSY potential, the partner potentials, the ground-state wave function, and then make some comments.

Example 3.5. Radial harmonic oscillator

$$\Phi(x) := \sqrt{\frac{m}{2}}\,\omega x - \frac{\hbar}{\sqrt{2m}}\frac{\eta}{x}, \quad \omega > 0,$$

$$V_\pm(x) = \frac{m}{2}\,\omega^2 x^2 + \frac{\hbar^2 \eta(\eta \pm 1)}{2mx^2} - \hbar\omega\left(\eta \mp \frac{1}{2}\right), \tag{3.54}$$

$$\phi_0^-(x) = Cx^\eta \exp\left\{-\frac{m\omega}{2\hbar}x^2\right\}, \quad |C|^2 = \left(\frac{m\omega}{\hbar}\right)^{\eta+1/2}\frac{2}{\Gamma(\eta + 1/2)}.$$

Obviously, for $\eta > 0$ SUSY is good. In particular, if we set $\eta = l + 1, l \in \mathbb{N}_0$, the two potentials $V_\pm$ are identical to those of the radial harmonic oscillator for fixed angular momentum l and $l + 1$, respectively. The fact that the two subspaces with fixed orbital angular momentum quantum number differing by one unit forming SUSY partners becomes transparent when we discuss the Pauli Hamiltonian for spherically symmetric potentials in section 8.3.

On the other hand, for $\eta \leqslant 0$ we have a broken SUSY. Despite the fact that $\eta \in (-1/2, 0]$ does lead to a square integrable ground-state wave function, it does not obey the boundary condition $\phi_0^-(0) = 0$. Because of this boundary condition the partner Hamiltonian associated with the above potentials $V_\pm$ are indeed self-adjoint. In fact, we have chosen a particular self-adjoint extension of the formal operators $H_\pm$ by equation (3.53). This is true for any $\eta \in \mathbb{R}$ because $\eta(\eta \pm 1) \geqslant -1/4$ [15]. In the broken case $\eta \leqslant 0$ we point out that for $\eta = -l$ we again recover (up to some constant) the same two radial harmonic oscillators with fixed angular momentum l and $l - 1$. Hence, the radial harmonic oscillator is a supersymmetric quantum system with good or broken SUSY, depending on the choice of the parameter η.

Example 3.6. The modified Pöschl–Teller problem

$$\Phi(x) := \frac{\hbar}{\sqrt{2m}}[\kappa \tanh x - \eta \coth x], \quad \kappa > 0,$$

$$V_\pm(x) = \frac{\hbar^2}{2m}\left[\frac{\eta(\eta \pm 1)}{\sinh^2 x} - \frac{\kappa(\kappa \pm 1)}{\cosh^2 x} + (\kappa - \eta)^2\right], \tag{3.55}$$

$$\phi_0^-(x) = C\frac{\sinh^\eta x}{\cosh^\kappa x}, \quad |C|^2 = \frac{2\Gamma(\kappa + 1/2)}{\Gamma(\eta + 1/2)\Gamma(\kappa - \eta)}.$$

The above potentials $V_\pm$ are called modified Pöschl–Teller potentials. Here SUSY is good if $0 < \eta < \kappa$. As in the above example the partner potentials are up to an additive constant invariant under the replacement $\eta \to -\eta$. What does change is the nature of SUSY, that is, good SUSY becomes broken and vice versa.

Example 3.7. Radial hydrogen atom

$$\Phi(x) := \frac{\sqrt{2m}}{\hbar} \frac{\alpha}{2\eta} - \frac{\hbar}{\sqrt{2m}} \frac{\eta}{x}, \quad \alpha, \eta > 0,$$

$$V_{\pm}(x) = -\frac{\alpha}{x} + \frac{\hbar^2 \eta(\eta \pm 1)}{2mx^2} + \frac{m\alpha^2}{2\eta^2 \hbar^2}, \tag{3.56}$$

$$\phi_0^-(x) = Cx^\eta \exp\left\{ -\frac{m\alpha}{\eta\hbar^2} x \right\}.$$

Note that because of our convention (3.20) we have to impose the condition $\eta > 0$. For $\eta = l + 1$ we recognise the radial hydrogen problem. See also section 8.3.3 where this system is derived from the three-dimensional Pauli Hamiltonian. Supersymmetric aspects of the one-dimensional hydrogen atom are discussed in [16]. There has been a long-lasting discussion about the spectral properties of the one-dimensional hydrogen atom in the literature. For clarification see [17].

3.6.3 Systems on a finite interval

As a last class of examples for the generalised Witten model we consider those which are defined on a finite interval $x \in \mathcal{M} = [a, b]$. Here, in contrast to the previous class, the momentum operator and hence the supercharges can be made self-adjoint if appropriate boundary conditions are specified at $x = a$ and b.

Example 3.8. Infinite square well and its SUSY partner
The first example we mention is the infinite square well of width π chosen, for convenience, to be symmetric about the origin:

$$\Phi(x) := \frac{\hbar}{\sqrt{2m}} \tan x, \quad x \in [-\pi/2, \pi/2],$$

$$V_+(x) = \frac{\hbar^2}{2m}\left(\frac{2}{\cos^2 x} - 1 \right), \quad V_-(x) = -\frac{\hbar^2}{2m}, \tag{3.57}$$

$$\phi_0^-(x) = \sqrt{2/\pi} \cos x,$$

with domain $\mathcal{H}^{\pm} := \{\phi \in L^2([-\pi/2, \pi/2]): \phi(-\pi/2) = 0 = \phi(\pi/2)\}$. Hence, the SUSY partner for the particle in a box represented by V_- is given by the symmetric Pöschl–Teller potential V_+. SUSY in this case is a good symmetry. Note that the square well V_- is shifted such that $E_0 = 0$. This model appears, for example, when studying static quarks in (1+1) dimensions interacting via an $SU(2)$ Yang–Mills field [18].

Example 3.9. The non-symmetric Pöschl–Teller oscillator

$$\Phi(x) := \frac{\hbar}{\sqrt{2m}}(\kappa \tan x - \eta \cot x), \quad \kappa > 0, \quad x \in [0, \pi/2],$$

$$V_{\pm}(x) = \frac{\hbar^2}{2m}\left[\frac{\kappa(\kappa \pm 1)}{\cos^2 x} + \frac{\eta(\eta \pm 1)}{\sin^2 x} - (\kappa + \eta)^2\right], \tag{3.58}$$

$$\phi_0^-(x) = C \cos^\kappa x \sin^\eta x, \quad |C|^2 = \frac{2\Gamma(\kappa + \eta + 1)}{\Gamma(\kappa + 1/2)\Gamma(\eta + 1/2)}$$

with domain $\mathcal{H}^{\pm} := \{\phi \in L^2([0, \pi/2]): \phi(0) = 0 = \phi(\pi/2)\}$. Obviously, the non-symmetric Pöschl–Teller oscillators with parameter set (κ, η) and $(\kappa - 1, \eta - 1)$ are SUSY partners. SUSY is good for $\eta > 0$ and broken for $\eta \leqslant 0$.

Example 3.10. The supersymmetric square well
Here we assume a vanishing SUSY potential in the configuration space $\mathcal{M} := [-1, 1]$ and consequently also the partner potentials vanish on $\mathcal{M}$

$$\Phi(x) := 0, \quad V_{\pm}(x) = 0. \tag{3.59}$$

The SUSY structure is introduced via the boundary conditions at $x = \pm 1$:

$$\begin{aligned}
\mathcal{H}^+ &:= \{\phi \in L^2([-1, 1]): \phi(-1) = 0 = \phi(1)\}, \\
\mathcal{H}^- &:= \{\phi \in L^2([-1, 1]): \phi'(-1) = 0 = \phi'(1)\}.
\end{aligned} \tag{3.60}$$

That is, for $H_+ = p^2/2m$ we choose Dirichlet conditions at $x = \pm 1$. Whereas, for $H_- = p^2/2m$ we take Neumann conditions at those points. These imply a good SUSY as the ground-state wave function is constant,

$$\phi_0^-(x) = 1/\sqrt{2}. \tag{3.61}$$

This example can be considered as the limiting case $\alpha \to \infty$ of equation (3.43), the supersymmetric anharmonic oscillator.

3.7 Problems

Problem 3.1. Real and complex supercharges of the Witten model

(a) Show that the real supercharges as defined in equation (3.1) can be represented in terms of the generalised annihilation operator A defined in equation (3.3) and its adjoint as follows:

$$Q_1 = \frac{1}{\sqrt{2}}\begin{pmatrix} 0 & A \\ A^\dagger & 0 \end{pmatrix}, \quad Q_2 = \frac{1}{\sqrt{2}}\begin{pmatrix} 0 & -iA \\ iA^\dagger & 0 \end{pmatrix}.$$

(b) Show that $[Q_1, Q_2] = iH \otimes \sigma_3$ and verify that the complex supercharges (3.2) are given by $Q = (Q_1 + iQ_2)/\sqrt{2}$ and $Q^\dagger = (Q_1 - iQ_2)/\sqrt{2}$.

(c) Show that the transformed supercharges (3.17) can be represented as follows:

$$\tilde{Q}_1 = \frac{1}{\sqrt{2}}\begin{pmatrix} 0 & -A^\dagger \\ -A & 0 \end{pmatrix}, \quad \tilde{Q}_2 = \frac{1}{\sqrt{2}}\begin{pmatrix} 0 & iA^\dagger \\ -iA & 0 \end{pmatrix}$$

and therefore the transformation $Q_i \to \tilde{Q}_i$ is equivalent to $A \to -A^\dagger$.

Problem 3.2. SUSY transformation for continuum states revisited

Following the discussion of section 3.5 consider now the case $\Phi_-^2 < E < \Phi_+^2$. That is, for $x \to -\infty$ the eigenfunction associated with energy eigenvalue E is still described by equation (3.38), however, equation (3.39) is now replaced by an exponentially decaying function

$$\phi_E^\pm(x) \sim T^\pm(E)\exp\{-\kappa_R(E)x\} \quad \text{for} \quad x \to \infty.$$

(a) Show that the inverse penetration length is given by

$$\kappa_R(E) = \frac{1}{\hbar}\sqrt{2m(\Phi_+^2 - E)}$$

and the corresponding amplitudes for the two partner potentials are related by

$$T^+(E) = \frac{\Phi_+ - \sqrt{\Phi_+^2 - E}}{\Phi_- + i\sqrt{E - \Phi_-^2}} T^-(E).$$

(b) Prove that $|T^+(E)|^2 < |T^-(E)|^2$ for $\Phi_-^2 < E < \Phi_+^2$.

Problem 3.3. Continuum states of the free particle and its SUSY partner

Consider example 3.3 of section 3.6.1 and show that the corresponding continuum states are given by

$$\phi_E^+(x) = \frac{1}{\sqrt{2\pi}}e^{ikx}, \quad \phi_E^-(x) = C(\tanh x - ik)e^{ikx},$$

where $E = \hbar^2(k^2 + 1)/2m$, $k \in \mathbb{R}$ and C is a proper normalisation constant.

Problem 3.4. General power law potentials on the half line

Consider a generalisation of examples 3.5 and 3.7 of section 3.6.2 defined by a SUSY potential of the form

$$\Phi(x) := \sqrt{\lambda}x^{a/2} - \frac{\hbar}{\sqrt{2m}}\frac{\eta}{x}, \quad a, \lambda, \eta > 0.$$

Show that SUSY is good and the partner potentials, respectively, the SUSY ground state, are given by

$$V_\pm(x) = \lambda x^a - \sqrt{\frac{\hbar^2 \lambda}{2m}} \left(2\eta \pm \frac{a}{2}\right) x^{(a-2)/2} + \frac{\hbar^2 \eta(\eta \pm 1)}{2mx^2},$$

$$\phi_0^-(x) = Cx^\eta \exp\left\{-\sqrt{\frac{2m\lambda}{\hbar^2}} \frac{2}{a+2} x^{(a+2)/2}\right\}.$$

Remark: For $\eta = l + 1$, and $a = 1$ the above partner potentials can be used to simulate a spherical symmetric linear quark-confining potential for large $r \equiv x \to \infty$ whose ground-state properties are exactly known [19]. This model is also used to study the spontaneous breakdown of PT symmetry and its relation to super-symmetry [20–22].

Problem 3.5. Spectral properties of infinite square well and its SUSY partner

Consider example 3.8 of section 3.6.3 and show the following.
 (a) The eigenvalues and eigenstates of H_- are given by

$$E_n = \frac{\hbar^2}{2m}(n^2 - 1), \quad \phi_{E_n}^-(x) = \sqrt{\frac{2}{\pi}} \sin(nx + n\pi/2), \quad n = 1, 2, 3, \dots.$$

 (b) The SUSY ground state $\phi_{E_1}^- \equiv \phi_0^-$ is annihilated by the operator A.
 (c) The eigenstates of H_+ are given by ($n = 2, 3, 4, \dots$)

$$\phi_{E_n}^+(x) = \sqrt{\frac{2}{\pi(n^2 - 1)}} (n \cos(nx + n\pi/2) + \tan x \sin(nx + n\pi/2)).$$

Problem 3.6. The square well with broken SUSY

As in example 3.10 of section 3.6.3 assume a vanishing SUSY potential on $\mathcal{M} = [-1, 1]$, that is, $H_\pm = p^2/2m$. However, now consider non-symmetric boundary conditions at $x = \pm 1$ by choosing the following Hilbert subspaces:

$$\mathcal{H}^+ := \{\phi \in L^2([-1, 1]): \phi'(-1) = 0 = \phi(1)\},$$
$$\mathcal{H}^- := \{\phi \in L^2([-1, 1]): \phi(-1) = 0 = \phi'(1)\}.$$

In other words, for $\mathcal{H}^+$ we assume a Neumann condition at $x = -1$ and a Dirichlet condition at $x = 1$, and vice versa for $\mathcal{H}^-$. Show that parity as well as SUSY are broken. This example may be understood as the $\alpha \to \infty$ limit of equation (3.46).

Problem 3.7. Free motion on the unit circle $\mathbb{S}^1$ and its SUSY structure

Consider the free Hamiltonian $H = -\hbar^2 \partial_x^2/2m$ on the unit circle with periodic boundary conditions, that is, $\mathcal{H} := \{\phi \in L^2(\mathbb{S}^1): \phi(x + 2\pi) = \phi(x)\}, x \in [0, 2\pi]$.

(a) Show that the eigenvalues and eigenstates are given by

$$E_n = \frac{\hbar^2 n^2}{2m}, \quad \phi_n(x) = \frac{e^{inx}}{\sqrt{2\pi}}, \quad n = 0, \pm 1, \pm 2, \pm 3, \ldots.$$

(b) Divide the Hilbert space into two disjunct subspaces describing a clockwise and counter clockwise motion on $\mathbb{S}^1$ as follows:

$$\mathcal{H}^- := \mathrm{span}\{\phi_n : n = 0, 1, 2, \ldots\}, \quad \mathcal{H}^+ := \mathrm{span}\{\phi_n : n = -1, -2, -3, \ldots\}.$$

What would be a suitable supercharge transforming between these subspaces?

(c) Show that $\phi_{E_n}^-(x) := \cos(nx)/\sqrt{\pi}$, $n \in \mathbb{N}_0$, and $\phi_{E_n}^+(x) := \sin(nx)/\sqrt{\pi}$, $n \in \mathbb{N}$, are also eigenfunctions of H and induce a grading of $\mathcal{H}$ into eigenspaces of the parity operator.

For more details, including also the Aharonov–Bohm effect, see [23, 24].

References

[1] Witten E 1981 Dynamical breaking of supersymmetry *Nucl. Phys.* **B188** 513–54
[2] Richtmyer R D 1978 *Principles of Advanced Mathematical Physics* vol I (New York: Springer)
[3] Reed M and Simon B 1980 *I Functional Analysis* (San Diego, CA: Academic)
[4] Gozzi E 1983 Ground-state wave-function 'representation' *Phys. Lett.* B **129** 432–6
[5] Jaffe A, Lesniewski A and Lewenstein M 1987 Ground state structure in supersymmetric quantum mechanics *Ann. Phys.* **178** 313–29
[6] Inomata A and Junker G 1993 Quasi-classical approach to path integrals in supersymmetric quantum mechanics *Lectures on Path Integration: Trieste 1991* ed H Cerdeira, S Lundqvist, D Mugnai, A Ranfagni, V Sa-yakanit and L S Schulman (Singapore: World Scientific) pp 460–82
[7] Thaller B 1992 *The Dirac Equation* (Berlin: Springer)
[8] Vogel J, Vogel E and Toepffer C 1985 Relaxation in a Fermi liquid *Ann. Phys.* **164** 463–94
[9] Cooper F, Khare A and Sukhatme U 1995 Supersymmetry and quantum mechanics *Phys. Rep.* **251** 267–385
[10] Jevicki A and Rodrigues J P 1984 Singular potentials and supersymmetry breaking *Phys. Lett.* B **146** 55–8
[11] Fuchs J 1986 Physical state conditions and supersymmetry breaking in quantum mechanics *J. Math. Phys.* **27** 349–53
[12] Shifman M A, Smilga A V and Vainshtein A I 1988 On the Hilbert space of supersymmetric quantum systems *Nucl. Phys.* B **299** 79–90
[13] Casahorran J 1991 A family of supersymmetric quantum mechanics models with singular superpotentials *Phys. Lett.* A **156** 425–8
[14] Panigrahi P K and Sukhatme U P 1993 Singular superpotentials in supersymmetric quantum mechanics *Phys. Lett.* A **178** 251–7
[15] Reed M and Simon B 1975 *Fourier Analysis, Self-Adjointness* (San Diego, CA: Academic)
[16] Sissakian A N, Ter-Antonyan V M, Pogosyan G S and Lutsenko I V 1990 Supersymmetry of a one-dimensional hydrogen atom *Phys. Lett.* A **143** 247–9

[17] Fischer W, Leschke H and Müller P 1995 The functional-analytic versus the functional-integral approach to quantum Hamiltonians: the one-dimensional hydrogen atom *J. Math. Phys.* **36** 2313–23

[18] Seeger M and Thies M 1998 QCD_{1+1} with static quarks as supersymmetric quantum mechanics *Phys. Rev.* D **58** 027701

[19] Inomata A and Junker G 2017 On the power-law duality in Feynman's path integral *Path Integration in Complex Dynamical Systems* (Leiden: Lorentz Center) invited talk

[20] Znojil M, Cannata F, Bagchi B and Roychoudhury R 2000 Supersymmetry without hermiticity within PT symmetric quantum mechanics *Phys. Lett.* B **483** 284–9

[21] Dorey P, Dunning C, Lishman A and Tateo R 2009 $\mathcal{PT}$ symmetry breaking and exceptional points for a class of inhomogeneous complex potentials *J. Phys. A: Math. Theor.* **42** 465302

[22] Dorey P, Dunning C and Tateo R 2001 Supersymmetry and the spontaneous breakdown of $\mathcal{PT}$ symmetry *J. Phys. A: Math. Gen.* **34** L391–400

[23] Correa F and Plyushchay M S 2006 Hidden supersymmetry in quantum bosonic systems *Ann. Phys.* **322** 2493–500

[24] Jakubský V, Nieto L-M and Plyushchay M S 2010 The origin of the hidden supersymmetry *Phys. Lett.* B **692** 51–6

IOP Publishing

Supersymmetric Methods in Quantum, Statistical and Solid State Physics
Enlarged and revised edition
Georg Junker

Chapter 4

One-dimensional quantum systems

The current chapter is devoted to the discussion of one-dimensional quantum systems. First, we will show that in principle any standard one-dimensional Schrödinger Hamiltonian may be put into the form of a Witten model. That is, one can find a family of SUSY potentials associated with the original quantum system [1, 2]. In a second step we will demonstrate that a particular invariance property of the SUSY potential, the so-called shape-invariance [3], will lead to exact results of the discrete eigenvalues and corresponding eigenfunctions. This shape-invariance is closely related to the factorisation method of Schrödinger, Infeld, and Hull [4, 5], which in fact goes back to Darboux [6]. The Darboux method can be utilised to construct new partner potentials, so-called conditionally exactly solvable potentials, where the eigenvalues and eigenfunctions of the associated Schrödinger Hamiltonian are obtained via SUSY transformations. We explicitly construct a family of SUSY partners for the linear harmonic oscillator. New creation and annihilation operators, motivated via SUSY transformations are also discussed, including their associated generalised coherent states. As an application of Witten's model in classical physics we will consider classical fields in $(1 + 1)$ dimensions. Here the SUSY structure naturally appears in the eigenvalue problem of the fluctuation operator around localised solutions called solitary waves. SUSY in combination with the shape-invariance condition can be utilised to construct stable as well as unstable field models for which the associated fluctuation equation exhibits exact solutions.

4.1 Supersymmetrisation of one-dimensional systems

In our previous discussion on the Witten model we have already seen that several well-known quantum Hamiltonians possess a SUSY structure according to definition 2.1. Here naturally the question arises: Which quantum systems exhibit a

doi:10.1088/2053-2563/aae6d5ch4

SUSY structure? This question has been considered by de Crombrugghe and Rittenberg [7] with the general result that SUSY, in particular the extended one with $N > 2$, imposes strong conditions on the Hamiltonian. For non-relativistic one-dimensional one-particle systems, however, it is always possible to bring them into the form of the Witten model [1, 2]. Here we will consider only systems of this type.

Suppose we are interested in the supersymmetric form of a given quantum problem characterised by the following standard Hamiltonian acting on $L^2(\mathbb{R})$,

$$H_V := \frac{p^2}{2m} + V(x), \tag{4.1}$$

where V is some continuous real-valued function such that H_V has a non-empty discrete spectrum which is below a possible continuous spectrum. Without loss of generality we choose as configuration space the Euclidean line, i.e. $\mathcal{M} = \mathbb{R}$. Suppose further that we want the Hamiltonian H_V to be identified (up to a constant) with the Hamiltonian H_- of the Witten model. The other choice H_+ can be treated similar to the procedure described below by replacing the SUSY potential Φ by $-\Phi$. In general, the ground-state energy of H_V cannot be expected to vanish or be strictly positive. Hence, in order to be able to accommodate a SUSY structure we have to put

$$H_- := H_V - \varepsilon, \tag{4.2}$$

where ε is an arbitrary constant energy shift sometimes called *factorisation energy* [8]. It is obvious that, if ε equals the ground-state energy of H_V, we will arrive at a good SUSY, whereas for values of ε which are below this ground-state energy we may obtain a broken SUSY. Finally, for ε above the ground-state energy of H_V the Hamiltonian H_- will have a negative eigenvalue which implies that we cannot obtain a well-defined SUSY structure.

For the moment let ε be an arbitrary real number. Then the SUSY potential we are looking for is determined by the generalised Riccati equation

$$\Phi^2(x) - \frac{\hbar}{\sqrt{2m}}\, \Phi'(x) = V(x) - \varepsilon. \tag{4.3}$$

Making the ansatz [8, 9]

$$\Phi_{\varepsilon,\lambda}(x) = -\frac{\hbar}{\sqrt{2m}} \frac{\mathrm{d}}{\mathrm{d}x} \ln\left[\rho_\varepsilon(x)\left(1 + \lambda \int_0^x \mathrm{d}z \rho_\varepsilon^{-2}(z)\right)\right]$$

$$= -\frac{\hbar}{\sqrt{2m}} \left(\frac{\rho_\varepsilon'(x)}{\rho_\varepsilon(x)} + \frac{\lambda \rho_\varepsilon^{-2}(x)}{1 + \lambda \int_0^x \mathrm{d}z \rho_\varepsilon^{-2}(z)} \right) \tag{4.4}$$

the non-linear Riccati equation reduces to a linear Schrödinger-like equation:

$$-\frac{\hbar^2}{2m} \rho_\varepsilon''(x) + V(x)\rho_\varepsilon(x) = \varepsilon \rho_\varepsilon(x). \tag{4.5}$$

Note that ρ_ε is not necessarily required to be square integrable. Hence, the parameter ε is indeed arbitrary. Let us remark that with ρ_ε

$$\tilde{\rho}_\varepsilon(x) := \rho_\varepsilon(x) \int_0^x \, dz \rho_\varepsilon^{-2}(z) \qquad (4.6)$$

is also a linear independent solution of equation (4.5) and hence any linear combination $\rho_\varepsilon + \lambda \tilde{\rho}_\varepsilon$, which is used in equation (4.4), will also yield such a solution.

With $\Phi_{\varepsilon,\lambda}$ as given in equation (4.4) we have obtained a two-parameter family of SUSY potentials which at least formally brings the original problem (4.1) into a SUSY Hamiltonian H_-. We are also able to find the corresponding partner Hamiltonian

$$H_+ := \frac{p^2}{2m} + \Phi_{\varepsilon,\lambda}^2(x) + \frac{\hbar}{\sqrt{2m}} \Phi_{\varepsilon,\lambda}'(x) = H_V - \varepsilon + \frac{2\hbar}{\sqrt{2m}} \Phi_{\varepsilon,\lambda}'(x), \qquad (4.7)$$

which may have energy levels identical to those of H_-. To make it more precise, let us discuss some particular ranges for the value of the parameter ε. For this we will denote the discrete eigenvalues of H_V by ε_n with the ordering $\varepsilon_n < \varepsilon_{n+1}$, $n \in \mathbb{N}_0$. We also limit our discussion to the special case $\lambda = 0$ for simplicity. A more general discussion is given in the original work by Sukumar [8]. In the present case, the zero-energy wave function ϕ_0^- as defined in equation (3.11) can be identified up to a normalisation constant C with ρ_ε,

$$\phi_0^-(x) = C\rho_\varepsilon(x), \qquad (4.8)$$

which is not required to be an element of the Hilbert space. We consider the following cases.

(a) $\varepsilon < \varepsilon_0$. In this case it is well known from Sturmian theory [9] that ρ_ε and therefore ϕ_0^- will have no zeros and will not be normalisable. It is, however, possible that $1/\rho_\varepsilon$ is normalisable. Or in other words, changing the minus sign in front of equation (4.4) to a plus sign may give rise to a normalisable $\phi_0^+ = C/\rho_\varepsilon$. This in essence leads to case (b) discussed below with Φ replaced by $-\Phi$. We will discuss both cases in more detail in section 4.3 below.

In the general case, where neither ρ_ε nor $1/\rho_\varepsilon$ is normalisable, equation (4.4) gives rise to a well-defined (note that $\lambda = 0$ and ρ_ε has no zeros) SUSY potential with broken SUSY. That is, the partner Hamiltonians H_- and H_+ have identical spectra. Consequently, the Hamiltonian

$$\tilde{H}_V := H_V + \frac{2\hbar}{\sqrt{2m}} \Phi_{\varepsilon,0}'(x) \qquad (4.9)$$

will have the same spectrum as H_V. Thus by supersymmetrisation with a factorisation energy ε below the ground-state energy of a given Hamiltonian H_V one can find another Hamiltonian $\tilde{H}_V$ with an identical spectrum but, of course, different eigenfunctions. These eigenfunction are related by the SUSY transformation (3.9).

(b) $\varepsilon = \varepsilon_0$. Now $C\rho_\varepsilon = \phi_0^-$ is normalisable and hence, the SUSY potential (4.4) gives rise to a good SUSY. The two partner Hamiltonians H_- and H_+ are essential iso-spectral, that is, the ground-state energy of H_- vanishes and all other eigenvalues $E_n = \varepsilon_n - \varepsilon$, $n \in \mathbb{N}$, of H_- coincide with that of H_+. As a consequence, the Hamiltonian $\tilde{H}_V$, as defined above, has a spectrum which coincides with the set $\{\varepsilon_1, \varepsilon_2, ...\}$ of energy eigenvalues of the original Hamiltonian H_V without its ground-state energy ε_0. The SUSY potential is represented by the ground-state wave function [10] of H_-, respectively, H_V:

$$\Phi(x) \equiv \Phi_{\varepsilon_0, 0}(x) = -\frac{\hbar}{\sqrt{2m}} \frac{(\phi_0^-)'(x)}{\phi_0^-(x)} = -\frac{\hbar}{\sqrt{2m}} \frac{\mathrm{d}}{\mathrm{d}x} \ln \phi_0^-(x). \tag{4.10}$$

(c) $\varepsilon_{n-1} < \varepsilon \leqslant \varepsilon_n$, $n = 1, 2, 3, ...$ Here from Sturmian theory we know that ρ_ε has n distinct zeros (of multiplicity one) and hence the SUSY potential (4.4) will have singularities at those points. Here the superalgebra will only be valid on a formal level and one has to be careful about a proper definition of the domains for the operators in question. In particular, the supercharges (3.1) cannot be self-adjoint because the Hamiltonian $H_- = H_V - \varepsilon$ has n strictly negative eigenvalues implying complex eigenvalues for the super-charges (3.1). The formal Hamiltonian H_+ will in general not be essential iso-spectral to H_- [11–14].

The generic power-law potential as an example for case (a)

As an illustrative example for the case $\varepsilon < \varepsilon_0$ let us consider the anharmonic-oscillator potential

$$V(x) := \frac{\hbar^2}{2m} x^n, \qquad n \in \mathbb{N}. \tag{4.11}$$

The generalised Riccati equation (4.3) reduces in this case to the usual one if we set the factorisation energy ε to zero, which indeed corresponds to case (a) above as obviously $\varepsilon_0 > 0$. The ansatz (4.4) with $\lambda = 0$ then allows us to express ρ_0 in terms of an arbitrary solution of Bessels differential equation denoted by Z_ν [15]:

$$\Phi(x) = -\frac{\hbar}{\sqrt{2m}} \frac{\rho_0'(x)}{\rho_0(x)}, \qquad \rho_0(x) := \sqrt{x} Z_{\frac{1}{n+2}}\left(\frac{2i}{n+2} x^{(n+2)/2}\right). \tag{4.12}$$

This example shows that a rather simple problem can lead to a complicated SUSY potential if supersymmetrisation according to case (a) is chosen. Hence, the natural choice for the factorisation energy is the ground-states energy of the Hamiltonian (4.1) with the SUSY potential expressed in terms of the ground-state wave function as in equation (4.10).

A non-trivial example for case (b)

The relation (4.10) displays for the case of good SUSY, that is, $\varepsilon = \varepsilon_0$, an interesting relation between the zeros of the SUSY potential and the extrema of the ground-state wave function. To be explicit, these extrema, or more precisely the point with vanishing slope, are given by the zeros of the SUSY potential. These zeros are in general not identical to the minima or maxima of the full potential V but coincide with the minima of Φ^2. This is rather surprising, because one naively expects the maxima (local minima) of the ground-state wave function to be located at the minima (local maxima) of the potential V.

To investigate this a bid further, let us consider the following SUSY potential:

$$\Phi(x) := \frac{\hbar}{\sqrt{2m}}(x - a \tanh x), \qquad a > 0. \tag{4.13}$$

Obviously, for $a \leqslant 1$ this SUSY potential has a single zero at $x_0 = 0$ which is also the position of the maximum of the ground-state wave function

$$\phi_0^-(x) = C \exp\{-x^2/2\}\cosh^a x. \tag{4.14}$$

However, for $a > 1$ the SUSY potential (4.13) has two more zeros, $\Phi(x_{\pm 1}) = 0$, located symmetrically about the first zero $x_0 = 0$ which now becomes the position of a local minimum of ϕ_0^-. The two non-trivial zeros $x_{\pm 1}$ are maxima of the ground-state wave function and are determined by the non-trivial solutions of $x_{\pm 1} = a \tanh(x_{\pm 1})$ with $x_{+1} = -x_{-1} > 0$. Graphs of the SUSY potential and the ground-state wave function for various values of a are given in figure 4.1. The functional dependence of the zeros x_i on the parameter a is shown in figure 4.2. It is clearly visible that the single maximum of ϕ_0^- for $a < 1$ bifurcates at $a = 1$ into two maxima at positions $x_{\pm 1}$. In addition, the trivial zero $x_0 = 0$ is now the location of the local minimum of ϕ_0^-. This bifurcation signals the onset of tunnelling in the full potential V_-. The two partner potentials are given by

$$V_{\pm}(x) = \frac{\hbar^2}{2m}[x^2 + a(a \pm 1)\tanh^2 x - 2ax \tanh x \mp (a - 1)] \tag{4.15}$$

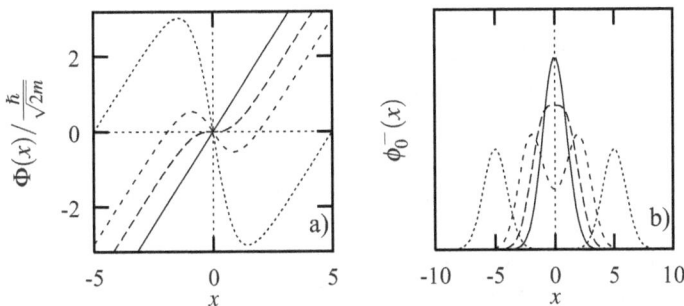

Figure 4.1. (a) The SUSY potential (4.13) for parameter values $a = 0$ (——), 1 (– – –), 2 (- - - -), and 5 (······). (b) The ground-state wave function (4.14) for the same parameters as in (a).

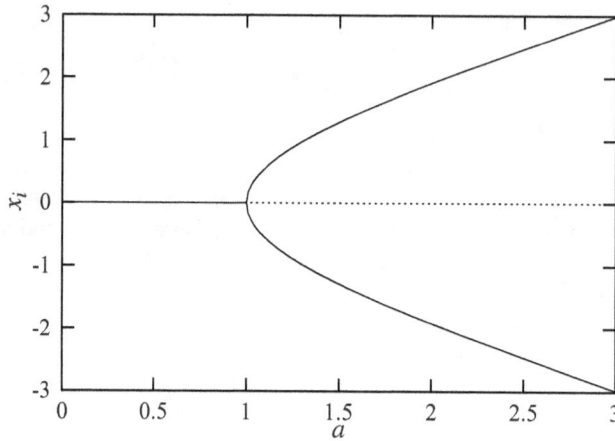

Figure 4.2. The locations x_i of the maxima (——) and minima (·····) of the ground-state wave function (4.14) as a function of the parameter a.

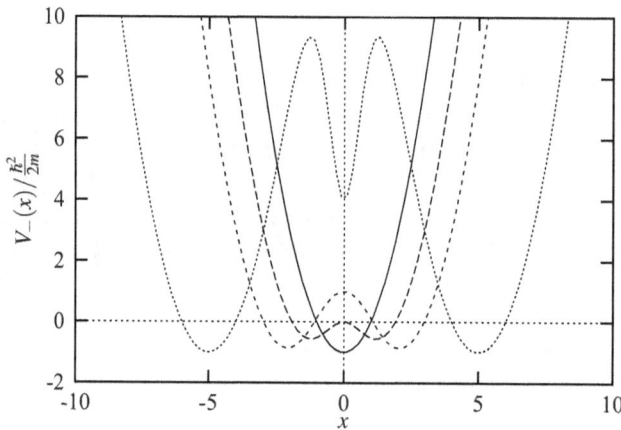

Figure 4.3. The potential V_- of equation (4.15) with parameters as in figure 4.1.

and are shown in figures 4.3 and 4.4. In particular, we note that $V_-(0) = \frac{\hbar^2}{2m}(a-1) > 0$ for $a > 1$ but the ground-state energy is identical zero for all values of a due to good SUSY. Hence, $a = 1$ characterises indeed the onset of tunnelling. We also point out that for $a > \frac{1}{2}(3 + \sqrt{5})$ the potential V_- has a local minimum at $x = 0$ where the ground-state wave function also has a local minimum. Therefore, a local minimum of a given potential does not necessary imply a local maximum in the ground-state wave function.

Finally, we note that this example demonstrates that SUSY can also be used for the discussion of tunnelling problems. In fact, the tunnelling splitting for the lowest energy eigenvalues is given by the lowest eigenvalue of H_+ if SUSY is good. This

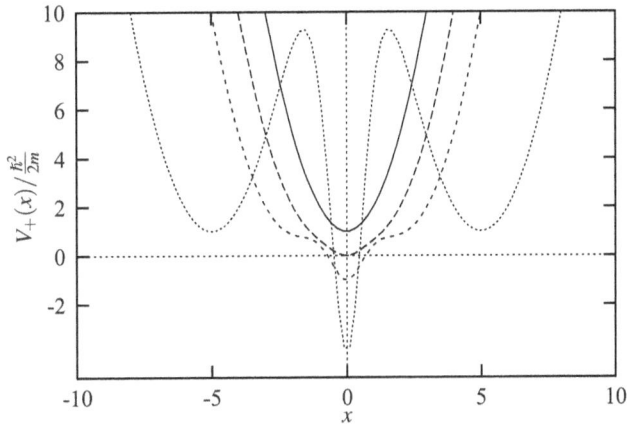

Figure 4.4. The potential V_+ of equation (4.15) with parameters as in figure 4.1.

eigenvalue can, for example, easily be obtained using the instanton method [16–18], a perturbative method [17, 19, 20], or a WKB-like method [21].

We have seen in the above discussion that for a factorisation energy $\varepsilon = \varepsilon_0$ it is always possible to find for a given Hamiltonian H_V an associated partner Hamiltonian $\tilde{H}_V$ given in equation (4.9) which is essential iso-spectral to H_V. It is obvious that by repetition of this supersymmetrisation procedure one can construct families of essential iso-spectral Hamiltonians. This method, in fact, was developed in the nineteenth century and is known as Darboux's transformation method. We will revisit this method in more detail in section 4.3.

4.2 Shape-invariance and exact solutions

Following the above discussion we have seen that, in principle, one can construct a hierarchy of potentials whose corresponding Hamiltonians are pairwise essential iso-spectral. In fact, one only needs to know the ground-state energy ε_0 of H_V and the corresponding ground-state wave function ϕ_0^- in order to obtain the partner Hamiltonian $\tilde{H}_V$. Repeating this process for the construction of a hierarchy of pairwise essential iso-spectral Hamiltonians one now needs also the ground-state properties of $\tilde{H}_V$, etc. In other words, in finding a family of such Hamiltonians a complete knowledge of the discrete spectrum of the starting Hamiltonian is required and nothing is gained.

In this section we want to show that under a certain condition called *shape-invariance* the full information on the discrete spectrum of the starting Hamiltonian is not required and can be obtained in a straightforward way using the SUSY transformation.

Let us assume that we are given a SUSY potential for good SUSY. We further assume that this SUSY potential depends on some set of parameters collectively denoted by a_0. We will explicitly denote the dependence on these parameters in the form $\Phi(a_0, x)$. Then the corresponding partner potentials are given by

$$V_{\pm}(a_0, x) = \Phi^2(a_0, x) \pm \frac{\hbar}{\sqrt{2m}} \frac{\partial}{\partial x} \Phi(a_0, x). \tag{4.16}$$

Definition 4.1. The partner potentials $V_{\pm}(a_0, x)$ are called *shape-invariant* if they are related by

$$\boxed{V_+(a_0, x) = V_-(a_1, x) + R(a_1),} \tag{4.17}$$

where a_1 is a new set of parameters uniquely determined from the old set a_0 via the mapping $F: a_0 \mapsto a_1 = F(a_0)$ and the residual term $R(a_1)$ is independent of the variable x.

In other words, shape-invariance implies that the partner potential $V_+(a_0, x)$ can, after subtracting the constant $R(a_1)$, be interpreted as a new partner potential $V_-(a_1, x)$ associated with a new SUSY potential $\Phi(a_1, x)$.

As an example we mention

$$\Phi(a_0, x) := \frac{\hbar}{\sqrt{2m}} a_0 \tanh x, \qquad a_0 > 0. \tag{4.18}$$

The corresponding partner potentials read

$$V_{\pm}(a_0, x) = \frac{\hbar^2}{2m} \left[a_0^2 - \frac{a_0(a_0 \mp 1)}{\cosh^2 x} \right] \tag{4.19}$$

and are shape-invariant because of the relation

$$V_+(a_0, x) = V_-(a_0 - 1, x) + \frac{\hbar^2}{2m} \left[a_0^2 - (a_0 - 1)^2 \right]. \tag{4.20}$$

One immediately reads off the function F and R,

$$a_1 = F(a_0) = a_0 - 1, \qquad R(a_1) = \frac{\hbar^2}{2m} \left[a_0^2 - a_1^2 \right], \tag{4.21}$$

and the ground-state wave function

$$\phi_0^-(a_0, x) = C \cosh^{-a_0} x. \tag{4.22}$$

Let us now assume that we have a SUSY potential $\Phi(a_0, x)$ generating shape-invariant partner potentials with a new parameter set a_1 such that $\Phi(a_1, x)$ is also a good SUSY potential. We further assume that the mapping $F: a_{s-1} \mapsto a_s = F(a_{s-1})$ may be iterated n times leading to a family of SUSY potentials $\Phi(a_s, x)$, $s = 0, 1, 2, \ldots, n$, all with good SUSY. Then the discrete energy eigenvalues and corresponding wave functions of the original Hamiltonian $H_0 := \frac{p^2}{2m} + V_-(a_0, x)$ are given by [3, 22, 23]

$$E_n = \sum_{s=1}^{n} R(a_s),$$

$$\phi_n^-(a_0, x) = \prod_{s=0}^{n-1} \left(\frac{A^{\dagger}(a_s)}{[E_n - E_s]^{1/2}} \right) \phi_0^-(a_n, x),$$

$$\phi_0^-(a_n, x) = C \exp\left\{ -\frac{\sqrt{2m}}{\hbar} \int_0^x dz\; \Phi(a_n, z) \right\},$$

(4.23)

where the product of the operators

$$A^{\dagger}(a_s) := -\frac{\hbar}{\sqrt{2m}} \frac{\partial}{\partial x} + \Phi(a_s, x)$$

(4.24)

is ordered such that $A^{\dagger}(a_s)$ stands to the left of $A^{\dagger}(a_{s+1})$. The validity of relations (4.23) is an obvious consequence of the SUSY transformation (3.9) which is illustrated in figure 4.5. Hence, shape-invariance is a sufficient condition for obtaining complete and exact information about the spectral properties of the bound states of the family of Hamiltonians

$$H_s := -\frac{\hbar^2}{2m} \frac{\partial^2}{\partial x^2} + V_-(a_s, x) + E_s, \qquad s = 0, 1, 2, \dots, n.$$

(4.25)

Note that the energy eigenfunctions

$$H_s \phi_{n-s}^-(a_s, x) = E_n \phi_{n-s}^-(a_s, x), \qquad n \geqslant s,$$

(4.26)

of this family of Hamiltonians are related by

$$\phi_{n-s}^-(a_s, x) = \frac{A^{\dagger}(a_s)}{\sqrt{E_n - E_s}} \phi_{n-(s+1)}^-(a_{s+1}, x).$$

(4.27)

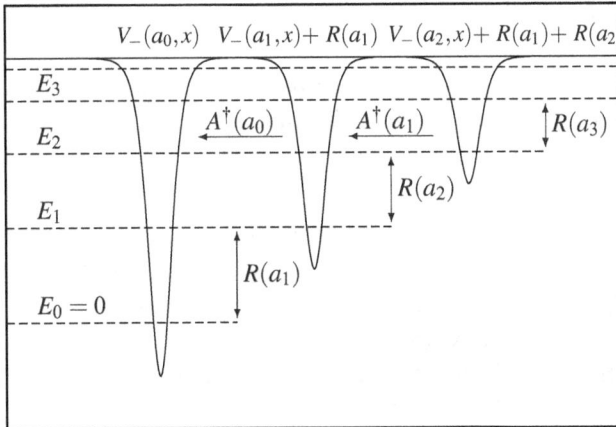

Figure 4.5. An illustration of a family of shape-invariant potentials. The arrows indicate the relations (4.23) and (4.27).

It should be noted that despite the fact that shape-invariance is a sufficient condition it is not necessary for the solvability of the eigenvalue problem of Schrödinger's equation [22].

4.2.1 An explicit example

As an example let us briefly discuss the family associated with the SUSY potential (4.18). Note that from equation (4.21) follows

$$a_s = a_{s-1} - 1 = a_0 - s, \qquad R(a_s) = \frac{\hbar^2}{2m}\left[a_{s-1}^2 - a_s^2\right], \qquad (4.28)$$

and therefore

$$\phi_0^-(a_n, x) = \phi_0^-(a_0 - n, x) = C \cosh^{n-a_0} x \qquad (4.29)$$

will be normalisable for all $n \in \mathbb{N}_0$ with $n < a_0$. The eigenvalues are also immediately found:

$$E_n = \frac{\hbar^2}{2m}\sum_{s=1}^{n}\left(a_{s-1}^2 - a_s^2\right) = \frac{\hbar^2}{2m}[a_0^2 - (a_0 - n)^2], \qquad n = 0, 1, 2, \dots < a_0. \quad (4.30)$$

Finally, the energy eigenfunctions are given by

$$\phi_n^-(a_0, x) = C_n\left(-\frac{\partial}{\partial x} + a_0 \tanh x\right) \times \dots$$
$$\times\left(-\frac{\partial}{\partial x} + (a_0 - n - 1)\tanh x\right)\cosh^{n-a_0} x, \qquad (4.31)$$

where C_n are normalisation constants determined by the ground-state normalisation C

$$C_n := C\prod_{s=0}^{n-1}[(a_0 - s)^2 - (a_0 - n)^2]^{-1/2}, \qquad n = 1, 2, 3, \dots < a_0. \qquad (4.32)$$

Let us mention that for the special case $a_0 \in \mathbb{N}$ the partner potentials (4.19) are reflectionless. This immediately follows from example 3.1 discussed in section 3.6.1 as for the last step $n = a_0 - 1$ one arrives at the free particle and its SUSY partner.

4.2.2 Comparison with the factorisation method

Despite the fact that originally shape-invariance was believed [3] to be a new hidden symmetry which allows for an exact solution of eigenvalue problems, it is equivalent to a similar condition already known from the factorisation method [24, 25]. The factorisation method has been developed by Schrödinger [4, 26, 27] and Infeld and Hull [5, 28, 29] as an algebraic method for solving the stationary Schrödinger equation in one dimension [30, 31]. See also the review [32]. Its close connection with SUSY quantum mechanics [1, 2] is obvious. In fact, it is amusing to note that already in 1967 Joseph [33], in his search for self-adjoint ladder operators (we now call them supercharges), found quite naturally that the Hilbert space has to be graded into two subspaces (he called them the 'a' and 'b' subspaces). He also noted that the Hamiltonian together with these self-adjoint ladder operators close an algebra containing anticommutators, that is, a superalgebra.

Let us make the connection between the factorisation method and shape-invariance of SUSY quantum mechanics transparent. Writing the shape-invariance condition in the form

$$\Phi^2(a_{s+1}, x) - \Phi^2(a_s, x) - \frac{\hbar}{\sqrt{2m}}[\Phi'(a_{s+1}, x) + \Phi'(a_s, x)] = -R(a_{s+1}) \qquad (4.33)$$

its equivalence to the condition (3.1.2) of Infeld and Hull becomes obvious if we identify the functions $k(x, s)$ and $L(s)$ in [5] as follows:

$$\Phi(a_s) =: -\frac{\hbar}{\sqrt{2m}} k(x, s), \qquad R(a_{s+1}) =: \frac{\hbar^2}{2m}[L(s + 1) - L(s)]. \qquad (4.34)$$

Note that because of $R(a_{s+1}) > 0$ we arrive at the class I type of factorisation of Infeld and Hull. This is due to our convention that the ground state is an eigenstate of H_-. Otherwise, we would arrive at the class II type of factorisation. In order to complete our comparison let us note that from equation (4.34) follows

$$E_n = \sum_{s=1}^{n} R(a_s) = \frac{\hbar^2}{2m}[L(n) - L(0)], \qquad (4.35)$$

which is identical to theorem IV of Infeld and Hull ($n = l + 1$), and

$$\phi_0^-(a_n, x) = C \exp\left\{\int_0^x \mathrm{d}z\, k(z, n)\right\}, \qquad (4.36)$$

which agrees with equation (2.7.1) of Infeld and Hull [5]. Finally, the relation (4.27) coincides up to an overall minus sign with equation (2.7.2) in [5].

Hence, the shape-invariance condition does not lead to any new insight into the question: Why are some potentials exactly solvable? Shape-invariance is simply a restatement of the factorisation condition of Infeld and Hull [5]. In contrast, the solvability of these shape-invariant potentials can be related to an underlying dynamical Lie symmetry [34, 35]. For an embedding of SUSY into those dynamical Lie symmetries see Barut and Roy [36]. More general dynamical superalgebras are constructed by Baake et al [37]. Let us also mention the recent work by Oikonomou [38] who was able to relate shape-invariance and the above mentioned Lie symmetry to a Z_3-graded symmetry structure.

The developing interest in SUSY quantum mechanics, however, has revived the important question: Which class of potentials are shape-invariant? This problem was attacked by Infeld and Hull by starting with a typical dependence of the energy eigenvalues on the quantum number n. Lévai [39] has reconsidered this problem in the

Table 4.1. List of known SUSY potentials giving rise to shape-invariant partner potentials.

SUSY potential $\Phi(x)/\frac{\hbar}{\sqrt{2m}}$	Config. space[a]	Parameter range for good SUSY[b]	Partner potentials $V_\pm(x)/\frac{\hbar^2}{2m}$
$A\tanh x + B/\cosh x$	$\mathbb{R}$	$A>0$	$A^2 + \dfrac{B^2 - A(A\mp 1) + B(2A\mp 1)\sinh x}{\cosh^2 x}$
$A\coth x - B/\sinh x$	$\mathbb{R}^+$	$B>A>0$	$A^2 + \dfrac{B^2 + A(A\mp 1) - B(2A\pm 1)\cosh x}{\sinh^2 x}$
$-A\cot x + B/\sin x$	$[0,\pi]$	$A>B>0$	$-A^2 + \dfrac{B^2 + A(A\pm 1) - B(2A\mp 1)\cos x}{\sin^2 x}$
$A\tan x - B\cot x$	$[0,\pi/2]$	$A>0,\ B>0^c$	$-(A+B)^2 + \dfrac{A(A\pm 1)}{\cos^2 x} + \dfrac{B(B\pm 1)}{\sin^2 x}$
$A\tanh x - B\coth x$	$\mathbb{R}^+$	$A>B>0^c$	$(A-B)^2 - \dfrac{A(A\mp 1)}{\cosh^2 x} + \dfrac{B(B\pm 1)}{\sinh^2 x}$
$A\tanh x + B/A$	$\mathbb{R}$	$A>B\geqslant 0$	$A^2 + \dfrac{B^2}{A^2} - \dfrac{A(A\mp 1)}{\cosh^2 x} + 2B\tanh x$
$-A\coth x + B/A$	$\mathbb{R}^+$	$B>A>0$	$A^2 + \dfrac{B^2}{A^2} + \dfrac{A(A\pm 1)}{\sinh^2 x} - 2B\coth x$
$-A\cot x + B/A$	$[0,\pi]$	$A>0$	$-A^2 + \dfrac{B^2}{A^2} + \dfrac{A(A\pm 1)}{\sin^2 x} - 2B\cot x$
$Ax - B/x$	$\mathbb{R}^+$	$A>0,\ B>0^c$	$-A(2B\mp 1) + A^2 x^2 + \dfrac{B(B\pm 1)}{x^2}$
$-A/x + B/A$	$\mathbb{R}^+$	$A>0,\ B>0$	$\dfrac{B^2}{A^2} - \dfrac{2B}{x} + \dfrac{A(A\pm 1)}{x^2}$
$-Ae^{-x} + B$	$\mathbb{R}$	$A>0,\ B>0$	$B^2 + A^2 e^{-2x} - A(2B\mp 1)e^{-x}$
$Ax + B$	$\mathbb{R}$	$A>0$	$(Ax+B)^2 \pm A$

[a] For $x \in \mathbb{R}^+$, $x \in [0, \pi/2]$, and $x \in [0, \pi]$ we impose Dirichlet boundary conditions on the wave functions at $x = 0$, $x = 0, \pi/2$, and $x = 0, \pi$, respectively.
[b] With our convention that the ground state is an eigenstate of H_-.
[c] These examples belong to class 2 of Gendensteîn and will give rise to a broken SUSY potential if B is replaced by $-B$.

light of SUSY $x \in \mathbb{R}^+$, $x \in [0, \pi/2]$, and $x \in [0, \pi]$ quantum mechanics. Although his approach was rather different from that of Infeld and Hull, he essentially arrived at the same potentials. See table 4.1.

However, inspired by this work Wittmer and Weiss [40] were able to find a shape-invariant generalisation of the harmonic-oscillator potential based on the Hongler–Zheng model [41, 42]. Another new class of shape-invariant potentials has been found by Khare and Sukhatme [43]. Unfortunately, this latter class of potentials cannot be expressed in closed form, that is, in terms of known functions. There exist only series representations which are even not guaranteed to be convergent on the real axis. Actually, these series are believed to have a finite radius of convergence [44]. Hence, the possibility of one or more singularities on the real axis cannot be excluded. In fact, the claimed properties of these potentials (being bounded from below as well as from above because of reflectionlessness and still having an infinite number of bound states) indicates that these potentials do not lead to a Hamiltonian with a well-defined domain. In other words, SUSY may only be a formal symmetry algebra. See also our comments on singular SUSY potentials made in section 4.1 case (c). For a detailed discussion see also [45].

The concept of shape-invariance has also been found to be valuable in connection with exact solutions of so-called discrete quantum mechanics which are related to multi-particle systems of Calogero and Sutherland and their generalisations [46]. Another example which shows the success of this concept is its application to systems having a position-dependent effective mass [47]. See also section 9.5 for an explicit example of a position-dependent band structure in a semiconductor system.

As mentioned before the factorisation method is closely related to an underlying Lie algebra which is denoted by $G(a,b)$ in the excellent book by Miller [34]. Its close connection to potential algebras discussed in the context of SUSY quantum mechanics can be found in [48].

Finally, for completeness, we also present the classification scheme of shape-invariant potentials as given by Gendenshteîn [3]:

Class 1:

$$\Phi_1(x) := \frac{\hbar}{\sqrt{2m}}\left(af_1(x) + b\right) \quad \text{with}$$

$$f_1'(x) = pf_1^2(x) + qf_1(x) + r,$$

$$V_\pm^{(1)}(x) = \frac{\hbar^2}{2m}\left[a(a \pm p)f_1^2(x) + a(2b \pm q)f_1(x) + (b^2 \pm ar)\right].$$

(4.37)

Class 2:

$$\Phi_2(x) := \frac{\hbar}{\sqrt{2m}}\left(af_2(x) + b/f_2(x)\right) \quad \text{with}$$

$$f_2'(x) = pf_2^2(x) + q,$$

$$V_\pm^{(2)}(x) = \frac{\hbar^2}{2m}\left[a(a \pm p)f_2^2(x) + \frac{b(b \mp q)}{f_2^2(x)} + 2ab \pm (aq - bp)\right].$$

(4.38)

Class 3:

$$\Phi_3(x) := \frac{\hbar}{\sqrt{2m}}\left(a + b\sqrt{pf_3^2(x) + q}\right)/f_3(x) \quad \text{with}$$

$$f_3'(x) = \sqrt{pf_3^2(x) + q},$$

$$V_\pm^{(3)}(x) = \frac{\hbar^2}{2m}\left[\frac{a^2 + pq(b \mp 1)}{f_3^2(x)} + \frac{\sqrt{pf_3^2(x) + q}}{f_3^2(x)}a(2b \mp 1) + b^2 p\right].$$

(4.39)

Here $a, b, p, q, r \in \mathbb{R}$ are arbitrary potential parameters. Depending on the values of these parameters SUSY will be good or broken. Let us note that the potentials belonging to class 2 possess the additional reparameterisation invariances:

$$\left.\begin{array}{l} a \to -(a \pm p) \\ b \to -(b \mp q) \end{array}\right\} \Rightarrow V_\pm^{(2)}(x) \to V_\pm^{(2)}(x) + \text{const.}$$

(4.40)

That is, in both cases the full potentials are only shifted by an additional constant. This reparametrisation implies the following changes in the SUSY potential:

$$a \rightarrow -(a \pm p) \Rightarrow \Phi_2(x) \rightarrow -(a \pm p)f_2(x) + b/f_2(x) \qquad (4.41)$$

$$b \rightarrow -(b \mp q) \Rightarrow \Phi_2(x) \rightarrow af_2(x) - (b \mp q)/f_2(x). \qquad (4.42)$$

The particular form of Φ_2 shows that if the parameters originally have been chosen such that SUSY is good, after reparametrisation SUSY will be broken [49–52]. Examples which belong to this class are the radial harmonic oscillator (section 3.6.1 example 1) and the Pöschl–Teller oscillators (section 3.6.2 example 3.7 and section 3.6.3 example 3.9).

4.3 The Darboux transformation method

The method being discussed here goes back to a work by Darboux in 1882 [6] and considers two Hamilton operators $H_\pm$ acting on some function space $\mathcal{H}$. In addition one assumes the existence of a linear operator $A: \mathcal{H} \mapsto \mathcal{H}$ obeying the relation

$$H_+A = AH_-. \qquad (4.43)$$

Let us further assume that the spectral properties of H_+ are known, that is, we know its eigenvalues E_n and eigenstates $|\phi_n^+\rangle \in \mathcal{H}$,

$$H_+|\phi_n^+\rangle = E_n|\phi_n^+\rangle. \qquad (4.44)$$

For simplicity and without loss of generality we assume a purely discrete spectrum. As a consequence of condition (4.43) we can now solve the eigenvalue problem for H_-. Considering the adjoint version of equation (4.43)

$$A^\dagger H_+ = H_-A^\dagger \qquad (4.45)$$

it is obvious that

$$|\phi_n^-\rangle := C_n A^\dagger |\phi_n^+\rangle \neq 0 \qquad (4.46)$$

is an eigenstate of H_- with the same eigenvalue E_n,

$$H_-|\phi_n^-\rangle = E_n|\phi_n^-\rangle, \qquad (4.47)$$

with C_n being the normalisation constant.

Clearly, states $|\phi_n^+\rangle$ which are annihilated by $A^\dagger$, that is, $|\phi_n^+\rangle \in \ker A^\dagger \subset \mathcal{H}$, are not allowed and hence the associated eigenvalues, in general, do not belong to the spectrum of H_-. On the other hand H_- may have additional eigenstates belonging to the kernel of A which is obvious from equation (4.43).

In summary, using the Darboux condition (4.43), sometimes also called the intertwining relation, one is able to obtain from the known solutions of the eigenvalue problem of H_+ those of a new problem H_-. This approach is very generic and thus not only applicable to quantum mechanical eigenvalue problems, but also to more general problems characterisable via ordinary differential equations [9]. Hence, the Darboux method is applied in many areas of physics.

Let us mention that an alternative to the Darboux method presented here is due to Abraham and Moses [53] and is based on a theorem of Gel'fand and Levitan [54, 55]. However, in the context of SUSY quantum mechanics they are equivalent [56, 57]. See also the work by Sukumar and others [2, 8, 58, 59]. For some historical remarks see [25, 60]. Such methods of constructing essential iso-spectral Hamiltonians are successfully applied, for example, in a SUSY variant of the variational method [61], in the inverse scattering method [62, 63] and in quantum cosmology [64]. Applications of higher order Darboux transformations to time-dependent Schrödinger operators and to the Fokker–Planck equation can, for example, be found in [65, 66]. For its application to position-dependent mass problems see [47]. A recent overview is given by Andrianov and Ioffe [67] and references therein. The Darboux method has also inspired the development of so-called conditionally exactly solvable potentials [68, 69] and was successfully applied to the construction of complex potentials exhibiting real eigenvalues [70]. We will discuss this to some extent in the following sections.

4.3.1 Modelling of conditionally exactly solvable potentials

As an explicit application of the Darboux method related to SUSY quantum mechanics let us consider a pair of one-dimensional Schrödinger Hamiltonians of the standard form, $\partial_x := \partial/\partial x$,

$$H_\pm = -\frac{\hbar^2}{2m}\partial_x^2 + V_\pm(x), \qquad \mathcal{H} = L^2(\mathbb{R}). \tag{4.48}$$

As a general ansatz for the intertwining operator A in equation (4.43) let us assume the form

$$A = \sum_{k=0}^{N} f_k(x)\partial_x^k, \tag{4.49}$$

where the functions $f_k : \mathbb{R} \mapsto \mathbb{R}$, $k = 0, 1, 2, \ldots, N$, are twice continuously differentiable. This ansatz is then inserted into the relation (4.43). Bringing all differential operators to the right and then equating the coefficient functions for each power of ∂_x results in coupled differential equations for the unknown functions f_k. These are in general highly non-trivial. Only for the highest power $N + 2$ one finds that $f_N(x)$ is actually arbitrary and therefore, without loss of generality one may fix this to a constant, $f_N(x) = \hbar/\sqrt{2m}$ for convenience.

Let us consider the simplest non-trivial case which is $N = 1$. Then the ansatz $A = (\hbar/\sqrt{2m})\partial_x + \Phi(x)$, i.e. $\Phi = f_0$, results in two coupled differential equations for Φ,

$$V_-(x) = V_+(x) - \frac{2\hbar}{\sqrt{2m}}\Phi'(x),$$

$$\frac{\hbar}{\sqrt{2m}}V_-'(x) + \Phi(x)V_-(x) = -\frac{\hbar^2}{2m}\Phi''(x) + \Phi(x)V_+(x).$$

(4.50)

Inserting the first one into the second equation and integration results in the Riccati equation

$$\Phi^2(x) + \frac{\hbar}{\sqrt{2m}}\Phi'(x) = V_+(x) - \varepsilon, \quad \varepsilon \in \mathbb{R},$$

(4.51)

where ε is an arbitrary constant of integration, which *a priori* could be complex but will be restricted to be real for later reasons. The non-linear Riccati equation can be linearised with the ansatz $\Phi(x) = (\hbar/\sqrt{2m})u'(x)/u(x)$ and leads to a Schrödinger-like equation:

$$\left(-\frac{\hbar^2}{2m}\partial_x^2 + V_+(x)\right)u(x) = \varepsilon u(x).$$

(4.52)

Here it must be noted that u is not restricted to be square integrable. In terms of a solution of equation (4.52) the potential V_- reads

$$V_-(x) = \frac{\hbar^2}{m}\left(\frac{u'(x)}{u(x)}\right)^2 - V_+(x) + 2\varepsilon.$$

(4.53)

Obviously, a well-defined potential requires ε to be real and u to have no zeros on the real axis to avoid singularities. Hence, we conclude that ε should be real and be bounded by the ground-state energy E_0 of H_+, $\varepsilon \leqslant E_0$. Otherwise the solution of equation (4.52) would have a zero.

Let us look at the kernel of $A^\dagger$ to see if H_+ may have an eigenvalue which does not belong to H_-. The condition $A^\dagger|\phi_\varepsilon^+\rangle = 0$ reads in the coordinate representation $\frac{d}{dx}\phi_\varepsilon^+(x) = \frac{u'(x)}{u(x)}\phi_\varepsilon^+(x)$ which results in the obvious solution $\phi_\varepsilon^+(x) = Cu(x)$, with C being a normalisation factor. As u was required to be nodeless the only normalisable solution of equation (4.52) would be the ground state of H_+ and hence $\varepsilon = E_0$. We will explicitly exclude this by requiring $\varepsilon < E_0$ from now on and therefore all eigenvalues of H_+ also belong to the spectrum of H_-.

The second case we need to consider is the kernel of A, that is, we need to look into solutions of $A|\phi_\varepsilon^-\rangle = 0$, which results in the condition $\frac{d}{dx}\phi_\varepsilon^-(x) = -\frac{u'(x)}{u(x)}\phi_\varepsilon^-(x)$ and has the solution

$$\phi_\varepsilon^-(x) = \frac{C}{u(x)}.$$

(4.54)

Compare this with case (a) discussed in section 4.1. Hence, in the case that u diverges fast enough for $x \to \pm\infty$ such that ϕ_ε^- becomes square integrable the Hamiltonian H_- possesses an additional eigenvalue ε and we have

$$\text{spec}H_- = \{\varepsilon, E_0, E_1, ...\} = \{\varepsilon\} \cup \text{spec}H_+. \tag{4.55}$$

The corresponding eigenstates for the eigenvalues E_n directly follow from equation (4.46) and explicitly read

$$\phi_n^-(x) = \frac{\hbar}{\sqrt{2m(E_n - \varepsilon)}} \left(\frac{u'(x)}{u(x)} \phi_n^+(x) - \phi_n^{+\prime}(x) \right). \tag{4.56}$$

We leave it as an exercise to the reader (see problem 4.2 below) to show that the normalisation constant in equation (4.46) is given by $|C_n|^{-2} = E_n - \varepsilon > 0$. Hence, when starting out with a Hamiltonian H_+ whose eigenvalues E_n and corresponding eigenfunctions ϕ_n^+ are known, one can construct new potentials (4.53) for which one immediately knows the spectral properties of the associated Hamiltonian H_- given by equations (4.54), (4.55), and (4.56). As certain conditions such as $\varepsilon < E_0$ and u being nodeless are involved, such newly constructed potentials have been called *conditionally exactly solvable potentials* [68, 69].

Before discussing an explicit example let us note that ε may be identified with the factorisation energy discussed in section 4.1. In fact, expressing the two potentials in terms of the function Φ,

$$V_\pm(x) = \Phi^2(x) \pm \frac{\hbar}{\sqrt{2m}} \Phi'(x) + \varepsilon \tag{4.57}$$

reveals the close relation with the Witten model of SUSY quantum mechanics where Φ plays the role of the SUSY potential. Indeed, we can identify the partner Hamiltonian as follows:

$$H_+ = AA^\dagger + \varepsilon, \qquad H_- = A^\dagger A + \varepsilon. \tag{4.58}$$

4.4 A SUSY-partner family for the harmonic oscillator

As an explicit example for the above discussion let us consider the one-dimensional harmonic oscillator

$$V_+(x) = \frac{m}{2} \omega^2 x^2. \tag{4.59}$$

The corresponding general solution of equation (4.52) for the harmonic-oscillator case is given by Kummer's confluent hypergeometric function [71],

$$u(x) = e^{-x^2/2} \left[\alpha \, _1F_1\left(\frac{1 - 2\varepsilon}{4}, \frac{1}{2}; x^2 \right) + \beta x \, _1F_1\left(\frac{3 - 2\varepsilon}{4}, \frac{3}{2}; x^2 \right) \right]. \tag{4.60}$$

For simplicity we use here units such that $\hbar = \omega = m = 1$, that is, the coordinate x is measured in units of $\lambda = \sqrt{\hbar/m\omega}$ and energies are in units of $\hbar\omega$. It is evident that the general discussion above is independent of an overall factor in the solution u and therefore we may fix the parameter α in equation (4.60) to unity, $\alpha = 1$, without loss of generality. However, the second parameter β is not arbitrary as only solutions u

without zeros are admissible. A corresponding condition can be obtained [69] by looking at the large z behaviour of u and require that to be positive. This leads to the condition

$$|\beta| < \beta_c(\varepsilon) := 2\frac{\Gamma(3/4 - \varepsilon/2)}{\Gamma(1/4 - \varepsilon/2)}. \tag{4.61}$$

Note that the eigenfunctions and eigenvalues of the harmonic oscillator are given by

$$E_n = n + \frac{1}{2}, \qquad n = 0, 1, 2, \ldots, \tag{4.62}$$

and

$$\phi_n^+(x) = \sqrt{\frac{1}{\sqrt{\pi}\, 2^n n!}}\, H_n(x) \exp\{-x^2/2\}, \tag{4.63}$$

respectively, with H_n denoting the Hermite polynomial of order n. Hence, we have in addition to (4.61) the restriction $\varepsilon < \frac{1}{2}$. The potential V_- explicitly reads in terms of the solution of (4.60)

$$V_-(x) = \left[\left(\frac{u'(x)}{u(x)}\right)^2 - \frac{1}{2}x^2 + 2\varepsilon\right]. \tag{4.64}$$

The spectral properties of the corresponding Hamiltonian H_- are given by the ground state

$$\phi_\varepsilon^-(x) = \frac{Ce^{x^2/2}}{{}_1F_1\!\left(\frac{1-2\varepsilon}{4}, \frac{1}{2}; x^2\right) + \beta x\, {}_1F_1\!\left(\frac{3-2\varepsilon}{4}, \frac{3}{2}; x^2\right)} \tag{4.65}$$

with ground-state energy $\varepsilon < 1/2$. The excited energy eigenvalues are given by equation (4.62) and the corresponding eigenfunctions read, $n = 0, 1, 2, \ldots,$

$$\phi_n^-(x) = \frac{e^{-x^2/2}}{[\sqrt{\pi}\, 2^{n+1} n!(n + 1/2 - \varepsilon)]^{1/2}}\left[H_{n+1}(x) + \left(\frac{u'(x)}{u(x)} - x\right)H_n(x)\right]. \tag{4.66}$$

In figure 4.6 we show the family of potentials V_- for the symmetric case $\beta = 0$ and for the range $-\frac{5}{2} < \varepsilon < \frac{1}{2}$. For values of ε close to 1/2 we see two deep potential wells where, as for large negative ε, the potential V_- has a single deep dip at the origin and for large x values is essentially identical to V_+. For an illustration of V_- in the non-symmetric case $\beta \neq 0$ we refer to the work by Sukumar [62]. It should also be noted that we can in principle allow for complex values of $\beta \in \mathbb{C}/\{[-\infty, -\beta_c(\varepsilon)] \cup [\beta_c(\varepsilon), \infty]\}$. In that case H_- is no longer Hermitian, however, it still has real eigenvalues bounded from below [70]. In particular, for a purely imaginary $\beta \in i\mathbb{R}$ the potential obeys the symmetry $V_-(x) = V_-^*(-x)$ and hence H_- is invariant under so-called PT transformations. These parity-time-symmetric Hamiltonians have attracted considerable

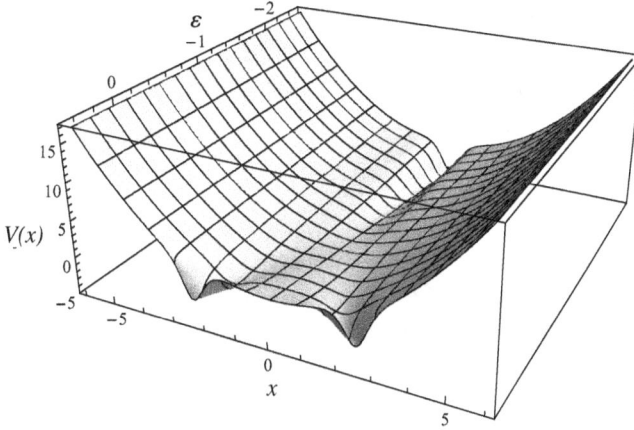

Figure 4.6. The family of conditionally exactly solvable potentials $V_-(x)$ related to the harmonic oscillator (4.59) for the symmetric case $\beta = 0$ in equation (4.60). Units are such that $\hbar = m = \omega = 1$.

attention in recent years [72, 73]. For a discussion and various graphs of V_- for complex β we refer the reader to the work [70].

For a complete discussion of conditionally exactly solvable potentials related to all shape-invariant potentials as listed in table 4.1 see the work by Junker and Roy [68, 69, 74].

4.4.1 Generalised annihilation and creation operators, non-linear algebras

For the harmonic oscillator $H_+ = a^\dagger a + \frac{1}{2}$ corresponding to potential (4.59) the standard annihilation and creation operators are given by

$$a = \frac{1}{\sqrt{2}}(\partial_x + x), \qquad a^\dagger = \frac{1}{\sqrt{2}}(-\partial_x + x) \tag{4.67}$$

and obey the well-known Heisenberg algebra

$$[H_+, a] = -a, \qquad [H_+, a^\dagger] = a^\dagger, \qquad [a, a^\dagger] = 1. \tag{4.68}$$

These operators act on the harmonic-oscillator eigenstates ϕ_n^+ as lowering and raising operators:

$$a\phi_n^+(x) = \sqrt{n}\,\phi_{n-1}^+(x), \qquad a^\dagger\phi_n^+(x) = \sqrt{n+1}\,\phi_{n+1}^+(x). \tag{4.69}$$

Hence, the question arises if similar annihilation and creation operators can be defined for H_-. For this let us define the generalised operators

$$B := A^\dagger a A, \qquad B^\dagger = A^\dagger a^\dagger A. \tag{4.70}$$

Obviously, the ground state $\phi_\varepsilon^- \in \ker A$ is annihilated by both operators, that is,

$$B\phi_\varepsilon^-(x) = 0 = B^\dagger \phi_\varepsilon^-(x). \tag{4.71}$$

For the excited states ϕ_n^- it is easy to show that

$$B\phi_{n+1}^-(x) = \sqrt{(E_n - \varepsilon)(n+1)(E_{n+1} - \varepsilon)}\,\phi_n^-(x)$$
$$B^\dagger\phi_n^-(x) = \sqrt{(E_{n+1} - \varepsilon)(n+1)(E_n - \varepsilon)}\,\phi_{n+1}^-(x) \tag{4.72}$$
$$B\phi_0^-(x) = 0.$$

Hence the generalised operators (4.70) indeed act as annihilation and creation operators on the eigenstates of H_- except for the ground state (4.71) which remains isolated. The algebra obeyed by these operators, however, is no longer a standard Heisenberg algebra but rather becomes a non-linear deformed Heisenberg algebra. It explicitly reads

$$[H_-, B] = -B, \qquad [H_-, B^\dagger] = B^\dagger, \qquad [B^\dagger, B] = 3H_-^2 - 4\varepsilon H_- + \varepsilon^2. \tag{4.73}$$

This quadratic algebra belongs to the so-called W_2-class of algebras. It may be viewed as a polynomial deformation of the $su(1, 1)$ Lie algebra [75–77].

4.4.2 Generalised coherent states

The canonical coherent states may be defined as eigenstates of the annihilation operator, that is,

$$a|\alpha\rangle = \alpha|\alpha\rangle, \qquad \alpha \in \mathbb{C}. \tag{4.74}$$

Utilising the relation $|\phi_n^+\rangle = (a^\dagger)^n/\sqrt{n!}\,|\phi_0^+\rangle$ we find

$$\langle\phi_n^+|\alpha\rangle = \frac{\alpha^n}{\sqrt{n!}}\langle\phi_0^+|\alpha\rangle. \tag{4.75}$$

Together with $|\langle\phi_0^+|\alpha\rangle|^2 = \exp\{-|\alpha|^2/2\}$ we arrive at the representation

$$|\alpha\rangle = e^{-|\alpha|^2/2}\sum_{n=0}^\infty \frac{\alpha^n}{\sqrt{n!}}|\phi_n^+\rangle = \exp\{-|\alpha|^2/2 + \alpha a^\dagger\}|\phi_0^+\rangle. \tag{4.76}$$

The canonical coherent states are overcomplete in the sense that

$$\langle\alpha|\beta\rangle = \exp\{-|\alpha|^2/2 - |\beta|^2/2 + \alpha^*\beta\} \tag{4.77}$$

and provide the resolution of unity as follows:

$$\int_\mathbb{C} \frac{\mathrm{d}^2\alpha}{\pi}|\alpha\rangle\langle\alpha| = 1. \tag{4.78}$$

The last relation is easily shown using the polar coordinate representation in the form $\alpha = \sqrt{x}\,e^{i\varphi}$ together with $\mathrm{d}^2\alpha/\pi = \mathrm{d}x\mathrm{d}\varphi/2\pi$ and $\int_0^\infty \mathrm{d}x\, x^n e^{-x} = n!$. Let us stress that this resolution of unity is an essential requirement in the definition of general coherent states [78]. The proof of the existence of a positive measure such as $\mathrm{d}^2\alpha/\pi$ in the case of canonical coherent states is usually non-trivial and closely related to the inverse momentum problem. This is the reconstruction of a probability distribution, i.e. a measure, out of its known momenta. We will perform this process now for the family of the SUSY partners of the harmonic oscillator.

Let us start with the definition of generalised non-linear coherent states as eigenstates of the generalised annihilation operator B obeying the non-linear algebra (4.73),

$$B|\mu\rangle = \mu|\mu\rangle \quad \mu \in \mathbb{C}. \tag{4.79}$$

As the ground states ϕ_ε^- of H_- are isolated we will construct these coherent states over the excited states $\{|\phi_n^-\rangle\}_{n \in \mathbb{N}_0}$ only. That is, we make the ansatz

$$|\mu\rangle = \sum_{n=0}^{\infty} c_n \mu^n |\phi_n^-\rangle. \tag{4.80}$$

Using the (4.79) and the first relation in equation (4.72) we arrive at the recursion relation

$$c_{n+1} = c_n \left[\left(n + \frac{1}{2} - \varepsilon \right)(n+1)\left(n + \frac{3}{2} - \varepsilon \right) \right]^{-1/2}. \tag{4.81}$$

Hence, we can express the coefficients c_n in terms of c_0 and the Pochhammer symbol $(z)_n := \Gamma(z+n)/\Gamma(z)$ as follows:

$$c_n = c_0 \left[n! \left(\frac{1}{2} - \varepsilon \right)_n \left(\frac{3}{2} - \varepsilon \right)_n \right]^{-1/2}. \tag{4.82}$$

The remaining coefficient $c_0 = c_0(|\mu|^2)$ is determined via the normalisation condition $\langle \mu|\mu \rangle = 1$ and may be expressed in terms of a generalised hypergeometric function

$$c_0^{-2}(|\mu|^2) = \sum_{n=0}^{\infty} \frac{|\mu|^{2n}}{n!} \frac{1}{\left(\frac{1}{2} - \varepsilon \right)_n \left(\frac{3}{2} - \varepsilon \right)_n} = {}_0F_2\left(\frac{1}{2} - \varepsilon, \frac{3}{2} - \varepsilon; |\mu|^2 \right). \tag{4.83}$$

The generalised hypergeometric function is defined in terms of a power series [79]:

$$_pF_q(\alpha_1, \alpha_2, \ldots, \alpha_p; \gamma_1, \gamma_2, \ldots, \gamma_q; z) := \sum_{n=0}^{\infty} \frac{(\alpha_1)_n(\alpha_2)_n \cdots (\alpha_p)_n}{(\gamma_1)_n(\gamma_2)_n \cdots (\gamma_q)_n} \frac{z^n}{n!}.$$

As expected these non-linear coherent states are overcomplete and obey the obvious relation

$$\langle \mu|\nu \rangle = \frac{{}_0F_2\left(\frac{1}{2} - \varepsilon, \frac{3}{2} - \varepsilon; \mu^*\nu \right)}{\sqrt{{}_0F_2\left(\frac{1}{2} - \varepsilon, \frac{3}{2} - \varepsilon; |\mu|^2 \right) {}_0F_2\left(\frac{1}{2} - \varepsilon, \frac{3}{2} - \varepsilon; |\nu|^2 \right)}}. \tag{4.84}$$

However, the derivation of the resolution of unity is less trivial. In order to construct a proper measure on the complex μ-plane we make an ansatz which splits off the isolated subspace spanned by $|\phi_\varepsilon^-\rangle$,

$$|\phi_\varepsilon^-\rangle\langle\phi_\varepsilon^-| + \int_C d^2\rho(\mu^*, \mu)|\mu\rangle\langle\mu| = 1. \tag{4.85}$$

It is evident that the measure ρ only depends on the modulus $|\mu|$ and hence with the further substitution $\mu = \sqrt{x}\,e^{i\varphi}$ and

$$d^2\rho(\mu^*, \mu) = d\varphi dx \frac{\sigma(x)}{2\pi c_0^2(x)} \tag{4.86}$$

we obtain the condition

$$\int_0^\infty dx\, \sigma(x)x^n = \Gamma(n+1)\frac{\Gamma\left(\dfrac{1}{2} - \varepsilon + n\right)\Gamma\left(\dfrac{3}{2} - \varepsilon + n\right)}{\Gamma\left(\dfrac{1}{2} - \varepsilon\right)\Gamma\left(\dfrac{3}{2} - \varepsilon\right)}. \tag{4.87}$$

This yet unknown probability density σ is defined via equation (4.87) by its moments. Here it shall be noted that the integral in equation (4.87) represents a so-called Mellin transformation [80, 81]. Hence, the inverse Mellin transformation allows us to calculate σ. It turns out that in the current case this inverse Mellin transformation reduces to the integral representation of Meijer's G-function [79] which in turn is closely related to the generalised hypergeometric function. Hence, we have the explicit form, see figure 4.7,

$$\sigma(x) = \frac{1}{\Gamma\left(\dfrac{1}{2} - \varepsilon\right)\Gamma\left(\dfrac{3}{2} - \varepsilon\right)} G_{03}^{30}\left(x \middle| 0, -\frac{1}{2} - \varepsilon, \frac{1}{2} - \varepsilon\right). \tag{4.88}$$

Let us consider a few properties of these generalised non-linear coherent states. First let us consider the time evolution generated by H_-. This turns out to be rather simple as

$$|\mu(t)\rangle := \exp\{-itH_-/\hbar\}|\mu\rangle = e^{-i\omega t/2}|\mu e^{-i\omega t}\rangle. \tag{4.89}$$

For convenience we have used here the explicit ω dependence. It is also instructive to look into the dynamics of some observables. Let us consider the two self-adjoint operators

$$X := \frac{1}{2}(B + B^\dagger), \qquad P := \frac{\omega}{2i}(B - B^\dagger), \tag{4.90}$$

which may be interpreted as generalised position and momentum operators, respectively. From the above result follows that their corresponding expectation values perform a harmonic motion,

$$\langle\mu(t)|X|\mu(t)\rangle = \mu\cos(\omega t), \qquad \langle\mu(t)|P|\mu(t)\rangle = -\mu\omega\sin(\omega t). \tag{4.91}$$

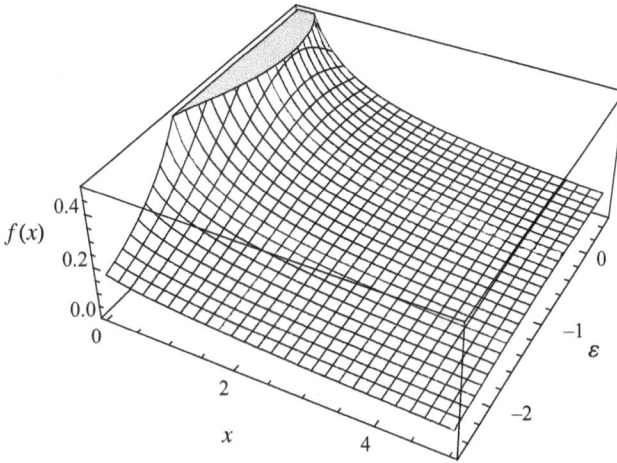

Figure 4.7. The density $f(x) = \sigma(x)/c_0^2(x)$ of the resolution of unity (4.86) of the generalised non-linear coherent states.

We may also consider the uncertainty relation for these observables. Their variance in an arbitrary state ψ is defined by

$$\langle \Delta X \rangle_\psi^2 := \langle \psi | X^2 | \psi \rangle - \langle \psi | X | \psi \rangle^2, \qquad \langle \Delta P \rangle_\psi^2 := \langle \psi | P^2 | \psi \rangle - \langle \psi | P | \psi \rangle^2 \qquad (4.92)$$

and obey the general uncertainty relation

$$\langle \Delta X \rangle_\psi^2 \langle \Delta P \rangle_\psi^2 \geqslant \frac{1}{4} |\langle \psi | [X, P] | \psi \rangle|^2. \qquad (4.93)$$

We leave it as an exercise to the reader to show that the generalised coherent states $|\mu\rangle$ are minimum uncertainly states in the sense that they obey the equality sign in the above relation.

Let us conclude that the above program executed for the linear harmonic oscillator, following the original work [68, 69, 82], can also be performed for the radial harmonic oscillator, that is, for a potential of the form $V_+(x) = x^2/2 + \gamma(\gamma + 1)/2x^2$ on the positive half line $\mathbb{R}^+$. Here unbroken as well as broken SUSY can be studied leading to a cubic algebra [68, 74]. The associated non-linear coherent states have been discussed in [83].

4.5 Classical fields in (1 + 1) dimensions

So far we have considered quantum mechanical models realising a SUSY structure and leading to exact results. However, Schrödinger-like differential equations also appear in many other areas of physics. Here we will considered classical fields in (1 + 1) dimensions, that is, scalar fields defined on the real line and evolving in time. These field theoretic models are of particular interest as they can exhibit finite-energy solutions such as the soliton solution and the kink solution of the sine-Gordon and the ϕ^4-model, respectively. With the concept of shape-invariance it is possible to construct new stable and unstable field models with finite-energy configurations having an exactly solvable fluctuation equation. The discussion here will be limited to the construction of stable field models. For the case of unstable field models see [84].

The scalar fields to be discussed here are real-valued functions $\phi: \mathbb{R} \times \mathbb{R} \to \mathbb{R}$, $(x, t) \mapsto \phi(x, t)$ with vanishing variations at infinity, that is, $\partial_x \phi \to 0$ and $\partial_t \phi \to 0$ for $x, t \to \pm\infty$. The corresponding Lagrange density is defined as

$$\mathcal{L}(\partial\phi, \phi) := \frac{1}{2}(\partial_t\phi)^2 - \frac{1}{2}(\partial_x\phi)^2 - U(\phi) = \frac{1}{2}(\partial_\mu\phi)(\partial^\mu\phi) - U(\phi) \qquad (4.94)$$

with a real-valued field potential U bounded from below, i.e. $U \geqslant 0$. Prominent examples are $U(\phi) = \frac{1}{2}\phi^2$, $U(\phi) = 1 + \cos\phi$, and $U(\phi) = \frac{1}{2}(1 - \phi^2)^2$, which describe the well-known Klein–Gordon, sine-Gordon, and ϕ^4-models, respectively. The classical equation of motion follows from the Euler–Lagrange equation and explicitly reads

$$\ddot{\phi} - \phi'' = -\frac{\partial U}{\partial\phi}. \qquad (4.95)$$

From now on we use the dot and the prime to indicate partial differentiation with respect to time-variable t and space-variable x, respectively. It can be shown that solutions of this equation lead to a conservation of the energy functional

$$E[\phi] := \int_\mathbb{R} dx \left[\frac{1}{2}\dot{\phi}^2 + \frac{1}{2}\phi'^2 + U(\phi)\right], \qquad (4.96)$$

that is, $dE[\phi]/dt = 0$. If, in addition, we assume that $U(\phi) \to 0$ as $x \to \pm\infty$ such solutions lead to a finite-energy functional. Remember that we also assumed $\dot{\phi} \to 0$ and $\phi' \to 0$ for $x \to \pm\infty$. These localised classical solutions, obeying the boundary conditions

$$\phi_\pm := \lim_{x \to \pm\infty} \phi(x, t) \qquad \text{with} \qquad U(\phi_\pm) = 0, \qquad (4.97)$$

are called solitary waves if they can be put into the form $\phi(x, t) = f(x - vt)$ and can be obtained from static, i.e. time-independent, solutions $\phi_s = \phi_s(x)$ of equation (4.95) via a Lorentz-boost. For such static solutions the equation of motion (4.95) reduces to the equation

$$\frac{1}{2}\phi_s'^2(x) = U(\phi_s(x)). \qquad (4.98)$$

The constant solutions $\phi_\pm$ are the classical vacua of the field theory.

The objective of this section is to construct field models, i.e. find appropriate field potentials U, which lead to such static finite-energy solution which are stable under small fluctuations. Hence we consider small fluctuations ψ about the static solution,

$$\phi(x) = \phi_s(x) + \psi(x) \qquad \text{with} \qquad \psi(x) \to 0 \quad \text{as} \quad x \to \pm\infty, \qquad (4.99)$$

leading to a fluctuation in the energy functional $E[\phi] \approx E[\phi_s] + \delta E[\psi]$ with lowest order in ψ being of the form

$$\delta E[\psi] := \frac{1}{2} \int_{\mathbb{R}} \mathrm{d}x\, \psi(x) \left[-\partial_x^2 + U''(\phi_s(x)) \right] \psi(x). \tag{4.100}$$

The stability condition $\delta E[\psi] \geqslant 0$ implies that the eigenvalues of the stability operator

$$H := -\partial_x^2 + U''(\phi_s(x)) \tag{4.101}$$

are non-negative. More explicitly, let us assume we know the spectral properties of H, that is, its eigenvalues μ_n and eigenfunctions ψ_n,

$$H\psi_n = \mu_n \psi_n, \tag{4.102}$$

then the fluctuations ψ can be written as $\psi(x) = \sum_n a_n \psi_n(x)$ with $a_n = \int \mathrm{d}x\, \psi_n^*(x)\psi(x)$ resulting in

$$\delta E[\psi] = \frac{1}{2} \sum_n \mu_n |a_n|^2. \tag{4.103}$$

Hence, stability indeed requires all eigenvalues to be non-negative, $\mu_n \geqslant 0$. In fact, it is easy to show that the lowest eigenvalue $\mu_0 = 0$ and the corresponding zero mode is given by $\psi_0(x) = C\phi_s'(x)$. This is due to the translation invariance of the energy functional. It is obvious that equation (4.96) is invariant under translations $\phi_s(x) \to \phi_s(x + \delta x)$. On the other side a translation corresponds to a fluctuation along the zero mode ψ_0 as $\phi_s(x + \delta x) \simeq \phi_s(x) + \delta x \phi_s'(x) = \phi_s(x) + (\delta x/C)\psi_0(x)$. In other words, the fluctuation operator H may be considered as the SUSY-partner Hamiltonian H_- of Witten's model with units $2m = \hbar = 1$ and a SUSY potential denoted by W:

$$H \equiv H_- = -\partial_x^2 + W^2(x) - W'(x). \tag{4.104}$$

The idea for constructing field models with an exactly solvable fluctuation equation is now reduced to choosing an appropriate SUSY potential W leading to an exactly solvable SUSY Hamiltonian. This, for example, can be one belonging to the class of the known shape-invariant SUSY potentials or one of the conditionally exactly solvable systems.

The programme would be as follows. Select the SUSY potential W, calculated the zero mode $\psi_0(x) = N \exp\left\{ -\int \mathrm{d}x\, W(x) \right\}$, integrate to find the static solution $\phi_s(x) = \int \mathrm{d}x\, \psi_0(x)/C$, and use relation (4.98) to find the corresponding field potential by eliminating x. In case the resulting field potential is *a priori* only defined in a finite range limited by the vacuum solutions $\phi_\pm$ it may be analytically continued to all values for ϕ. The main problem in this programme is to find explicit analytic expressions for the integrals involved.

As an example, let us consider $W(x) = \tanh(x)$ which is the SUSY partner of the free particle (cf. example 3.3 in section 3.6.1). The zero mode is given by $\psi_0(x) = \sqrt{2}/\cosh x$ and with the choice $C = 1/\sqrt{2}$ the static solution reads

$\phi_s(x) = 2\arcsin(\tanh x)$. Now we utilise relation (4.98) to find an explicit expression for the field potential

$$U(\phi_s(x)) = \frac{1}{2}\phi_s'^2(x) = \frac{2}{\cosh^2 x} = 2(1 - \tanh^2 x)$$

$$= 2\left(1 - \sin^2\frac{\phi_s}{2}\right) = 1 + \cos\phi_s. \tag{4.105}$$

Here we have made use of the relation $\sin\phi_s/2 = \tanh x$ to eliminate x. Analytic continuation beyond the values $\phi_\pm = \pm\pi$ then leads to the well-known sine-Gordon model:

$$U(\phi) = 1 + \cos\phi. \tag{4.106}$$

This programme has systematically been executed in [84] for all shape-invariant SUSY potentials on the real line. For example, the SUSY potential $W(x) = 2\tanh x$ results in the known ϕ^4-model but for $W(x) = 4\tanh x$ a new explicit field potential was found in addition to others related to the δ- and Morse potential. In addition, also unstable field models were considered.

4.6 Problems

Problem 4.1. Shape-invariance of the symmetric Pöschl–Teller potential

In analogy to section 4.2.1 consider a family of SUSY potentials given by

$$\Phi(a_0, x) := \frac{\hbar}{\sqrt{2m}} a_0 \tan x, \qquad x \in [-\pi/2, \pi/2], \qquad a_0 > 0.$$

Show that this leads to shape-invariant partner potentials and calculate the spectral properties of the associated SUSY Hamiltonians.

Problem 4.2. Alternative approach to conditionally exactly solvable potentials

Consider the Witten model with a generic SUSY potential W, that is, a pair of Hamiltonians $H_\pm = -\frac{1}{2}\partial_x^2 + V_\pm(x)$ with $V_\pm(x) = \frac{1}{2}W^2(x) \pm \frac{1}{2}W'(x)$, $\hbar = m = 1$. Furthermore assume that $W(x) = \Phi(x) + g(x)$ where Φ is a SUSY potential belonging to a shape-invariant system.

(a) Show that for

$$g^2(x) + 2\Phi(x)g(x) + g'(x) = b, \qquad b = \text{const.}$$

the potential V_+ becomes a shape-invariant, hence, exactly solvable potential. That is,

$$V_+(x) = \frac{1}{2}[\Phi^2(x) + \Phi'(x) + b]$$

and the partner potential reads

$$V_-(x) = \frac{1}{2}[\Phi^2(x) - \Phi'(x) - 2g'(x) + b].$$

(b) Use the ansatz $g(x) = \frac{u'(x)}{u(x)} - \Phi(x)$ to derive a Schrödinger-like equation for u and compare this with equation (4.52). Which conditions must be met by the parameter b and the function u to allow for an exact solution of the eigenvalue problem of H_-?

(c) Discuss explicitly the example $\Phi(x) = x$ and compare this with the results of section 4.4.

See [69] for a complete discussion for all shape-invariant cases listed in table 4.1.

Problem 4.3. The harmonic-oscillator SUSY partner for $\varepsilon = -5/2$ and $\beta = 0$

Consider the special member $\varepsilon = -5/2$ and $\beta = 0$ of the family of SUSY partners of the harmonic oscillator discussed in section 4.4. This case corresponds to the solution of the Schrödinger-like equation (4.52) given by $u(x) = -(4x^2 + 2)\exp\{x^2/2\}$. Show that the corresponding partner potential explicitly reads

$$V_-(x) = \frac{x^2}{2} + \frac{8x^2}{2x^2 + 1} + \left(\frac{4x}{2x^2 + 1}\right)^2 - 5.$$

Problem 4.4. A family of SUSY partners for the free particle

Following section 4.4 consider the case of a free particle, that is, $V_+(x) = 0$ and show that for this case the general solution of (4.52) reads $u(x) = e^{\kappa x} + \beta e^{-\kappa x}$, where $\varepsilon = -\hbar^2\kappa^2/2m$. What are the allowed values for the parameter β and ε? Show that the family of partner potentials of the free particle explicitly reads

$$V_-(x) = \frac{\hbar^2\kappa^2}{m}\left[\left(\frac{e^{\kappa x} - \beta e^{-\kappa x}}{e^{\kappa x} + \beta e^{-\kappa x}}\right)^2 - 1\right].$$

Derive the eigenfunction of the corresponding Hamiltonian H_- associated with the eigenvalues $E_k = \hbar^2 k^2/2m$ with $k \in \mathbb{R}$. Why is this class of potentials reflectionless?

Problem 4.5. Generalised coherent states minimising the uncertainty relation

Prove that for the generalised coherent states (4.79) the equals sign is valid in the uncertainty relation (4.93). That is,

$$\langle \Delta X \rangle_\mu^2 \langle \Delta P \rangle_\mu^2 = \frac{1}{4}|\langle \mu |[X, P]|\mu\rangle|^2.$$

Problem 4.6. Partition function and internal energy of the HO SUSY-partner family

Consider the SUSY-partner Hamiltonians of the harmonic oscillator discussed in section 4.4 but shifted by the factorisation energy ε. That is, $\text{spec}H_+ = \left\{ n + \frac{1}{2} - \varepsilon : n \in \mathbb{N}_0 \right\}$ with $\varepsilon < 1/2$ and $\text{spec}H_- = \{0\} \cup \text{spec}H_+$. Calculate the partition functions $Z_\pm(\beta) := \text{Tr}\, e^{-\beta H_\pm}$, the associated internal energies $U_\pm(\beta) := -\partial_\beta \ln Z_\pm(\beta)$ and prove the relation (2.68), that is, $U_-(\beta)Z_-(\beta) = U_+(\beta)Z_+(\beta)$.

Problem 4.7. The double quadratic field model via SUSY

Consider the SUSY potential $W(x) = \text{sgn}\, x$, where $\text{sgn}\, x$ denotes the step function, i.e. $\text{sgn}\, x = x/|x|$ for $x \neq 0$ and $\partial_x \text{sgn}\, x = 2\delta(x)$. Execute the programme discussed in section 4.5 and show that this SUSY potential induces a field potential of the form

$$U(\phi) = \frac{1}{2}(1 - |\phi|)^2.$$

References

[1] Andrianov A A, Borisov N V and Ioffe M V 1984 The factorization method and quantum systems with equivalent energy spectra *Phys. Lett.* A **105** 19–22

[2] Sukumar C V 1985 Supersymmetry, factorisation of the Schrödinger equation and a Hamiltonian hierarchy *J. Phys. A: Math. Gen.* **18** L57–61

[3] Gendenshteîn L É 1983 Derivation of exact spectra of the Schrödinger equation by means of supersymmetry *JETP Lett.* **38** 356–8

[4] Schrödinger E 1940 A method of determining quantum-mechanical eigenvalues and eigenfunctions *Proc. R. Irish Acad.* A **46** 9–16

[5] Infeld L and Hull T E 1951 The factorization method *Rev. Mod. Phys.* **23** 21–68

[6] Darboux G 1882 Sur une proposition relative aux équations linéaires *C. R. Acad. Sci.* **94** 1456–9

[7] de Crombrugghe M and Rittenberg V 1983 Supersymmetric quantum mechanics *Ann. Phys.* **151** 99–126

[8] Sukumar C V 1985 Supersymmetric quantum mechanics of one-dimensional systems *J. Phys. A: Math. Gen.* **18** 2917–36

[9] Ince E L 1956 *Ordinary Differential Equations* (New York: Dover)

[10] Gozzi E 1983 Ground-state wave-function 'representation' *Phys. Lett.* B **129** 432–6

[11] Jevicki A and Rodrigues J P 1984 Singular potentials and supersymmetry breaking *Phys. Lett.* B **146** 55–8

[12] Fuchs J 1986 Physical state conditions and supersymmetry breaking in quantum mechanics *J. Math. Phys.* **27** 349–53

[13] Shifman M A, Smilga A V and Vainshtein A I 1988 On the Hilbert space of supersymmetric quantum systems *Nucl. Phys.* B **299** 79–90

[14] Falomir H and Pisani P A G 2005 Self-adjoint extensions and SUSY breaking in supersymmetric quantum mechanics *J. Phys. A: Math. Gen.* **38** 4665–83

[15] Magnus W, Oberhettinger F and Soni R P 1966 *Formulas and Theorem for the Special Functions of Mathematical Physics* 3rd edn (Berlin: Springer)

[16] Gildner E and Patrascioiu A 1977 Pseudoparticle contributions to the energy spectrum of a one-dimensional system *Phys. Rev.* D **16** 423–30

[17] Salomonson P and van Holten J W 1982 Fermionic coordinates and supersymmetry in quantum mechanics *Nucl. Phys.* B **196** 509–31

[18] Kaul R K and Mizrachi L 1989 On non-perturbative contributions to vacuum energy in supersymmetric quantum mechanical models *J. Phys. A: Math. Gen.* **22** 675–85

[19] Keung W-Y, Kovacs E and Sukhatme U 1988 Supersymmetry and double-well potentials *Phys. Rev. Lett.* **60** 41–4

[20] Gangopadhyaya A, Panigrahi P K and Sukhatme U P 1993 Supersymmetry and tunneling in an asymmetric double well *Phys. Rev.* A **47** 2720–4

[21] Giler S, Kosiński P, Rembieliński J and Maślanka P 1989 Ground-state energy calculation in broken SUSYQM by a WKB-like method *J. Phys. A: Math. Gen.* **22** 647–61

[22] Cooper F, Ginocchio J N and Khare A 1987 Relationship between supersymmetry and solvable potentials *Phys. Rev.* D **36** 2458–73

[23] Roy B, Roy P and Roychoudhury R 1991 On solutions of quantum eigenvalue problems: a supersymmetric approach *Fortschr. Phys.* **89** 211–58

[24] Baumgartner B 1991 Perturbations of supersymmetric systems in quantum mechanics *Recent Developments in Quantum Mechanics, Mathematical Physics Studies Nr. 12* ed A Boutet de Monvel, P Dita, G Nenciu and R Purice (Dordrecht: Kluwer) pp 195–208

[25] Grosse H 1991 Supersymmetric quantum mechanics *Recent Developments in Quantum Mechanics, Mathematical Physics Studies Nr. 12* ed A Boutet de Monvel, P Dita, G Nenciu and R Purice (Dordrecht: Kluwer) pp 299–327

[26] Schrödinger E 1941 Further studies on solving eigenvalue problems by factorization *Proc. R. Irish Acad.* A **46** 183–206

[27] Schrödinger E 1941 The factorization of the hypergeometric equation *Proc. R. Irish Acad.* A **47** 53–4

[28] Infeld L 1941 On a new treatment of some eigenvalue problems *Phys. Rev.* **59** 737–47

[29] Hull T E and Infeld L 1948 The factorization method, hydrogen intensities, and related problems *Phys. Rev.* **74** 905–9

[30] Green H S 1966 *Quantenmechanik in algebraischer Darstellung* (Berlin: Springer)

[31] deLange O L and Raab R E 1991 *Operator Methods in Quantum Mechanics* (Oxford: Clarendon)

[32] Mielnik B and Rosas-Ortiz O 2004 Factorization: little or great algorithm? *J. Phys. A: Math. Gen.* **37** 10007–35

[33] Joseph A 1967 Self-adjoint ladder operators (I) *Rev. Mod. Phys.* **39** 829–37

[34] Miller W Jr 1968 *Lie Theory and Special Functions* (New York: Academic) pp 267–76

[35] Olshanetsky M A and Perelomov A M 1983 Quantum integrable systems related to Lie symmetries *Phys. Rep.* **94** 313–404

[36] Barut A O and Roy P 1992 The embedding of supersymmetry into the dynamical groups *Group Theory in Physics AIP Conference Proceedings* vol 266 ed A Frank, T Seligman and K B Wolf (New York: American Institute of Physics) pp 248–54

[37] Baake M, Delbourgo R and Jarvis P D 1991 Models for supersymmetric quantum mechanics *Aust. J. Phys.* **44** 353–62

[38] Oikonomou V K 2014 A relation between Z^3-graded symmetry and shape invariant supersymmetric systems *J. Phys. A: Math. Theor.* **47** 1–17

[39] Lévai G 1989 A search for shape-invariant solvable potentials *J. Phys. A: Math. Gen.* **22** 689–702

[40] Wittmer S and Weiss U 1992 A shape-invariant generalization of the harmonic oscillator in supersymmetric quantum mechanics *University of Stuttgart preprint* unpublished

[41] Hongler M-O and Zheng W M 1983 Exact results for the diffusion in a class of asymmetric bistable potentials *J. Math. Phys.* **24** 336–40

[42] Zheng W M 1984 The Darboux transformation and solvable double-well potential models for Schrödinger equations *J. Math. Phys.* **25** 88–90

[43] Khare A and Sukhatme U P 1993 New shape-invariant potentials in supersymmetric quantum mechanics *J. Phys. A: Math. Gen.* **26** L901–4

[44] Barclay D T, Dutt R, Gangopadhyaya A, Khare A, Pagnamenta A and Sukhatme U P 1993 New exactly solvable Hamiltonians: shape invariance and self-similarity *Phys. Rev.* A **48** 2786–97

[45] Cooper F, Khare A and Sukhatme U 1995 Supersymmetry and quantum mechanics *Phys. Rep.* **251** 267–385

[46] Odake S and Sasaki R 2005 Shape invariant potentials in discrete quantum mechanics *J. Nonlinear Math. Phys. Supp.* **1 12** 507–21

[47] Gangulyand A and Nieto L M 2007 Shape-invariant quantum Hamiltonian with position-dependent effective mass through second-order supersymmetry *J. Phys. A: Math. Theor.* A **40** 7265–81

[48] Gangopadhyaya A, Mallow J V and Sukhatme U P 1998 Translational shape invariance and the inherent potential algebra *Phys. Rev.* A **58** 4287–92

[49] Inomata A and Junker G 1993 Quasi-classical approach to path integrals in supersymmetric quantum mechanics *Lectures on Path Integration: Trieste 1991* ed H Cerdeira, S Lundqvist, D Mugnai, A Ranfagni, V Sa-yakanit and L S Schulman (Singapore: World Scientific) pp 460–82

[50] Inomata A and Junker G 1993 Quasi-classical approach in supersymmetric quantum mechanics *Proceedings of International Symposium on Advanced Topics of Quantum Physics* ed J Q Liang, M L Wang, S N Qiao and D C Su (Beijing: Science Press) pp 61–74

[51] Suparmi A 1992 Semi-classical quantization rules in supersymmetric quantum mechanics *PhD Thesis* State University of New York, Albany

[52] Inomata A, Junker G and Suparmi A 1993 Remarks on semiclassical quantization rule for broken SUSY *J. Phys. A: Math. Gen.* **26** 2261–4

[53] Abraham P B and Moses H E 1980 Changes in potentials due to changes in the point spectrum: anharmonic oscillators with exact solutions *Phys. Rev.* A **22** 1333–40

[54] Gel'fand I M and Levitan B M 1951 On the determination of a differential equation from its spectral function *Am. Math. Soc. Transl.* **1** 253–304

[55] Faddeyev L D 1963 The inverse problem in the quantum theory of scattering *J. Math. Phys.* **4** 72–104

[56] Nieto M M 1984 Relationship between supersymmetry and the inverse method in quantum mechanics *Phys. Lett.* B **145** 208–10

[57] Sinha A and Roy P 2004 Comment on The Darboux transformation and algebraic deformations of shape-invariant potentials *J. Phys. A: Math. Gen.* **37** 8401–4

[58] Luban M and Pursey D L 1986 New Schrödinger equations for old: inequivalence of Darboux and Abraham–Moses constructions *Phys. Rev.* D **33** 431–6

[59] Pursey D L 1986 New families of isospectral Hamiltonians *Phys. Rev.* D **33** 1048–55

[60] Rosu H C 1999 Short survey of Darboux transformations *Symmetries in Quantum Mechanics and Quantum Optics* ed A Ballesteros *et al* (Burgos: Serv. de Publ. Univ. Burgos) pp 301–15

[61] Gozzi E, Reuter M and Thacker W D 1993 Variational methods via supersymmetric techniques *Phys. Lett.* A **183** 29–32

[62] Sukumar C V 1985 Supersymmetric quantum mechanics and the inverse scattering method *J. Phys. A: Math. Gen.* **18** 2937–55

[63] Baye D and Sparenberg J-M 1994 Most general form of phase-equivalent radial potentials for arbitrary modifications of the bound spectrum *Phys. Rev. Lett.* **73** 2789–92

[64] García A, Guzmàn W, Sabido M and Socorro J 2006 Iso-spectral potentials and inflationary quantum cosmology *Int. J. Theor. Phys.* **45** 2483–96

[65] Cannata F, Ioffe M, Junker G and Nishnianidze D 1999 Intertwining relations of non-stationary Schrödinger operators *J. Phys. A: Math. Gen.* **32** 3583–98

[66] Song D-Y and Klauder J R 2003 Generalization of the Darboux transformation and generalized harmonic oscillators *J. Phys. A: Math. Gen.* **36** 8673–84

[67] Andrianov A A and Ioffe M V 2012 Nonlinear supersymmetric quantum mechanics: concepts and realizations *J. Phys. A: Math. Theor.* **45** 503001

[68] Junker G and Roy P 1997 Conditionally exactly solvable problems and non-linear algebras *Phys. Lett.* A **232** 155–61

[69] Junker G and Roy P 1998 Conditionally exactly solvable potentials: a supersymmetric construction method *Ann. Phys.* **270** 155–77

[70] Cannata F, Junker G and Trost J 1998 Schrödinger operators with complex potential but real spectrum *Phys. Lett.* A **246** 219–26

[71] Galindo A and Pascual P 1990 *Quantum Mechanics I* (Berlin: Springer)

[72] Bender C M and Boettcher S 1998 Real spectra in non-Hermitian Hamiltonians having PT symmetry *Phys. Rev. Lett.* **80** 5243–6

[73] Bender C M 2016 PT symmetry in quantum physics: from a mathematical curiosity to optical experiments *Europhys. News* **47** 17–20

[74] Junker G and Roy P 1998 Supersymmetric construction of exactly solvable potentials and nonlinear algebras *Phys. Atomic Nucl.* **61** 1736–43

[75] Roček M 1991 Representation theory of the nonlinear $SU(2)$ algebra *Phys. Lett.* B **255** 554–7

[76] Karassiov V P 1994 G-invariant polynomial extensions of Lie algebras in quantum many-body physics *J. Phys. A: Math. Gen.* **27** 153–65

[77] Katriel J and Quesne C 1996 Recursively minimally deformed oscillators *J. Math. Phys.* **37** 1650–61

[78] Klauder J R and Skagerstam B-S 1985 *Coherent States: Applications in Physics and Mathematical Physics* (Singapore: World Scientific)

[79] Erdélyi A, Magnus W, Oberhettinger F and Tricomi F G 1953 *Higher Transcedental Functions* vol 1 (New York: McGraw-Hill)

[80] Erdélyi A, Magnus W, Oberhettinger F and Tricomi F G 1954 *Table of Integral Transforms* vol 1 (New York: McGraw-Hill)

[81] Poularikas A (ed) 2000 *The Transforms and Applications Handbook* 2nd edn (Boca Raton, FL: CRC Press)

[82] Cannata F, Junker G and Trost J 1998 Solvable potentials, non-linear algebras, and associated coherent states *Particles, Fields, and Gravitation* ed J Rembieliński (New York: American Institute of Physics) pp 209–18

[83] Junker G and Roy P 1999 Non-linear coherent states associated with conditionally exactly solvable problems *Phys. Lett.* A **257** 113–9

[84] Junker G and Roy P 1997 Construction of (1 + 1)-dimensional field models with exactly solvable fluctuation equations about classical finite-energy configurations *Ann. Phys.* **256** 302–19

IOP Publishing

Supersymmetric Methods in Quantum, Statistical and Solid State Physics

Enlarged and revised edition
Georg Junker

Chapter 5

Supersymmetric classical mechanics

In this chapter we will discuss the classical version of the supersymmetric Witten model. Supersymmetric classical systems are special cases of so-called *pseudoclassical* models. These are classical systems with Cartesian and Grassmannian degrees of freedom, which have first been discussed by Berezin and Marinov [1]. The expression pseudoclassical mechanics has been introduced in 1976 by Casalbuoni [2] and is now an established notion [3] describing classical systems which in addition to their usual commuting bosonic (Cartesian) also have anticommuting fermionic (Grassmannian) degrees of freedom. Pseudoclassical mechanics is of particular interest because of its ability to describe spin-degrees of freedom on a classical level [4–9]. However, there exist alternatives to describe spin in a classical way, see, for example, [10, 11].

5.1 Pseudoclassical models

The construction of pseudoclassical models has systematically been discussed by Casalbuoni [12]. Here we will study the properties of the simplest non-trivial example which allows for an interaction between bosonic and fermionic degrees of freedom. It is characterised by the pseudoclassical Lagrangian ($m = 1$):

$$L_0(\dot{x}, x, \dot{\psi}, \psi, \dot{\bar{\psi}}, \bar{\psi}) := \frac{1}{2}\dot{x}^2 - V_1(x) + \frac{i}{2}(\bar{\psi}\dot{\psi} - \dot{\bar{\psi}}\psi) - V_2(x)\bar{\psi}\psi. \tag{5.1}$$

In the above x denotes a bosonic degree of freedom and ψ and $\bar{\psi}$ denote independent fermionic degrees of freedom. The potentials V_1 and V_2 are differentiable functions of x only.

Let us first discuss the nature of the bosonic and fermionic degrees of freedom. They are not represented by ordinary c-numbers but by even and odd elements of a Grassmann algebra. In the present case this algebra may be constructed via two generators denoted by ψ_0 and $\bar{\psi}_0$ which obey the following relations:

$$\{\psi_0, \bar{\psi}_0\} = 0, \qquad \psi_0^2 = 0 = \bar{\psi}_0^2. \tag{5.2}$$

The overbar denotes the adjoint Grassmann number. For an arbitrary element of the Grassmann algebra

$$B := a_1 + a_2\psi_0 + a_3\bar{\psi}_0 + a_4\bar{\psi}_0\psi_0, \qquad a_i \in \mathbb{C}, \tag{5.3}$$

its adjoint is defined as

$$\bar{B} := a_1^* + a_2^*\bar{\psi}_0 + a_3^*\psi_0 + a_4^*\bar{\psi}_0\psi_0. \tag{5.4}$$

Here, the mapping $\psi \mapsto \bar{\psi}$ is linear and involutive. Even elements of this algebra are those which commute with all Grassmann numbers of the form (5.3). The bosonic degree of freedom x is, therefore, expected to be of the form

$$x = x_1 + x_2\bar{\psi}_0\psi_0, \tag{5.5}$$

where $x_1, x_2 \in \mathbb{R}$ which assures the reality of the bosonic variable $\bar{x} = x$. As x is a real and even Grassmann number, as given in equation (5.5), so are the potentials V_i:

$$V_i(x) = V_i(x_1) + V_i'(x_1)x_2\bar{\psi}_0\psi_0, \qquad i = 1, 2, \tag{5.6}$$

which follows formally from the Taylor expansion noting that higher-order terms in $x_2\bar{\psi}_0\psi_0$ vanish due to the property (5.2). On the other hand, the fermionic degrees of freedom are odd elements of the Grassmann algebra

$$\psi = a\psi_0 + b\bar{\psi}_0, \qquad \bar{\psi} = a^*\bar{\psi}_0 + b^*\psi_0, \tag{5.7}$$

with $a, b \in \mathbb{C}$. For more details about Grassmann numbers and Grassmann algebras we refer to [13–16].

Note that here we will limit ourselves to a Grassmann algebra with two generators. This is sufficient to discuss all basic aspects of pseudoclassical and supersymmetric classical models. For a discussion within higher-order Grassmann algebras we refer to [17, 18] and to [19] for a more generic approach.

5.2 A supersymmetric classical model

Now we will specialise the above Lagrangian (5.1) by setting

$$V_1(x) := \frac{1}{2}\,\Phi^2(x), \qquad V_2(x) := \Phi'(x), \tag{5.8}$$

where, as in the sense above,

$$\Phi(x) = \Phi(x_1) + \Phi'(x_1)x_2\bar{\psi}_0\psi_0. \tag{5.9}$$

It will be shown later that Φ is up to a constant factor $\sqrt{2}$ identical to the SUSY potential in Witten's model.

The special class of systems we are now dealing with is characterised by a Lagrangian of the form

$$L(\dot{x}, x, \dot{\psi}, \psi, \dot{\bar{\psi}}, \bar{\psi}) := \frac{1}{2}\dot{x}^2 - \frac{1}{2}\Phi^2(x) + \frac{i}{2}(\bar{\psi}\dot{\psi} - \dot{\bar{\psi}}\psi) - \Phi'(x)\bar{\psi}\psi. \qquad (5.10)$$

This Lagrangian characterises pseudoclassical systems being supersymmetric. Indeed, this model can be derived via a supersymmetrisation of a field theory in (0+1) dimensions. By supersymmetrisation we mean the extension of the (0+1)-dimensional space spanned by the time variable t to a superspace spanned by $(t, \theta, \bar{\theta})$ where θ and $\bar{\theta}$ are odd Grassmann variables. The superspace formulation has been introduced by Salam and Strathdee [20]. For the properties of superspaces see, for example, [14–16].

The general procedure has been outlined by Nicolai [21] and is basically a special case of the supersymmetrisation of (3+1)-dimensional field theories [22, 23]. For more details see section 6.5 in the book by Kalk and Soff [24]. In essence, one first arrives at a Lagrangian density $\mathcal{L}$ over the superspace $(t, \theta, \bar{\theta})$ being invariant under SUSY transformations, which consist basically of translations generated by $\partial/\partial t$, $\partial/\partial\theta$, and $\partial/\partial\bar{\theta}$. The 'effective' Lagrangian (5.10) is then obtained via integration, $L = \int d\theta d\bar{\theta}\, \mathcal{L}$, and characterises a system being invariant under the following SUSY transformation of the fields:

$$\begin{aligned}
x &\mapsto x + \delta x, & \delta x &:= \bar{\varepsilon}\psi + \bar{\psi}\varepsilon, \\
\psi &\mapsto \psi + \delta\psi, & \delta\psi &:= -(i\dot{x} + \Phi(x))\varepsilon, \\
\bar{\psi} &\mapsto \bar{\psi} + \delta\bar{\psi}, & \delta\bar{\psi} &:= (i\dot{x} - \Phi(x))\bar{\varepsilon},
\end{aligned} \qquad (5.11)$$

where ε and $\bar{\varepsilon}$ denote 'infinitesimal' versions of the Grassmann variables θ and $\bar{\theta}$. The invariance of such systems under the above SUSY transformation becomes obvious by noting that equation (5.11) implies

$$L \mapsto L + \delta L, \qquad \delta L = \frac{1}{2}\frac{d}{dt}[(\dot{x} - i\Phi)\bar{\varepsilon}\psi + (\dot{x} + i\Phi)\bar{\psi}\varepsilon], \qquad (5.12)$$

which gives rise to a Lagrangian $L + \delta L$ being gauge-equivalent to the original Lagrangian L. In other words, $L + \delta L$ and L characterise the same pseudoclassical system.

5.3 The classical dynamics

Despite the fact that these pseudoclassical and supersymmetric classical models were known around the time when SUSY quantum mechanics was introduced, it has taken almost two decades to observe [25, 26] that the solutions of the corresponding classical equations of motion can be derived in a rather explicit form and have some interesting properties. Indeed, similar to the integrability of standard one-dimensional classical systems which is due to the fact that the energy is a constant of motion, one-dimensional pseudoclassical system also allow for an explicit integration of their equations of motion.

Let us start with the equations of motion derived from equation (5.10):

$$\dot{\psi} = -i\Phi'(x)\psi, \qquad \dot{\bar{\psi}} = i\Phi'(x)\bar{\psi}, \qquad (5.13)$$

$$\ddot{x} = -\Phi(x)\Phi'(x) - \Phi''(x)\bar{\psi}\psi. \tag{5.14}$$

The first-order differential equations for the fermionic degrees of freedom can immediately be integrated. With initial conditions $\psi_0 := \psi(0)$ and $\bar{\psi}_0 := \bar{\psi}(0)$ integration of equation (5.13) gives

$$\psi(t) = \psi_0 \exp\left\{-\mathrm{i}\int_0^t \mathrm{d}\tau\, \Phi'(x(\tau))\right\}, \qquad \bar{\psi}(t) = \bar{\psi}_0 \exp\left\{\mathrm{i}\int_0^t \mathrm{d}\tau\, \Phi'(x(\tau))\right\}, \tag{5.15}$$

where x denotes the (yet unknown) solution of equation (5.14). Note that we choose the initial values of the fermionic degrees of freedom as generators of the underlying Grassmann algebra. This is a rather natural choice when restricting the discussion to a two-dimensional Grassmann algebra. Let us note that the solutions (5.15) imply that $\bar{\psi}(t)\psi(t) = \bar{\psi}_0\psi_0$ is a constant and, therefore, equation (5.14) simplifies to

$$\ddot{x} = -\Phi(x)\Phi'(x) - \Phi''(x)\bar{\psi}_0\psi_0. \tag{5.16}$$

As the SUSY potential Φ is of the form (5.9), the bosonic degree of freedom x is necessarily of the form (5.5). That is,

$$x(t) = x_{\mathrm{qc}}(t) + q(t)\bar{\psi}_0\psi_0, \tag{5.17}$$

where x_{qc} and q are real-valued functions of time. We will call x_{qc}, which is the so-called body of x, the *quasi-classical* solution in order to differentiate it from the full pseudoclassical solution x which contains also the soul- or $\bar{\psi}_0\psi_0$-term and in general is an even-Grassmann-valued function of time. It is also worth mentioning that in equation (5.15) one may replace $x(\tau)$ by $x_{\mathrm{qc}}(\tau)$ because of equations (5.17) and (5.2). Hence, we have

$$\psi(t) = \psi_0 \exp\left\{-2\mathrm{i}\varphi[x_{\mathrm{qc}}]\right\}, \qquad \bar{\psi}(t) = \bar{\psi}_0 \exp\left\{2\mathrm{i}\varphi[x_{\mathrm{qc}}]\right\}, \tag{5.18}$$

where we have introduced the *fermionic phase*

$$\boxed{\varphi[x] := \frac{1}{2}\int_0^t \mathrm{d}\tau\, \Phi'(x(\tau)),} \tag{5.19}$$

a functional which we will revisit in our quasi-classical approximation to the quantum propagator for the Witten model. In fact, it will turn out that this is a so-called *topological phase* and is closely related to the Witten index.

Now let us study the solution for the bosonic degree of freedom. As in standard classical mechanics we make use of the energy conservation, that is, we multiply equation (5.16) with $\dot{x}$ and integrate, resulting in the conservation of the energy

$$\mathcal{E} = \frac{1}{2}\dot{x}^2 + \frac{1}{2}\Phi^2(x) + \Phi'(x)\bar{\psi}_0\psi_0, \tag{5.20}$$

which is a constant even Grassmann number. The ansatz (5.17) together with $\mathcal{E} =: E + F\bar{\psi}_0\psi_0$ (here $E \geqslant 0$ and $F \in \mathbb{R}$ are constants denoting the body and the soul of $\mathcal{E}$, respectively) results in

$$\dot{x}_{qc}^2 = 2E - \Phi^2(x_{qc}), \tag{5.21}$$

$$\dot{q} = \frac{1}{\dot{x}_{qc}}\left[F - \Phi'(x_{qc}) - \Phi(x_{qc})\Phi'(x_{qc})q\right]. \tag{5.22}$$

The first equation is the standard Newton equation for the one-dimensional Cartesian degree of freedom x_{qc} in a potential of the form $\frac{1}{2}\Phi^2$ with energy E. The second equation, which determines $q(t)$, can also be integrated if $\dot{x}_{qc} \not\equiv 0$, and yields [25]

$$q(t) = \frac{\dot{x}_{qc}(t)}{\dot{x}_{qc}(0)}q(0) + \dot{x}_{qc}(t)\int_0^t d\tau \frac{F - \Phi'(x_{qc}(\tau))}{2E - \Phi^2(x_{qc}(\tau))}. \tag{5.23}$$

Note that $\dot{x}_{qc} \equiv 0$ implies $E = 0$. Hence the result is only valid for $E > 0$. The case $E = 0$ is discussed separately at the end of this section. In the above, $q(0)$ is a constant of integration. Again we find, as for the fermionic degrees of freedom, that $q(t)$ is expressible in terms of the quasi-classical solution $x_{qc}(t)$ determined by equation (5.21). Let us note that the singularity of the integral in equation (5.23) near the turning points of the quasi-classical path is precisely canceled by its prefactor as $\dot{x}_{qc}(t)$ vanishes at those points. Hence, $q(t)$ remains finite for all $t \geqslant 0$. Let us also note that even for the initial condition $q(0) = 0$ we have in general $q(t) \neq 0$ for $t > 0$. In other words, even assuming the pseudoclassical solution initially to be real, $x(0) \in \mathbb{R}$, it will in general become a Grassmann-valued quantity. It is only in the special case $\Phi'(x) = F = $ const., that is, for a linear SUSY potential Φ, where a real $x(0)$ remains real forever.

Let us now discuss some properties of the quasi-classical solution x_{qc}. The equation of motion (5.21) for the quasi-classical path can be obtained from a 'quasi-classical' Lagrangian defined by

$$L_{qc} := \frac{1}{2}\dot{x}^2 - \frac{1}{2}\Phi^2(x) = \frac{1}{2}(\dot{x} \pm i\Phi(x))^2 \mp i\Phi(x)\dot{x}. \tag{5.24}$$

Here $x \in \mathbb{R}$ will denote the usual real-valued degree of freedom. The second equality above shows that this Lagrangian is gauge-equivalent to

$$\tilde{L}_{qc}^\pm := \frac{1}{2}(\dot{x} \pm i\Phi(x))^2 \tag{5.25}$$

as $L_{qc} = \tilde{L}_{qc}^\pm \mp i\frac{d}{dt}\Phi(x)$. As an aside we note that the above complex gauge transformations $L_{qc} \to \tilde{L}_{qc}^\pm$ become real by using Euclidean instead of real time. The canonical momenta obtained from the two Lagrangians $\tilde{L}_{qc}^\pm$ are

$$\xi^{\pm} := \frac{\partial \tilde{L}^{\pm}_{\text{qc}}}{\partial \dot{x}} = \dot{x} \pm i\Phi(x) = (\xi^{\mp})^* \tag{5.26}$$

and, surprisingly, coincide with the generators of the SUSY transformation (5.11) of the fermionic degrees of freedom:

$$\delta\psi = -i\xi^{-}\bar{\varepsilon}, \qquad \delta\bar{\psi} = i\xi^{+}\varepsilon. \tag{5.27}$$

In fact, with these canonical momenta one can construct classical supercharges:

$$Q := \frac{i}{\sqrt{2}}\xi^{-}\bar{\psi}, \qquad \bar{Q} = -\frac{i}{\sqrt{2}}\xi^{+}\psi. \tag{5.28}$$

It is easily verified with the help of equations (5.13) and (5.14) that these supercharges are constants of motion. Indeed, from equation (5.25) we derive the equations of motion

$$\dot{\xi}^{\pm} = \mp i\Phi'(x_{\text{qc}})\xi^{\pm}, \tag{5.29}$$

which are identical in form with those for the fermionic degrees of freedom. Obviously, the solutions read

$$\xi^{\pm}(t) = \xi^{\pm}(0)\exp\{\mp 2i\varphi[x_{\text{qc}}]\} \tag{5.30}$$

and explicate the conservation of the supercharges (5.28). It is also obvious that the conserved energy E of the quasi-classical solution can be expressed in terms of these canonical momenta, $E = \frac{1}{2}\xi^{+}\xi^{-}$. As a consequence we have the relation

$$\xi^{\pm}/\sqrt{2E} = \left(\xi^{\mp}/\sqrt{2E}\right)^{-1} \tag{5.31}$$

valid for $E > 0$.

For $E = 0$ the quasi-classical solutions are given by $x_{\text{qc}}(t) = x_k$, where x_k are the zeros of the potential Φ, $\Phi(x_k) = 0$. These are precisely the classical ground states for good SUSY. SUSY will be broken on the classical level if the SUSY potential Φ does not have zeros, because then $E \geqslant \frac{1}{2}\Phi^2(x) > 0$ for all $x \in \mathbb{R}$. See also figure 5.2. Note that the solution (5.23) for q may not be used in this case. Actually, because of $\dot{x}_{\text{qc}} = \Phi(x_k) = 0$ the equation of motion corresponding to equation (5.22) reads

$$F = \Phi'(x_{\text{qc}}) = \Phi'(x_k). \tag{5.32}$$

As a consequence, the solutions of the fermionic degrees of freedom are given by

$$\psi(t) = \psi_0 e^{-itF}, \qquad \bar{\psi}(t) = \bar{\psi}_0 e^{-itF}, \tag{5.33}$$

that is, they perform a harmonic motion with frequency F. The equation of motion (5.16) reduces to that of a forced harmonic oscillator with frequency F for q.

Finally, let us make two remarks. First we note, that the above procedure for solving the dynamics of the supersymmetric classical system (5.10) can also be used to derive [26] solutions for the more general pseudoclassical system (5.1). In a second remark we point out that the gauge transformations $L_{qc} \to \tilde{L}_{qc}^{\pm}$ are the classical analogues of the so-called Nicolai map [27, 28]. This map characterises a transformation of bosonic fields of a SUSY quantum field theory. It has the interesting property that the full bosonic action is mapped into a free action of boson fields. Note that the quasi-classical Lagrangians $\tilde{L}_{qc}^{\pm}$ are quadratic in their canonical momenta $\xi^{\pm}$. For an extensive review on the Nicolai map see Ezawa and Klauder [29] and section 7.3.

Let us now discuss in detail the fermionic phase defined in equation (5.19). This discussion will be presented in a separate section because of its fundamental role played in the quasi-classical approximation considered in section 6.1.2.

5.4 Discussion of the fermionic phase

According to its definition (5.19) the fermionic phase in equation (5.18) is a functional of the quasi-classical path. For simplicity let us assume that the quasi-classical solution starts with positive velocity $\dot{x}_{qc}(0) > 0$, hence $E > 0$, and t is sufficiently small such that $\dot{x}_{qc}(\tau) > 0$ for all $0 \leqslant \tau \leqslant t$. In this case the fermionic phase can easily be calculated using the equation of motion (5.21)

$$\varphi[x_{qc}] = \frac{1}{2} \int_{x_{qc}(0)}^{x_{qc}(t)} dx \frac{\Phi'(x)}{\sqrt{2E - \Phi^2(x)}} = \frac{1}{2}[a(x'') - a(x')], \tag{5.34}$$

where, for convenience, we have introduced the abbreviations

$$a(x) := \arcsin \frac{\Phi(x)}{\sqrt{2E}} \in \left[-\frac{\pi}{2}, \frac{\pi}{2}\right], \qquad x'' := x_{qc}(t), \qquad x' := x_{qc}(0). \tag{5.35}$$

If we further assume, without loss of generality, that we initially start at a zero of the SUSY potential, that is, $\Phi(x') = 0$, the fermionic solution ψ takes the simple form

$$\psi(t) = \psi_0 \exp\left\{-i \arcsin \frac{\Phi(x'')}{\sqrt{2E}}\right\} = \psi_0 \left[\sqrt{1 - \frac{\Phi^2(x'')}{2E}} - i \frac{\Phi(x'')}{\sqrt{2E}}\right]$$

$$= \psi_0 \left[\dot{x}_{qc}(t) - i\Phi(x_{qc}(t))\right]/\sqrt{2E}. \tag{5.36}$$

Hence, the fermionic degrees of freedom are expressible in terms of the canonical momenta $\xi^{\pm}$ of the quasi-classical Lagrangians $\tilde{L}_{qc}^{\pm}$ as expected from relation (5.30):

$$\psi(t) = \frac{1}{\sqrt{2E}} \xi^-(t)\psi_0 = \frac{\psi_0}{2E} \xi^+(0)\xi^-(t),$$

$$\bar{\psi}(t) = \frac{1}{\sqrt{2E}} \xi^+(t)\bar{\psi}_0 = \frac{\bar{\psi}_0}{2E} \xi^-(0)\xi^+(t). \tag{5.37}$$

These are the finite versions of the infinitesimal SUSY transformation (5.27). Note that because of the assumption $\Phi(x') = 0$ we have $\xi^+(0) = \xi^-(0)$ and therefore $\xi^{\pm}(0) = 1/\sqrt{2E}$.

Let us now consider a more general quasi-classical solution which also passes through turning points. These are points where $\dot{x}_{qc}$ vanishes. Without loss of generality we also assume that Φ^2 has one global minimum and diverges for $x \to \pm\infty$. See figure 5.1 for a typical shape of $\frac{1}{2}\Phi^2$. In other words, we assume the quasi-classical solution to be a bounded periodic motion about this minimum. For the SUSY potential this assumption implies that it has at most one zero which, however, may have a multiplicity larger than one. Typical SUSY potentials with good and broken SUSY at the classical as well as quantum level [30] are shown in figures 5.2(a)–(c).

Being a one-dimensional system we can rather explicitly discuss the possible quasi-classical paths for a given energy $E > 0$. It is convenient [31] to consider the following four classes. Without loss of generality, we may assume $x'' > x'$, see figure 5.1.

(1) Paths which leave x' to the right and reach x'' from the left:

$$\dot{x}_{qc}(0) > 0, \quad \dot{x}_{qc}(t) > 0.$$

(2) Paths which leave x' to the left and reach x'' from the left:

$$\dot{x}_{qc}(0) < 0, \quad \dot{x}_{qc}(t) > 0.$$

(3) Paths which leave x' to the right and reach x'' from the right:

$$\dot{x}_{qc}(0) > 0, \quad \dot{x}_{qc}(t) < 0.$$

(4) Paths which leave x' to the left and reach x'' from the right:

$$\dot{x}_{qc}(0) < 0, \quad \dot{x}_{qc}(t) < 0.$$

Within each class the paths are uniquely characterised by the number of complete cycles they perform before arriving at x'' at time t. Each of these full cycles contributes a term $a(x_R) - a(x_L)$ to the fermionic phase, where x_R and x_L are the quasi-classical right and left turning points of the periodic motion with energy E:

$$\Phi^2(x_R) = 2E = \Phi^2(x_L). \tag{5.38}$$

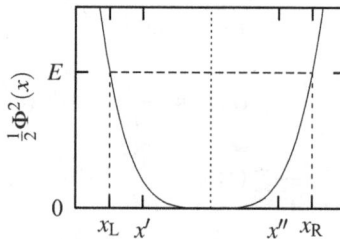

Figure 5.1. A typical shape of the potential $\frac{1}{2}\Phi^2$. The quasi-classical motion for a given energy E starts at $x_{qc}(0) = x'$ and reaches at time t the position $x_{qc}(t) = x''$. During its motion the left and right turning points x_L and x_R may be visited several times.

Thus for a quasi-classical path belonging to class (i), $i = 1, 2, 3, 4$, and containing $k \in \mathbb{N}_0$ complete cycles we obtain [32, 33]

$$\varphi[x_{\text{qc}}] \equiv \varphi_k^{(i)} = \varphi_0^{(i)} + k[a(x_{\text{R}}) - a(x_{\text{L}})], \qquad (5.39)$$

where

$$
\begin{aligned}
\varphi_0^{(1)} &:= \frac{1}{2}[a(x'') - a(x')], \\
\varphi_0^{(2)} &:= \frac{1}{2}[a(x'') + a(x')] - a(x_{\text{L}}), \\
\varphi_0^{(3)} &:= a(x_{\text{R}}) - \frac{1}{2}[a(x'') + a(x')], \\
\varphi_0^{(4)} &:= a(x_{\text{R}}) - a(x_{\text{L}}) - \frac{1}{2}[a(x'') - a(x')].
\end{aligned}
\qquad (5.40)
$$

The explicit values for $a(x_{\text{R}})$ and $a(x_{\text{L}})$ depend on the shape of the SUSY potential Φ. The turning-point condition (5.38) has the two possible solutions:

(a) $\Phi(x_{\text{R}}) = -\Phi(x_{\text{L}}) = \pm\sqrt{2E}$. This situation only occurs when SUSY is good on the classical as well as on the quantum level (see figure 5.2(a)). We have $a(x_{\text{R}}) = -a(x_{\text{L}}) = \pm\frac{\pi}{2}$ and thus

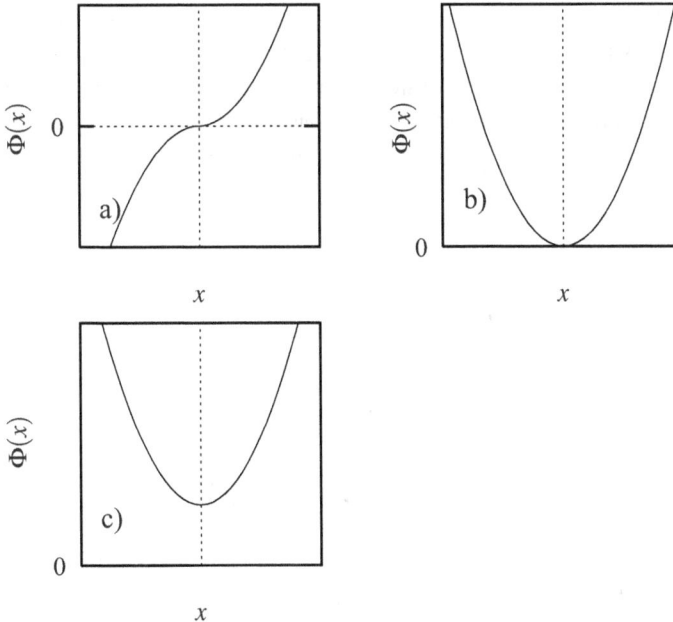

Figure 5.2. The three qualitative different shapes for the SUSY potential. In case (a) SUSY will be good on the classical as well as on the quantum level. Case (b) displays a situation where SUSY is good on the classical level but will be broken at the quantum level due to quantum fluctuations. The last case (c) shows a SUSY potential where SUSY is already broken on the classical level.

$$\varphi_k^{(i)} = \varphi_0^{(i)} \pm k\pi. \tag{5.41}$$

(b) $\Phi(x_R) = \Phi(x_L) = \pm\sqrt{2E}$. This is the situation where SUSY is broken on the quantum level. Note that classically SUSY may still be good (see figures 5.2(b) and (c)). Here we find $a(x_R) = a(x_L) = \pm\frac{\pi}{2}$ and consequently

$$\varphi_k^{(i)} = \varphi_0^{(i)}. \tag{5.42}$$

Note that because of the factor two in the exponent of equation (5.18) the classical fermionic solutions do not require a separate discussion for these two cases. Even more interestingly, they are identical within each class of paths. Denoting with $\psi^{(i)}(t)$ the solution for the class (i) we have

$$\left.\begin{aligned} \psi^{(1)}(t) &= \psi_0 \exp\{-i(a(x'') - a(x'))\} \\ \psi^{(2)}(t) &= -\psi_0 \exp\{-i(a(x'') + a(x'))\} \\ \psi^{(3)}(t) &= -\psi_0 \exp\{i(a(x'') + a(x'))\} \\ \psi^{(4)}(t) &= \psi_0 \exp\{i(a(x'') - a(x'))\} \end{aligned}\right\} = \frac{\psi_0}{2E}\xi^+(0)\xi^-(t) \tag{5.43}$$

and similar expressions for $\bar{\psi}^{(i)}(t)$. See also equation (5.37).

Here we emphasise that the fermionic phase, in contrast to the fermionic solutions, distinguishes between the cases of good and broken SUSY on the quantum level. Although it is a purely classical quantity, the fermionic phase along the quasi-classical path provides some information about the nature of the SUSY of the corresponding quantum system. To be more precise, because of our assumption that $\frac{1}{2}\Phi^2$ forms a single-well potential (see figure 5.1) the Witten index may be put into the following form [34]:

$$\Delta = \frac{1}{\pi}[a(x_R) - a(x_L)]. \tag{5.44}$$

Hence, the fermionic phase accumulates with each complete cycle of the quasi-classical motion an additional phase $\Delta \pi$:

$$\varphi_k^{(i)} = \varphi_0^{(i)} + k\Delta\pi. \tag{5.45}$$

This obviously coincides in the case of good SUSY ($\Delta = \pm 1$) with equation (5.41) and for broken SUSY ($\Delta = 0$) with equation (5.42).

5.5 Quantisation

So far the discussion of this chapter has been devoted to the classical dynamics of supersymmetric systems. Now we want to demonstrate that the quantised version of the model characterised by the Lagrangian (5.10) is identical to the Witten model. There are at least two different approaches for the quantisation of a classical dynamical system. These are the standard canonical quantisation procedure and the path-integral approach of Feynman [35]. We will sketch both approaches for illustrative purposes. More detailed discussions can be found in [36–39].

5.5.1 Canonical approach

Let us begin with the canonical quantisation. For this, one first has to derive from Lagrangian (5.10) the corresponding classical Hamiltonian. For this we need the canonical momenta

$$p := \frac{\partial L}{\partial \dot{x}} = \dot{x}, \qquad \pi := \frac{\partial L}{\partial \dot{\psi}} = -\frac{i}{2}\,\bar{\psi}, \qquad \bar{\pi} := \frac{\partial L}{\partial \dot{\bar{\psi}}} = -\frac{i}{2}\,\psi. \tag{5.46}$$

Here we observe that the fermionic momenta π and $\bar{\pi}$ are proportional to the fermionic degrees of freedom. As a consequence, we have the second-class constraints

$$\chi_1 := \pi + \frac{i}{2}\,\bar{\psi} = 0, \qquad \chi_2 := \bar{\pi} + \frac{i}{2}\,\psi = 0. \tag{5.47}$$

This is a typical feature of all pseudoclassical systems [12]. Obviously, the phase space is six-dimensional. However, the dynamics takes place in a four-dimensional submanifold defined by equation (5.47). In order to respect these constraints we have to introduce two odd (Grassmann-valued) Lagrange multipliers λ_1, λ_2 and arrive at the so-called total Hamiltonian [40, 41]

$$H_T := \frac{p^2}{2} + \frac{\Phi^2(x)}{2} + \Phi'(x)\bar{\psi}\psi + \left(\pi + \frac{i}{2}\bar{\psi}\right)\lambda_1 + \left(\bar{\pi} + \frac{i}{2}\psi\right)\lambda_2. \tag{5.48}$$

The corresponding equations of motion read (note the extra minus sign for the odd Grassmann variables)

$$\dot{x} = \frac{\partial H_T}{\partial p}, \qquad \dot{\psi} = -\frac{\partial H_T}{\partial \pi}, \qquad \dot{\bar{\psi}} = -\frac{\partial H_T}{\partial \bar{\pi}},$$

$$\dot{p} = -\frac{\partial H_T}{\partial x}, \qquad \dot{\pi} = -\frac{\partial H_T}{\partial \psi}, \qquad \dot{\bar{\pi}} = -\frac{\partial H_T}{\partial \bar{\psi}}, \tag{5.49}$$

which are equivalent to the Euler–Lagrange equations derived from L. In fact, using the consistency conditions $\dot{\chi}_1 = 0 = \dot{\chi}_2$ [40, 41], which guarantee that the constraints (5.47) are respected during the time evolution, one finds for the Lagrange multipliers $\lambda_1 = i\Phi'\psi$, $\lambda_2 = \bar{\lambda}_1 = -i\Phi'\bar{\psi}$. Inserting this result in H_T we obtain

$$H_T = \frac{1}{2}p^2 + \frac{1}{2}\Phi^2(x) + i\Phi'(x)(\bar{\psi}\bar{\pi} - \psi\pi). \tag{5.50}$$

In this representation the total Hamiltonian obviously gives rise to the correct equations of motion (5.13)–(5.14). Here let us note that often the classical Hamiltonian

$$H := \frac{1}{2} p^2 + \frac{1}{2} \Phi^2(x) + \Phi'(x) \bar{\psi} \psi \tag{5.51}$$

is considered, which however, does not lead to the correct equations of motion. Actually, in Dirac's notion H_T and H are *weakly equal*. Two phase-space functions F and G are called weakly equal, $F \approx G$, if they are identical on the submanifold of the phase space which is defined by the constraints, see [40, p 12]. Such weak equalities are denoted by the sign $\approx$, that is,

$$H \approx H_T. \tag{5.52}$$

However, H and H_T lead to different flows in phase space. It is H_T which generates the correct phase-space flow staying on the submanifold of the phase space defined by the constraints (5.47). Because of these constraints the time evolution of an arbitrary phase-space function is determined by the Dirac bracket instead of the usual Poisson bracket:

$$\frac{dF}{dt} \approx \{F, H_T\}_{DB}, \tag{5.53}$$

where in the present case the Dirac bracket explicitly reads [39]

$$
\begin{aligned}
\{F, G\}_{DB} = {} & \frac{\partial F}{\partial x} \frac{\partial G}{\partial p} - \frac{\partial F}{\partial p} \frac{\partial G}{\partial x} \\
& + (-1)^{\deg F} \frac{1}{2} \left[\frac{\partial F}{\partial \psi} \frac{\partial G}{\partial \pi} + \frac{\partial F}{\partial \pi} \frac{\partial G}{\partial \psi} + \frac{\partial F}{\partial \bar{\psi}} \frac{\partial G}{\partial \bar{\pi}} + \frac{\partial F}{\partial \bar{\pi}} \frac{\partial G}{\partial \bar{\psi}} \right] \\
& + (-1)^{\deg F} i \left[\frac{\partial F}{\partial \psi} \frac{\partial G}{\partial \bar{\psi}} + \frac{\partial F}{\partial \bar{\psi}} \frac{\partial G}{\partial \psi} - \frac{1}{4} \frac{\partial F}{\partial \pi} \frac{\partial G}{\partial \bar{\pi}} - \frac{1}{4} \frac{\partial F}{\partial \bar{\pi}} \frac{\partial G}{\partial \pi} \right]
\end{aligned} \tag{5.54}
$$

with $\deg F = 0\,(1)$ if F is an even (odd) Grassmann-valued phase-space function. Note that the Dirac bracket has the following symmetry property:

$$\{F, G\}_{DB} = \begin{cases} \{G, F\}_{DB} & \text{for} \quad \deg G = 1 = \deg F \\ -\{G, F\}_{DB} & \text{else} \end{cases}. \tag{5.55}$$

Explicitly we have the following relations:

$$
\begin{aligned}
\{x, p\}_{DB} &= 1, \quad \{\psi, \pi\}_{DB} = -\frac{1}{2} = \{\bar{\psi}, \bar{\pi}\}_{DB}, \\
\{\psi, \bar{\psi}\}_{DB} &= -i, \quad \{\pi, \bar{\pi}\}_{DB} = \frac{i}{4},
\end{aligned} \tag{5.56}
$$

and all other Dirac brackets for the bosonic and fermionic variables vanish.

Quantisation of the system characterised by equation (5.48) can be achieved, see for example [41], by replacing the Grassmann variables by the corresponding operators for which we will use the same symbol. Simultaneously, the symmetric Dirac brackets are replaced by the (symmetric) anticommutator divided by (i$\hbar$). In contrast, the antisymmetric Dirac brackets are replaced by the (antisymmetric) commutator also divided by (i$\hbar$). Hence, the operators obey the algebra

$$[x, p] = i\hbar, \quad \{\psi, \pi\} = -\frac{i\hbar}{2} = \{\bar{\psi}, \bar{\pi}\}, \quad \{\psi, \bar{\psi}\} = \hbar, \quad \{\pi, \bar{\pi}\} = -\frac{\hbar}{4}. \quad (5.57)$$

Note that $\pi^2 = 0 = \bar{\pi}^2$ and $\psi^2 = 0 = \bar{\psi}^2$. The algebra satisfied by the fermionic operators is isomorphic to that obeyed by Pauli matrices. Therefore, we may choose the representation

$$\psi = \sqrt{\hbar}\,\sigma_-, \quad \bar{\psi} = \sqrt{\hbar}\,\sigma_+, \quad \pi = -\frac{i}{2}\sqrt{\hbar}\,\sigma_+, \quad \bar{\pi} = -\frac{i}{2}\sqrt{\hbar}\,\sigma_-. \quad (5.58)$$

Note that there are no other (inequivalent) irreducible representations [42]. Using this representation in equation (5.50) we finally arrive at Witten's quantum-mechanical Hamilton operator acting on $L^2(\mathbb{R}) \otimes \mathbb{C}^2$,

$$H_{\mathrm{T}} = \frac{1}{2}p^2 + \frac{1}{2}\Phi^2(x) + \frac{\hbar}{2}\Phi'(x)\sigma_3, \quad (5.59)$$

with unit mass, $m = 1$, and rescaled SUSY potential $\Phi/\sqrt{2}$. It is also worth mentioning that the ladder operators (3.3) are the quantised versions of the momenta $\xi^\pm$, that is,

$$A = \frac{i}{\sqrt{2}}\xi^-, \quad A^\dagger = -\frac{i}{\sqrt{2}}\xi^+ \quad (5.60)$$

and consequently the classical supercharges (5.28) become, after quantisation, the supercharges (3.2). Similarly, the relation $E = \frac{1}{2}\xi^+\xi^-$ is the classical analogue of the tree Hamiltonian $H_{\mathrm{tree}} = \frac{1}{2}\{A^\dagger, A\}$. Let us also mention, that the classical super-charges (5.28) and the classical Hamiltonian (5.48) obey the classical superalgebra

$$\{Q, \bar{Q}\}_{\mathrm{DB}} \approx -iH_{\mathrm{T}}, \quad \{Q, H_{\mathrm{T}}\}_{\mathrm{DB}} \approx 0, \quad \{\bar{Q}, H_{\mathrm{T}}\}_{\mathrm{DB}} \approx 0. \quad (5.61)$$

5.5.2 Path-integral approach

The second approach of quantising the pseudoclassical system (5.10) is based on Feynman's path integral [35, 43]. The basic proposition of Feynman is that the integral kernel of the quantum-mechanical time-evolution operator can be expressed as an integral over a certain set of continuous paths in (pseudo-)classical config-uration space, each of which is weighted with a phase given by the action calculated along these paths. For a later convenience, we will only consider the trace of the time-evolution operator $\exp\{-itH/\hbar\}$. Clearly, this requires a suitable regularisa-tion, for example, by letting the time t have a negative imaginary part, if the Hamiltonian H is bounded from below. In the case of a purely classical system described by a standard Lagrangian $L(x, \dot{x})$ Feynman's approach to construct H from L can then be written as

$$\text{Tr} \exp\{-itH/\hbar\} = \int_{x(t)=x(0)} \mathcal{D}[x] \exp\left\{\frac{i}{\hbar} \int_0^t d\tau \, L(x(\tau), \dot{x}(\tau))\right\}. \qquad (5.62)$$

Here $\int_{x(t)=x(0)} \mathcal{D}[x](\cdot)$ symbolises integration over all continuous paths $x: [0, t] \to \mathbb{R}$, which are periodic in the sense that $x(t) = x(0)$. More precise definitions of the Feynman path integration can be based, for example, on a theory of so-called pro-measures [44–47] or on a discrete time-lattice approach. The latter approach is equivalent to the Lie–Trotter formula [31, 35, 43, 48–53]. In the following we will perform only formal calculations for simplicity.

The path-integral approach is not only useful for quantising bosonic degrees of freedom but is also applied to field theories with fermionic degrees of freedom [13]. The formulation of fermionic path integrals is based on the integration rules for Grassmann variables (see, for example, [54, 55]):

$$\int d\theta \, 1 = 0, \qquad \int d\theta \, \theta = 1. \qquad (5.63)$$

From these follows the 'Gauss formula':

$$\int d\theta_1 \int d\bar{\theta}_1 \cdots \int d\theta_n \int d\bar{\theta}_n \exp\left\{\sum_{i,j=1}^{n} \bar{\theta}_i A_{ij} \theta_j\right\} = \det\left(A_{ij}\right) \qquad (5.64)$$

and its infinite-dimensional generalisation, the 'fermionic Gaussian-path-integral formula'

$$\int_{\psi(t)=\pm\psi(0)} \mathcal{D}[\psi] \int_{\bar{\psi}(t)=\pm\bar{\psi}(0)} \mathcal{D}[\bar{\psi}] \exp\left\{\int_0^t d\tau \int_0^t d\tau' \, \bar{\psi}(\tau) D(\tau, \tau') \psi(\tau')\right\}$$
$$= N_\pm \det\left(D(\tau, \tau')\right)_\pm. \qquad (5.65)$$

Again, this path integral is formal in the sense that the determinant on the right-hand side does in general not exist. For this, we have included a 'normalisation' constant $N_\pm$ to be chosen such that the right-hand side becomes meaningful. We also mention that in the above we consider periodic (+) as well as antiperiodic (−) boundary conditions for the fermionic fields. Clearly, the determinant on the right-hand side does also depend on the choice of the boundary conditions. This dependence is indicated by the subscript. In fact, the natural choice for fermionic degrees of freedom is the antiperiodic one [54]. However, we will consider both periodic and antiperiodic conditions in the following path-integral treatment [36, 37]:

$$Z_\pm(t) := \int_{x(t)=x(0)} \mathcal{D}[x] \int_{\psi(t)=\pm\psi(0)} \mathcal{D}[\psi] \int_{\bar{\psi}(t)=\pm\bar{\psi}(0)} \mathcal{D}[\bar{\psi}] \exp\left\{\frac{i}{\hbar} \int_0^t d\tau \, L\right\}, \qquad (5.66)$$

where

$$L := \frac{1}{2} \dot{x}^2 - \frac{1}{2} \Phi^2(x) + i\bar{\psi}\dot{\psi} - \Phi'(x)\bar{\psi}\psi \tag{5.67}$$

is a Lagrangian equivalent to equation (5.10) because of $\bar{\psi}(t)\psi(t) = \bar{\psi}(0)\psi(0)$. As the above Lagrangian is bilinear in the fermionic fields, the corresponding path integral is reduced to the problem of calculating the determinant of the kernel

$$D(\tau, \tau') := \left(i\frac{\partial}{\partial \tau} - \Phi'(x(\tau)) \right) \delta(\tau - \tau') \tag{5.68}$$

with appropriate boundary conditions. This problem is immediately reduced to the eigenvalue problem

$$\left(i\frac{\partial}{\partial \tau} - \Phi'(x(\tau)) \right) \psi_\lambda(\tau) = \lambda \, \psi_\lambda(\tau) \tag{5.69}$$

whose solution reads

$$\psi_\lambda(t) = \psi_\lambda(0)\exp\{-i\lambda t - 2i\varphi[x]\}, \tag{5.70}$$

where $\varphi[x]$ is the fermionic phase defined in equation (5.19). From the boundary condition $\psi_\lambda(t) = \pm\psi_\lambda(0)$ one obtains the eigenvalues

$$
\begin{aligned}
\lambda_n^- &:= \frac{1}{t}[(2n + 1)\pi - 2\varphi[x]] & \text{for} && \psi_\lambda(t) = -\psi_\lambda(0), \\
\lambda_n^+ &:= \frac{1}{t}[2n\pi - 2\varphi[x]] & \text{for} && \psi_\lambda(t) = +\psi_\lambda(0),
\end{aligned}
\tag{5.71}
$$

where $n \in \mathbb{Z}$. Consequently, the determinant is formally given by the infinite product

$$\det(D(\tau, \tau'))_\pm = \prod_{n \in \mathbb{Z}} \lambda_n^\pm, \tag{5.72}$$

which is not well defined. However, choosing the normalisation constants by

$$\frac{1}{N_-} := 2 \prod_{n \in \mathbb{Z}} (2n + 1)\frac{\pi}{t}, \quad \frac{1}{N_+} := 2i \prod_{n \in \mathbb{Z}} 2n\frac{\pi}{t} \tag{5.73}$$

and using the relations

$$\cos z = \prod_{k=0}^{\infty} \left(1 - \frac{4z^2}{(2k + 1)^2\pi^2} \right), \quad \sin z = z \prod_{k=1}^{\infty} \left(1 - \frac{z^2}{k^2\pi^2} \right) \tag{5.74}$$

we arrive at the result

$$Z_\pm(t) = z_-(t) \mp z_+(t), \tag{5.75}$$

where

$$
\begin{aligned}
z_{\pm}(t) &:= \int_{x(t)=x(0)} \mathcal{D}[x]\exp\left\{\frac{\mathrm{i}}{2\hbar}\left[\int_0^t \mathrm{d}\tau(\dot{x}^2 - \Phi^2(x))\right] \mp \mathrm{i}\varphi[x]\right\} \\
&= \int_{x(t)=x(0)} \mathcal{D}[x]\exp\left\{\frac{\mathrm{i}}{2\hbar}\int_0^t \mathrm{d}\tau[\dot{x}^2 - \Phi^2(x) \mp \hbar\Phi'(x)]\right\} \qquad (5.76) \\
&= \mathrm{Tr}\exp\left\{-\frac{\mathrm{i}}{\hbar} t\, H_{\pm}\right\}
\end{aligned}
$$

and

$$
H_{\pm} := \frac{1}{2}p^2 + \frac{1}{2}\Phi^2(x) \pm \frac{\hbar}{2}\Phi'(x) \qquad (5.77)
$$

are indeed the partner Hamiltonians of the Witten model with unit mass and rescaled SUSY potential $\Phi/\sqrt{2}$. As a result we find that antiperiodic boundary conditions for the fermion degrees of freedom lead to

$$
Z_-(t) = \mathrm{Tr}\exp\{-\mathrm{i}tH_{\mathrm{T}}/\hbar\}, \qquad (5.78)
$$

with H_T being the Witten Hamiltonian (5.59). In contrast to this, periodic boundary conditions give rise to

$$
Z_+(t) = \mathrm{Tr}\,[W\exp\{-\mathrm{i}tH_{\mathrm{T}}/\hbar\}], \qquad (5.79)
$$

which is related to the heat-kernel regularised Witten index by

$$
Z_+(t) = -\bar{\Delta}(\mathrm{i}t/\hbar). \qquad (5.80)
$$

Although this path-integral approach has been formal, it clearly demonstrates that the term proportional to $\hbar$ in the Hamiltonians (5.77) stems from taking all fermion loops into account. This has been the reason for introducing the notion tree Hamiltonian and (fermion-)loop correction in section 3.1.

5.6 Problems

Problem 5.1. SUSY invariance of the supersymmetric classical model

Based on the discussion of section 6.5 in the book by Kalka and Soff [24], show that the Lagrangian (5.10) transforms under the SUSY transformations defined in equation (5.11) according to equation (5.12).

Problem 5.2. Supersymmetric classical equations of motion

Derive the equations of motion (5.13) and (5.14) from the Lagrangian (5.10). Note that ∂_ψ is an odd Grassmann-valued operator.

Problem 5.3. The $E = 0$ solution of the equations of motion

Consider the special solution $x(t) = x_k + q(t)\bar{\psi}_0\psi_0$, where x_k is a zero of the SUSY potential, $\Phi(x_k) = 0$. Show that this corresponds to the case $E = 0$ and the relation (5.20) results in the algebraic relation (5.32).

Problem 5.4. Dirac brackets for Grassmann-valued degrees of freedom

Using the definition (5.54) of the Dirac brackets prove the explicit results given in equation (5.56).

References

[1] Berezin F A and Marinov M S 1975 Classical spin and Grassmann algebra *JETP Lett.* **21** 320–1
[2] Casalbuoni R 1976 On the quantization of systems with anticommuting variables *Nuovo Cim.* A **33** 115–25
[3] Freund P G O 1986 *Introduction to Supersymmetry Cambridge Monographs on Mathematical Physics* (Cambridge: Cambridge University Press)
[4] Casalbuoni R 1976 Relativity and supersymmetries *Phys. Lett.* B **62** 49–50
[5] Barducci A, Casalbuoni R and Lusanna L 1976 Supersymmetries and pseudoclassical relativistic electron *Nuovo Cim.* A **35** 377–99
[6] Brink L, Deser S, Zumino B, Di Veccia P and Howe P 1976 Local supersymmetry for spinning particles *Phys. Lett.* B **64** 435–8
[7] Berezin F A and Marinov M S 1977 Particle spin dynamics as the Grassmann variant of classical mechanics *Ann. Phys.* **104** 336–62
[8] Barducci A, Bordi F and Casalbuoni R 1981 Path integral quantization of spinning particles interacting with crossed external electromagnetic fields *Nuovo Cim.* B **64** 287–315
[9] Lasenby A, Doran C and Gull S 1993 Grassmann calculus, pseudoclassical mechanics and geometric algebra *J. Math. Phys.* **34** 3683–712
[10] Barut A O and Zanghi N 1984 Classical model of the Dirac electron *Phys. Rev. Lett.* **52** 2009–12
[11] Barut A O 1986 Electron theory and path integral formulation of quantum-electrodynamics from a classical action *Path Integrals from meV to MeV Bielefeld Encounters in Physics and Mathematics* vol 8 ed M C Gutzwiller, A Inomata, J R Klauder and L Streit (Singapore: World Scientific) pp 381–95
[12] Casalbuoni R 1976 The classical mechanics for Bose–Fermi systems *Nuovo Cim.* A **33** 389–431
[13] Berezin F A 1966 *The Method of Second Quantization* (New York: Academic)
[14] DeWitt B 1992 *Supermanifolds* 2nd edn (Cambridge: Cambridge University Press)
[15] Cornwell J F 1989 *Group Theory in Physics: Supersymmetries and Infinite-Dimensional Algebras* vol 3 (London: Academic)
[16] Constantinescu F and de Groote H F 1994 *Geometrische und algebraische Methoden der Physik: Supermannigfaltigkeiten und Virasoro-Algebren* (Stuttgart: Teubner)
[17] Heumann R and Manton N S 2000 Classical supersymmetric mechanics *Ann. Phys.* **284** 52–88

[18] Alonso Izquierdo A, González León M A, Mateos Guilarte J and de la Torre Mayado M 2003 Supersymmetry versus integrability in two-dimensional classical mechanics *Ann. Phys.* **308** 664–91

[19] Bruce A J, Grabowska K and Moreno G 2017 On a geometric framework for Lagrangian supermechanics *J. Geom. Mech.* **9** 411–37

[20] Salam A and Strathdee J 1974 Super-gauge transformations *Nucl. Phys.* **B76** 477–82

[21] Nicolai H 1976 Supersymmetry and spin systems *J. Phys. A: Math. Theor.* A **9** 1497–506

[22] Wess J and Zumino B 1974 Supergauge transformations in four dimensions *Nucl. Phys.* B **70** 39–50

[23] Wess J and Zumino B 1974 Supergauge invariant extension of quantum electrodynamics *Nucl. Phys.* B **78** 1–13

[24] Kalka H and Soff G 1997 *Supersymmetrie* (Stuttgart: Teubner)

[25] Junker G and Matthiesen S 1994 Supersymmetric classical mechanics *J. Phys. A: Math. Theor.* **27** L751–5

[26] Junker G and Matthiesen S 1995 Addendum: Pseudoclassical mechanics and its solution *J. Phys. A: Math. Theor.* A **28** 1467–8

[27] Nicolai H 1980 On a new characterization of scalar supersymmetric theories *Phys. Lett.* B **89** 341–6

[28] Nicolai H 1980 Supersymmetry and functional integration measures *Nucl. Phys.* B **176** 419–28

[29] Ezawa H and Klauder J R 1985 Fermions without fermions: the Nicolai map revisited *Prog. Theor. Phys.* **74** 904–15

[30] Witten E 1981 Dynamical breaking of supersymmetry *Nucl. Phys.* B **188** 513–54

[31] Schulman L S 1981 *Techniques and Applications of Path Integration* (New York: Wiley)

[32] Inomata A and Junker G 1993 *Quasi-Classical Approach to Path Integrals in Supersymmetric Quantum Mechanics* ed H Cerdeira, S Lundqvist, D Mugnai, A Ranfagni, V Sa-yakanit and L S Schulman (Singapore: World Scientific) 460–82

[33] Inomata A and Junker G 1994 Quasiclassical path-integral approach to supersymmetric quantum mechanics *Phys. Rev.* A **50** 3638–49

[34] Junker G, Matthiesen S and Inomata A 1996 *Classical and Quasi-Classical Aspects of Supersymmetric Quantum MechanicsSymmetry Methods in Physics* vol 1 ed A N Sissakian and G S Pogosyan (Dubna: Joint Institute for Nuclear Research) pp 290–7

[35] Feynman R P 1948 Space–time approach to non-relativistic quantum mechanics *Rev. Mod. Phys.* **20** 367–87

[36] Gildner E and Patrascioiu A 1977 Effect of fermions upon tunneling in a one-dimensional system *Phys. Rev.* D **16** 1802–4

[37] Cooper F and Freedman B 1983 Aspects of supersymmetric quantum mechanics *Ann. Phys.* **146** 262–88

[38] Nicolai H 1991 Supersymmetrische quantenmechanik *Phys. Blätter* **47** 387–92

[39] Matthiesen S 1995 Supersymmetrische klassische Mechanik und ihre Quantisierung *Diploma Thesis* University Erlangen-Nürnberg

[40] Dirac P A M 1964 *Lectures on Quantum Mechanics* (New York: Belfer Graduate School of Science)

[41] Henneaux M and Teitelboim C 1992 *Quantization of Gauge Systems* (Princeton, NJ: Princeton University Press)

[42] Lawson H B Jr. and Michelsohn M-L 1989 *Spin Geometry* (Princeton, NJ: Princeton University Press)

[43] Feynman R P and Hibbs A R 1965 *Quantum Mechanics and Path Integrals* (New York: McGraw-Hill)

[44] Albeverio S and Høegh-Krohn R J 1976 *Mathematical Theory of Feynman Path Integrals Lecture Notes in Mathematics* (Berlin: Springer)

[45] DeWitt-Morette C, Maheshwari A and Nelson B 1979 Path integration in non-relativistic quantum mechanics *Phys. Rep.* **50** 255–372

[46] Albeverio S and Brzeźniak Z 1995 Oscillatory integrals on Hilbert spaces and Schrödinger equation with magnetic fields *J. Math. Phys.* **36** 2135–56

[47] Cartier P and DeWitt-Morette C 1995 A new perspective on functional integration *J. Math. Phys.* **36** 2237–312

[48] Nelson E 1964 Feynman integrals and the Schrödinger equation *J. Math. Phys.* **5** 332–43

[49] Leschke H and Schmutz M 1977 Operator orderings and functional formulations of quantum and stochastic dynamics *Z. Phys.* B **27** 85–94

[50] Langouche F, Roekaerts D and Tirapegui E 1982 *Functional Integration and Semiclassical Expansion* (Dordrecht: Reidel)

[51] Exner P 1985 *Open Quantum Systems and Feynman Integrals* (Dordrecht: Reidel)

[52] Inomata A, Kuratsuji H and Gerry C C 1992 *Path Integrals and Coherent States of* SU*(2) and* SU*(1, 1)* (Singapore: World Scientific)

[53] Kleinert H 2009 *Path Integrals in Quantum Mechanics, Statistics, Polymer Physics, and Financial Markets* 5th edn (Singapore: World Scientific)

[54] Zinn-Justin J 2002 *Quantum Field Theory and Critical Phenomena* 4th edn (Oxford: Clarendon)

[55] Roepstorff G 1994 *Path Integral Approach to Quantum Physics* (Berlin: Springer)

[56] Misra S P 1991 *Introduction to Supersymmetry and Supergravity* (New York: Wiley)

IOP Publishing

Supersymmetric Methods in Quantum, Statistical and Solid State Physics
Enlarged and revised edition
Georg Junker

Chapter 6

Quasi-classical approximation

In this chapter we will consider a quasi-classical evaluation of the path integral for the Witten model. In contrast to the usual semi-classical evaluation, where one expands the action about the *classical* paths up to second order, we propose a modified approach by expanding the action about the *quasi-classical* paths. We arrive in the case of good SUSY at a quantisation condition which has previously been suggested by Comtet, Bandrauk, and Campbell [1]. For broken SUSY we find a modified form of this quantisation condition [2, 3]. A remarkable property of these two quasi-classical SUSY formulas is that they yield the exact discrete spectrum for all shape-invariant potentials.

6.1 The path-integral formalism

According to Feynman [4, 5] the kernel of the time-evolution operator generated by the standard Schrödinger Hamiltonian,

$$H := \frac{p^2}{2m} + V(x), \tag{6.1}$$

is expressible in terms of a path integral

$$\langle x''|\exp\{-itH/\hbar\}|x'\rangle = \int_{x(0)=x'}^{x(t)=x''} \mathcal{D}[x]\exp\left\{\frac{i}{\hbar}\int_0^t d\tau\, L(x(\tau), \dot{x}(\tau))\right\} \tag{6.2}$$

with the standard Lagrangian

$$L(x, \dot{x}) := \frac{m}{2}\dot{x}^2 - V(x). \tag{6.3}$$

Here, in contrast to the path integral in section 5.5.2, $\int_{x(0)=x'}^{x(t)=x''} \mathcal{D}[x](\cdot)$ symbolises integration over all continuous paths $x: [0, t] \rightarrow \mathbb{R}$ starting in $x' := x(0)$ and ending in $x'' := x(t)$.

From the integral kernel of the time-evolution operator one can obtain the Green's function, that is, the integral kernel of the resolvent of H via (complex) Laplace transformation:

$$\langle x''|(E - H)^{-1}|x'\rangle = \frac{1}{i\hbar} \int_0^\infty dt \, \langle x''|e^{-itH/\hbar}|x'\rangle \, e^{itE/\hbar}, \qquad \text{Im } E > 0. \qquad (6.4)$$

The real poles (in the complex E-plane) of this expression give rise to the discrete spectrum of H. The associated residues provide the corresponding normalised energy eigenfunctions.

6.1.1 The WKB approximation in the path integral

In the semi-classical approximation [6–8] one evaluates the path integral (6.2) using the method of stationary phase. That is, one first looks for all classical paths x_{cl} for which the action

$$S[x] := \int_0^t d\tau \left[\frac{m}{2} \dot{x}^2 - V(x) \right] \qquad (6.5)$$

is stationary, $\delta S[x_{\text{cl}}] = 0$, and obey the boundary conditions $x_{\text{cl}}(0) = x'$ and $x_{\text{cl}}(t) = x''$. Then one expands the action about these classical paths up to second order in $\eta(\tau) := x(\tau) - x_{\text{cl}}(\tau)$,

$$S[x] \simeq S[x_{\text{cl}}] + \int_0^t d\tau \left[\frac{m}{2} \dot{\eta}^2 - \frac{1}{2} V''(x_{\text{cl}}(\tau))\eta^2 \right], \qquad (6.6)$$

and thus arrives at

$$\langle x''|\exp\{-itH/\hbar\}|x'\rangle \simeq \sum_{x_{\text{cl}}}^{\text{fixed } t} F_V[x_{\text{cl}}]\exp\{iS[x_{\text{cl}}]/\hbar\} \qquad (6.7)$$

with the Fresnel-type path integral

$$F_V[x_{\text{cl}}] := \int_{\eta(0)=0}^{\eta(t)=0} \mathcal{D}[\eta]\exp\left\{ \frac{i}{\hbar} \int_0^t d\tau \left[\frac{m}{2} \dot{\eta}^2 - \frac{1}{2} V''(x_{\text{cl}}(\tau))\eta^2 \right] \right\}. \qquad (6.8)$$

In the above the symbol $\sum_{x_{\text{cl}}}^{\text{fixed } t}(\cdot)$ stands for the summation over all classical paths starting in x' and arriving after the fixed time t in x''. The remaining path integral (6.8) is easily calculated via the van Vleck–Pauli–Morette formula [9–11]:

$$F_V[x_{\text{cl}}] = \sqrt{\frac{i}{2\pi\hbar} \frac{\partial^2 S[x_{\text{cl}}]}{\partial x'' \partial x'}} = \frac{e^{i\pi/4}}{\sqrt{2\pi\hbar}} \left| \frac{\partial^2 S[x_{\text{cl}}]}{\partial x'' \partial x'} \right|^{1/2} \exp\{-i\mu[x_{\text{cl}}]\pi/2\}, \qquad (6.9)$$

where $\mu \in \mathbb{N}_0$ denotes the number of negative eigenvalues of the second variation of the classical action $S[x_{cl}]$. This integer-valued functional μ is usually called the Morse index and equals the number of points along x_{cl} which are conjugate to x' [7].

In a second step one performs the Laplace integration (6.4) using, again, the method of stationary phase. A detailed discussion of this calculation can be found in chapter 18 of Schulman's book [7], which is based on the original work of Gutzwiller [6]. The result reads

$$\langle x''|(E - H)^{-1}|x'\rangle \simeq \frac{1}{i\hbar}\sqrt{D_V(x', x'', E)}$$
$$\times \sum_{x_{cl}}^{\text{fixed } E} \exp\left\{\frac{i}{\hbar} W[x_{cl}] - i\nu[x_{cl}]\frac{\pi}{2}\right\}, \qquad (6.10)$$

where

$$D_V(x'', x', E) := \frac{m}{2} |(E - V(x''))(E - V(x'))|^{-1/2} \qquad (6.11)$$

and

$$W[x_{cl}] := S[x_{cl}] + Et = \int_{x_{cl}} dx\sqrt{2m(E - V(x))} \qquad (6.12)$$

is Hamilton's characteristic functional. Note that $t = t(E)$ is obtained from the relation $\partial S[x_{cl}]/\partial t = -E$ which follows from the stationarity condition in the evaluation of the time integration (6.4). The integral in equation (6.12) has to be taken along the classical path x_{cl}. In equation (6.10) the symbol $\sum_{x_{cl}}^{\text{fixed } E}(\cdot)$ denotes the summation over all classical paths from x' to x'' with a given fixed energy E. The integer-valued functional $\nu \in \mathbb{N}_0$ is called the Maslov index and equals the number of turning points along the classical path [6, 12].

If one finally assumes that the potential V has a single global minimum (see figure 6.1) the sum over all classical paths in equation (6.10) can be performed explicitly [7]. See also the following section. We arrive at the well-known

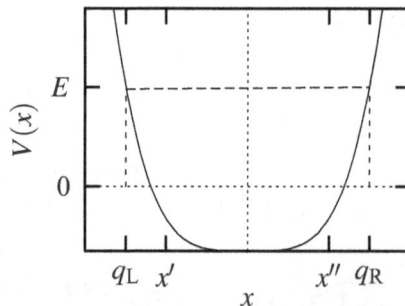

Figure 6.1. A single-well potential V and the corresponding classical paths for a fixed energy E starting in x' and arriving in x''. The left and right turning points q_L and q_R may be passed several times.

quantisation condition of Wentzel, Brillouin, and Kramers [13–16], the so-called WKB formula,

$$\int_{q_L}^{q_R} \mathrm{d}x \sqrt{2m(E - V(x))} = \hbar\pi(n + 1/2), \qquad n = 0, 1, 2, \dots , \qquad (6.13)$$

where q_L and q_R are the classical left and right turning points determined by $V(q_L) = E = V(q_R)$. The corresponding approximate energy eigenfunctions read

$$\phi_n(x) \simeq \sqrt{\frac{4m}{T_{E_n} p_{E_n}(x)}} \sin\left(\frac{1}{\hbar} \int_{q_L}^{x} \mathrm{d}z\, p_{E_n}(z) + \frac{\pi}{4}\right), \qquad (6.14)$$

where $p_E(x) := \sqrt{2(E - V(x))}$ is the magnitude of the classical momentum for a given energy E, $T_E := 2m \int_{q_L}^{q_R} \mathrm{d}x [p_E(x)]^{-1}$ is the period of the classical motion and E_n is the solution of the WKB formula (6.13) for a given $n \in \mathbb{N}_0$.

6.1.2 Quasi-classical modification for Witten's model

Let us now consider the case of Witten's model, where the potential appearing in the action (6.5) is given by one of the partner potentials $V_\pm$. That is, the two actions associated with Witten's partner Hamiltonians $H_\pm$ read

$$S^\pm[x] := \int_0^t \mathrm{d}\tau \left[\frac{m}{2} \dot{x}^2 - \Phi^2(x) \mp \frac{\hbar}{\sqrt{2m}} \Phi'(x)\right]. \qquad (6.15)$$

According to our discussion of section 5.5 we may split this actions into a tree part and a fermion-loop correction,

$$S^\pm[x] = S_{\text{tree}}[x] \mp \hbar\varphi[x], \qquad (6.16)$$

where

$$S_{\text{tree}}[x] := \int_0^t \mathrm{d}\tau \left[\frac{m}{2} \dot{x}^2 - \Phi^2(x)\right], \qquad \varphi[x] := \frac{1}{\sqrt{2m}} \int_0^t \mathrm{d}\tau\, \Phi'(x). \qquad (6.17)$$

Note that the above functional φ is identical to the fermionic phase (5.19). Here we have the original SUSY potential Φ, whereas in equation (5.19) the rescaled SUSY potential $\Phi/\sqrt{2}$ is used and the mass has been set to unity. The explicit appearance of $\hbar$ in front of the fermionic phase indicates that this part of the action stems from quantum corrections of Fermion loops. It is therefore reasonable to assume that the major contributions to the path integral will be supplied by paths which make the tree-action functional stationary. Those paths are in fact the quasi-classical paths introduced in section 5.3. Note, however, that the SUSY potential Φ will in general depend on Planck's constant as well. Actually, the natural units of Φ are $\hbar/\sqrt{2m}$. Hence, a formal power counting, as done by Comtet *et al* [1] and Eckhardt [2], has no *a priori* justification. Our suggested modification, which takes the quasi-classical paths and their quadratic fluctuations as dominant contributions to the path integral into account, is not based on such a formal power counting. It rather stems from our discussion of supersymmetric classical dynamics, which has already shown that the solutions of the classical equations of motion are governed by those quasi-classical paths. For this reason the present approach [3, 17, 18] has been called the *quasi-classical approximation* in order to distinguish it from the usual semi-classical approximation as discussed in section 6.1.1.

Therefore, let us expand the tree action up to second order in $\eta(\tau) := x(\tau) - x_{\mathrm{qc}}(\tau)$, that is,

$$S^{\pm}[x] \simeq S_{\mathrm{tree}}[x_{\mathrm{qc}}] \mp \hbar\varphi[x_{\mathrm{qc}}] + \int_0^t d\tau\left[\frac{m}{2}\,\dot{\eta}^2 - \frac{1}{2}(\Phi^2)''\big(x_{\mathrm{qc}}(\tau)\big)\eta^2\right]. \tag{6.18}$$

Then we again arrive at an approximate Fresnel-type path-integral representation for the kernel of the time-evolution operator associated with the SUSY partner Hamiltonians:

$$\langle x''|e^{-itH_{\pm}/\hbar}|x'\rangle \simeq \sum_{x_{\mathrm{cl}}}^{\mathrm{fixed}\ t} F_{\Phi^2}[x_{\mathrm{cl}}]\exp\left\{\frac{i}{\hbar}\,S_{\mathrm{tree}}[x_{\mathrm{qc}}] \mp i\varphi[x_{\mathrm{qc}}]\right\}, \tag{6.19}$$

where

$$F_{\Phi^2}[x_{\mathrm{cl}}] = \sqrt{\frac{i}{2\pi\hbar}\frac{\partial^2 S_{\mathrm{tree}}[x_{\mathrm{qc}}]}{\partial x''\partial x'}}\left|\frac{\partial^2 S_{\mathrm{tree}}[x_{\mathrm{qc}}]}{\partial x''\partial x'}\right|^{1/2}\exp\{-i\mu[x_{\mathrm{qc}}]\pi/2\}. \tag{6.20}$$

In essence, the calculation is identical to that of section 6.1.1 with the potential V replaced by Φ^2. The only difference is the additional fermionic phase-functional φ. Hence, we immediately arrive at the approximated Green's function

$$\langle x''|(E - H_{\pm})^{-1}|x'\rangle \simeq \frac{1}{i\hbar}\sqrt{D_{\Phi^2}(x', x'', E)}$$

$$\times \sum_{x_{\mathrm{qc}}}^{\mathrm{fixed}\ E} \exp\left\{\frac{i}{\hbar}\,W_{\mathrm{tree}}[x_{\mathrm{qc}}] \mp i\varphi[x_{\mathrm{qc}}] - i\nu[x_{\mathrm{qc}}]\frac{\pi}{2}\right\} \tag{6.21}$$

with

$$W_{\mathrm{tree}}[x_{\mathrm{qc}}] := \int_{x_{\mathrm{qc}}} dx\sqrt{2m(E - \Phi^2(x))}. \tag{6.22}$$

We will now explicitly perform the above path summation. To this end we assume that Φ^2 will have one single global minimum. See figure 6.2. Note that this does not necessarily imply that the full potentials $V_{\pm}$ are also of this type. In doing this summation we first classify, following section 5.4, the quasi-classical paths into the same four classes (i), $i = 1, 2, 3, 4$. All three phases appearing in equation (6.21) are functionals of the quasi-classical path. Hence, within each class (i) we can explicitly put them into the form

$$W_{\text{tree}}[x_{\text{cq}}] \equiv W_k^{(i)} = W_0^{(i)} + 2kw(x_R),$$
$$\nu[x_{\text{cq}}] \equiv \nu_k^{(i)} = \nu_0^{(i)} + 2k, \tag{6.23}$$
$$\varphi[x_{\text{cq}}] \equiv \varphi_k^{(i)} = \varphi_0^{(i)} + k[a(x_R) - a(x_L)],$$

where

$$w(x) := \int_{x_L}^{x} dz\sqrt{2m(E - \Phi^2(z))}, \qquad a(x) := \arcsin\frac{\Phi(x)}{\sqrt{E}}. \tag{6.24}$$

Note that the present definition of $a(x)$ differs from that in equation (5.35) because of the rescaling of the SUSY potential. As in section 5.4, the integer $k \in \mathbb{N}_0$ enumerates the number of complete cycles of a path within each class. The quantities for $k = 0$ are given in table 6.1 with the turning points $x_{L/R}$ to be derived from $\Phi^2(x_L) = E = \Phi^2(x_R)$. The path sum may be rewritten as

$$\sum_{x_{\text{qc}}}^{\text{fixed } E} (\cdot) = \sum_{i=1}^{4}\sum_{k=0}^{\infty}(\cdot) \tag{6.25}$$

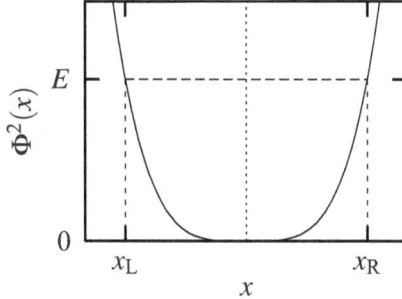

Figure 6.2. The single-well potential Φ^2 and the left and right turning points x_L and x_R of the quasi-classical motion for a given energy E.

Table 6.1. Phases of the $k = 0$ paths contributing to the resolvent kernel (6.21) within the quasi-classical approximation.

Class (i)	$W_0^{(i)}$	$\varphi_0^{(i)}$	$\nu_0^{(i)}$
(1)	$w(x'') - w(x')$	$\frac{1}{2}[a(x'') - a(x')]$	0
(2)	$w(x'') + w(x')$	$\frac{1}{2}[a(x'') + a(x')] - a(x_L)$	1
(3)	$2w(x_R) - w(x'') - w(x')$	$-\frac{1}{2}[a(x'') + a(x')] + a(x_R)$	1
(4)	$2w(x_R) - w(x'') + w(x')$	$-\frac{1}{2}[a(x'') + a(x')] + a(x_R) - a(x_L)$	2

and gives rise to a geometric series for the k-sum. Performing the latter summation we arrive at the approximate Green's function:

$$\langle x''|(E - H_\pm)^{-1}|x'\rangle \simeq \frac{1}{i\hbar} \frac{m}{\sqrt{p_E^{qc}(x'')p_E^{qc}(x')}}$$
$$\times \frac{\sum_{i=1}^{4} \exp\left\{\frac{i}{\hbar}W_0^{(i)} \mp i\varphi_0^{(i)} - i\nu_0^{(i)}\frac{\pi}{2}\right\}}{1 - \exp\left\{i\left[\frac{2}{\hbar}w(x_R) \mp (a(x_R) - a(x_L)) - \pi\right]\right\}},$$

(6.26)

where

$$p_E^{qc}(x) := \sqrt{2m(E - \Phi^2(x))}$$

(6.27)

is the magnitude of the quasi-classical momentum for a given energy E. From this resolvent kernel we may derive approximate energy eigenvalues and eigenfunctions.

6.2 Quasi-classical quantisation conditions

From the poles of equation (6.26) we obtain the following quasi-classical quantisation condition:

$$w(x_R) \equiv \int_{x_L}^{x_R} dx\, p_E^{qc}(x) = \hbar\pi\left(n + \frac{1}{2} \pm \frac{a(x_R) - a(x_L)}{2\pi}\right),$$

(6.28)

with $n \in \mathbb{N}_0$. This expression can be made more explicit by considering the turning-point condition of the quasi-classical paths, which implies $a(x_{R/L}) = \frac{\pi}{2}\mathrm{sgn}[\Phi(x_{R/L})]$:

$$\int_{x_L}^{x_R} dx\, p_E^{qc}(x) = \hbar\pi\left(n + \frac{1}{2} \pm \frac{1}{4}[\mathrm{sgn}\,\Phi(x_R) - \mathrm{sgn}\,\Phi(x_L)]\right).$$

(6.29)

This expression has first been obtained by Eckhardt [2] via Maslov theory based, however, on a very strong assumption. Eckhardt explicitly assumes that the $\hbar$ dependence of the SUSY potential Φ is such that $\lim_{\hbar \to 0}\Phi^2$ is well-defined. This, for example, is not the case for the radial hydrogen atom (3.56). In fact, for the natural choice of units $\Phi(x) = \frac{\hbar}{\sqrt{2m}}f(x)$, where f is independent of $\hbar$, Eckhardt's assumption covers the free-particle case only. The above path-integral approach, which is based on the stationary paths of the tree and not the full action, was first presented in [3]. In the following we will denote the solutions of this quantisation condition for a given $n \in \mathbb{N}_0$ by $E_n^\pm$ for $H_\pm$. For a discussion of the above quantisation condition (6.29) we consider the two possible cases $\Phi(x_L) = -\Phi(x_R)$ and $\Phi(x_L) = \Phi(x_R)$.

Case 1. $\Phi(x_L) = -\Phi(x_R) = \pm\sqrt{E}$

This case corresponds to a good SUSY. Note that we have assumed that Φ^2 is a continuous function with a single global minimum. Hence only the three cases shown in figure 5.2(a)–(c) are possible. The present case corresponds to the first case (a) and implies a good SUSY. We also note that because of our ground-state convention (3.20) only the case $\Phi(x_L) = -\Phi(x_R) = -\sqrt{E}$ may occur. We arrive at the quasi-classical quantisation conditions for good SUSY:

$$
\begin{aligned}
\int_{x_L}^{x_R} dx\sqrt{2m(E - \Phi^2(x))} &= \hbar\pi n \qquad \text{for } H_-, \\
\int_{x_L}^{x_R} dx\sqrt{2m(E - \Phi^2(x))} &= \hbar\pi(n + 1) \qquad \text{for } H_+.
\end{aligned}
\tag{6.30}
$$

Formula (6.30), which is sometimes referred to as the CBC formula, was first suggested by Comtet, Bandrauk, and Campbell [1] based on a formal WKB approach. They have assumed that Φ will be independent of $\hbar$.

Comparing the CBC formula (6.30) with the WKB formula (6.13) we note two differences. First, instead of the full potentials $V_\pm$ only Φ^2 appears in the integral on the left-hand side. Second, the extra term $\frac{1}{2}$ in the WKB formula, which stems from the Maslov indices, is in the case of H_- precisely cancelled by the contributions of the fermionic phase. In the case of H_+, however, these two contributions are equal and add up to unity. The CBC formula (6.30) has some remarkable properties.

(1) The quantisation condition (6.30) provides the exact ground-state energy for H_-. Note that for $n = 0$ one necessarily is led to $E \equiv E_0 = 0$. We have already shown in section 4.1, that the knowledge of the ground-state energy and wave function for a given Hamiltonian is needed in order to find the SUSY potential. Hence, the quasi-classical quantisation condition cannot provide an additional information about the ground state. Nevertheless, it is remarkable that the quasi-classical approach can reproduce the exact ground-state energy for good SUSY. This is a rather unexpected result.

(2) The presence of the fermionic phase has the consequence that the exact relation $E_{n+1}^- = E_n^+$ is also valid within the quasi-classical approximation. Note that the WKB formula (6.13) does in general not reproduce this relation.

(3) An immediate and obvious consequence of properties (1) and (2) is that the CBC formula will reproduce exact bound-state spectra for all shape-invariant potentials [1, 19]. This is in contrast to the WKB formula, which is not able to reproduce the exact spectra for those systems without any ad hoc modifications of the Langer type [20]. Only in the special case of the harmonic oscillator and the Morse oscillator is the WKB formula able to reproduce the exact bound-state spectrum. See also our discussion in section 6.4.1 below. Furthermore, it has been shown that for a wide class of shape-invariant potentials all higher-order corrections (in $\hbar$) to the CBC formula vanish [21, 22].

Case 2. $\Phi(x_L) = \Phi(x_R) = \pm\sqrt{E}$

This is the case where SUSY is broken (at the quantum level). See figure 5.2(a)–(c) cases (b) and (c). Here the contribution to the fermionic phase from the left turning point cancels that from the right turning point. The quantisation condition (6.29) reduces to

$$\int_{x_L}^{x_R} dx \sqrt{2m(E - \Phi^2(x))} = \hbar\pi\left(n + \frac{1}{2}\right) \quad \text{for } H_\pm. \qquad (6.31)$$

This formula, sometimes referred to as the EIJ formula, was first derived in [2, 3, 17, 18]. As in equation (6.30), instead of the full potential $V_\pm$ only Φ^2 appears in the integral on the left-hand side. Whereas, on the right-hand side it is identical to the WKB formula. The Maslov indices are not affected by the fermionic phase. Formula (6.31) also has some remarkable properties.

(1) The exact relation $E_n^- = E_n^+$ valid for broken SUSY is respected also by the quasi-classical expression (6.31). Again we note that the WKB formula (6.13) does in general not obey this relation.

(2) The surprising result, which was first noted in [3] is, that for those shape-invariant potentials for which the parameters can be chosen such that SUSY will be broken (see remark c in table 4.1), formula (6.31) does provide the exact bound-state spectrum. See also [17, 18, 23, 24] and our discussion in section 6.4.1. As for the CBC formula it can be shown that higher-order corrections to equation (6.31) vanish identically for some shape-invariant potentials [25].

It is interesting to note that equations (6.30) and (6.31) can be combined with the help of the Witten index (2.61), which in the case of the Witten model is given by $\Delta = \mp 1$ for good SUSY if the ground state belongs to $H_\pm$, and by $\Delta = 0$ for broken SUSY, respectively. See equation (3.30). Hence, we may write equations (6.30) and (6.31) in the compact form

$$\boxed{\int_{x_L}^{x_R} dx \sqrt{2m(E - \Phi^2(x))} = \hbar\pi\left(n + \frac{1}{2} \pm \frac{\Delta}{2}\right) \quad \text{for } H_\pm,} \qquad (6.32)$$

which is even independent of any convention about the ground state. The above expression displays an interesting interplay between the Witten and the Maslov index. We will refer to expression (6.32) as the *quasi-classical SUSY (qc-SUSY) approximation* in the following. The appearance of Witten's index in this formula is rather natural. Note that under the assumption we have made, that is, Φ^2 characterises a single-well potential (see figure 6.2) we may put expression (3.30) into the form

$$\Delta = \frac{1}{\pi}(a(x_R) - a(x_L)), \qquad x_R > x_L. \qquad (6.33)$$

This identification allows us to interpret the qc-SUSY approximation (6.32) as the pseudoclassical analogue of the Bohr–Sommerfeld quantisation condition [26]. In fact, let us consider the following pseudoclassical phase integral along one period of the quasi-classical motion (see equations (5.15) and (5.46))

$$\oint (\pi d\psi + \bar{\pi} d\bar{\psi}) = -(a(x_R) - a(x_L))[\bar{\psi}_0, \psi_0] = -\pi \Delta[\bar{\psi}_0, \psi_0] \tag{6.34}$$

and note that $[\bar{\psi}_0, \psi_0]$ is replaced by $\hbar\sigma_3$ upon quantisation (cf. equation (5.58)). Since, σ_3 has eigenvalues ± 1 in the subspace $\mathcal{H}^\pm$ the qc-SUSY formula (6.32) may formally be put into the form

$$\oint \left(p_E^{qc} dx + \pi d\psi + \bar{\pi} d\bar{\psi} \right) = 2\pi\hbar \left(n + \frac{1}{2} \right). \tag{6.35}$$

Let us now comment on the formal approaches of [1, 2]. In fact, assuming that Φ is independent of $\hbar$, the above expression (6.32) can be derived from the standard WKB formula (6.13) by simple Taylor expansion in $\hbar$ [23]. Replacing in the WKB formula (6.13) the potential V by the explicit form of $V_\pm$ we find

$$\begin{aligned}
\int_{q_L}^{q_R} dx \sqrt{\left(p_E^{qc}(x)\right)^2 \mp \frac{\hbar}{\sqrt{2m}} \Phi'(x)} &= \int_{x_L}^{x_R} dx \, p_E^{qc}(x) \\
&\pm \frac{\hbar}{2} \int_{x_L}^{x_R} dx \frac{\Phi'(x)}{\sqrt{E - \Phi^2(x)}} + O(\hbar^{3/2}),
\end{aligned} \tag{6.36}$$

where we have expanded the square root on the left-hand side in a Taylor series and kept only the first two terms. We have also replaced the classical turning points $V_\pm(q_{L/R}) = E$ by the quasi-classical turning points $\Phi^2(x_{L/R}) = E$. The second integral on the right-hand side can explicitly be evaluated and we will arrive at the expression (6.28) which is equivalent to equation (6.32). Hence, for a SUSY potential which is independent of $\hbar$ the quasi-classical SUSY approximation (6.32) is actually equivalent to the WKB formula in first order of $\hbar$. An example for such a case is the harmonic oscillator SUSY potential $\Phi(x) = \sqrt{m/2} \, \omega x$ for which the WKB as well as the CBC formula provide the exact spectrum. This, however, is not a typical but rather an exceptional case. Actually, whenever there is an intrinsic length scale in the underlying classical problem (for example $1/\alpha$ in $\Phi(x) = \hbar\alpha/\sqrt{2m} \tanh(\alpha x)$) then the energy has the unit $\hbar^2/2m$ if we measure distances in units of this length scale (i.e. $\alpha = 1$). Setting $\Phi(x) =: \frac{\hbar}{\sqrt{2m}} f(x)$ and $E =: \frac{\hbar^2}{2m} \varepsilon$ the stationary Schrödinger equation reads

$$\left(-\frac{\partial^2}{\partial x^2} + f^2(x) \pm f'(x) \right) \varphi(x) = \varepsilon \varphi(x). \tag{6.37}$$

The WKB and the quasi-classical SUSY formula lead to the following approximate quantisation conditions for the dimensionless eigenvalues ε, respectively,

$$\begin{aligned}
\int_{q_L}^{q_R} dx \sqrt{\varepsilon - f^2(x) \mp f'(x)} &= \pi \left(n + \frac{1}{2} \right), \\
\int_{x_L}^{x_R} dx \sqrt{\varepsilon - f^2(x)} &= \pi \left(n + \frac{1}{2} \pm \frac{\Delta}{2} \right).
\end{aligned} \tag{6.38}$$

Hence, the WKB and the quasi-classical SUSY approximations are indeed not equivalent.

6.3 Quasi-classical eigenfunctions

The approximate result for the resolvent kernel (6.26) also provides the quasi-classical wave functions to be obtained from the residues of its poles [3]. Again we will distinguish the two cases of good and broken SUSY.

Case 1. Good SUSY

Using the explicit forms of the quantities in table 6.1 we can perform the remaining sum in equation (6.26) at the poles $E_n^{\pm}$ determined by the CBC formula (6.30):

$$\text{Res}\langle x''|(E - H_+)^{-1}|x'\rangle|_{E=E_n^+} \simeq \frac{4m}{T_E^{\text{qc}}} \left. \frac{\sin\left(\dfrac{w(x')}{\hbar} - \dfrac{a(x')}{2}\right)}{\sqrt{p_E^{\text{qc}}(x')}} \frac{\sin\left(\dfrac{w(x'')}{\hbar} - \dfrac{a(x'')}{2}\right)}{\sqrt{p_E^{\text{qc}}(x'')}} \right|_{E=E_n^+},$$

$$\text{Res}\langle x''|(E - H_-)^{-1}|x'\rangle|_{E=E_n^-} \simeq \frac{4m}{T_E^{\text{qc}}} \left. \frac{\cos\left(\dfrac{w(x')}{\hbar} + \dfrac{a(x')}{2}\right)}{\sqrt{p_E^{\text{qc}}(x')}} \frac{\cos\left(\dfrac{w(x'')}{\hbar} + \dfrac{a(x'')}{2}\right)}{\sqrt{p_E^{\text{qc}}(x'')}} \right|_{E=E_n^-}, \tag{6.39}$$

where $T_E^{\text{qc}} := 2m \int_{x_{\text{L}}}^{x_{\text{R}}} dz [p_E^{\text{qc}}(z)]^{-1}$ is the period of the bounded quasi-classical motion for a given energy E. From these residues we read off the quasi-classical wave functions:

$$\boxed{\phi_n^+(x) \simeq \sqrt{\frac{4m}{T_{E_n^+}^{\text{qc}} p_{E_n^+}^{\text{qc}}(x)}} \sin\left(\frac{1}{\hbar} \int_{x_{\text{L}}}^x dz\, p_{E_n^+}^{\text{qc}}(z) - \frac{1}{2}\arcsin\frac{\Phi(x)}{\sqrt{E_n^+}}\right),} \tag{6.40}$$

$$\boxed{\phi_n^-(x) \simeq \sqrt{\frac{4m}{T_{E_n^-}^{\text{qc}} p_{E_n^-}^{\text{qc}}(x)}} \cos\left(\frac{1}{\hbar} \int_{x_{\text{L}}}^x dz\, p_{E_n^-}^{\text{qc}}(z) + \frac{1}{2}\arcsin\frac{\Phi(x)}{\sqrt{E_n^-}}\right).} \tag{6.41}$$

Note that $E_0^- = 0$ and, therefore, also $p_{E_0^-} = 0$. Hence, equation (6.41) is not defined for $n = 0$. We have to exclude the value $n = 0$ in equation (6.41). Actually, the ground-state wave function is already known exactly:

$$\phi_0^-(x) = C \exp\left\{-\frac{\sqrt{2m}}{\hbar} \int_{x_0}^x dz\, \Phi(z)\right\}. \tag{6.42}$$

In comparing the expressions (6.40) and (6.41) with the WKB result (6.14) we note that the latter has a constant phase shift $\pi/4$ whereas the SUSY wave functions contain x-dependent phase shifts stemming from the fermionic phase.

Case 2. Broken SUSY

In this case the residues of equation (6.26) explicitly read

$$
\mathrm{Res}\langle x''|(E - H_+)^{-1}|x'\rangle|_{E=E_n^+} \simeq \frac{4m}{T_E^{\mathrm{qc}}} \left. \frac{\cos\left(\dfrac{w(x')}{\hbar} - \dfrac{a(x')}{2}\right)}{\sqrt{p_E^{\mathrm{qc}}(x')}} \frac{\cos\left(\dfrac{w(x'')}{\hbar} - \dfrac{a(x'')}{2}\right)}{\sqrt{p_E^{\mathrm{qc}}(x'')}} \right|_{E=E_n^+},
$$

$$
\mathrm{Res}\langle x''|(E - H_-)^{-1}|x'\rangle|_{E=E_n^-} \simeq \frac{4m}{T_E^{\mathrm{qc}}} \left. \frac{\sin\left(\dfrac{w(x')}{\hbar} + \dfrac{a(x')}{2}\right)}{\sqrt{p_E^{\mathrm{qc}}(x')}} \frac{\sin\left(\dfrac{w(x'')}{\hbar} + \dfrac{a(x'')}{2}\right)}{\sqrt{p_E^{\mathrm{qc}}(x'')}} \right|_{E=E_n^-},
$$

$$\tag{6.43}$$

where now the poles at $E = E_n^+ = E_n^-$ are determined by the EIJ formula (6.31). Hence the corresponding quasi-classical wave functions read

$$
\boxed{\phi_n^+(x) \simeq \sqrt{\frac{4m}{T_{E_n^+}^{\mathrm{qc}} p_{E_n^+}^{\mathrm{qc}}(x)}} \cos\left(\frac{1}{\hbar} \int_{x_{\mathrm{L}}}^x \mathrm{d}z\, p_{E_n^+}^{\mathrm{qc}}(z) - \frac{1}{2}\arcsin\frac{\Phi(x)}{\sqrt{E_n^+}}\right),}
$$

$$\tag{6.44}$$

$$
\boxed{\phi_n^-(x) \simeq \sqrt{\frac{4m}{T_{E_n^-}^{\mathrm{qc}} p_{E_n^-}^{\mathrm{qc}}(x)}} \sin\left(\frac{1}{\hbar} \int_{x_{\mathrm{L}}}^x \mathrm{d}z\, p_{E_n^-}^{\mathrm{qc}}(z) + \frac{1}{2}\arcsin\frac{\Phi(x)}{\sqrt{E_n^-}}\right).}
$$

$$\tag{6.45}$$

As in the good-SUSY case we observe the additional x-dependent phase appearing in equations (6.44) and (6.45).

The wave functions as they stand (including the WKB wave function) are only valid in the quasi-classical (classical) allowed region $x_{\mathrm{L}} < x < x_{\mathrm{R}}$ ($q_{\mathrm{L}} < x < q_{\mathrm{R}}$). In the forbidden regions the trigonometric functions have to be replaced by appropriate decreasing exponential functions. As expected, these quasi-classical eigenfunctions are singular at the quasi-classical turning points x_{L} and x_{R}. For a discussion of regularised wave functions see [27–29].

6.4 Discussion of the results

6.4.1 Exactly soluble examples

As we have already mentioned above, the quasi-classical SUSY quantisation condition (6.32) provides the exact bound-state eigenvalues for all shape-invariant potentials. This is in contrast to the WKB formula (6.13), which in general requires an ad hoc replacement of the potential parameters in order to yield the exact eigenvalues for those shape-invariant potentials. As an example, let us consider the radial harmonic oscillator

$$V(r) = \frac{m}{2}\omega^2 r^2 + \frac{\hbar^2 l(l+1)}{2mr^2}, \qquad l = 0, 1, 2, \ldots, \qquad r > 0. \tag{6.46}$$

Using the WKB approximation, it is known [20] that only upon the Langer modification $l(l+1) \rightarrow (l + \frac{1}{2})^2$ the WKB formula

$$\int_{q_L}^{q_R} dr \sqrt{2m\left(\tilde{E} - \frac{m}{2}\omega^2 r^2 - \frac{\hbar^2\left(l + \frac{1}{2}\right)^2}{2mr^2}\right)} = \hbar\pi(n + 1/2) \tag{6.47}$$

will give rise to the exact spectrum (see problem 6.1 below)

$$\tilde{E}_n = \hbar\omega(2n + l + 3/2). \tag{6.48}$$

Let us now consider the SUSY potential

$$\Phi(r) := \sqrt{\frac{m}{2}}\,\omega r - \frac{\hbar(l+1)}{\sqrt{2m}\,r}, \tag{6.49}$$

which leads to a good SUSY. The associated partner potentials are

$$V_+(r) = \frac{m}{2}\,\omega^2 r^2 + \frac{\hbar^2(l+1)(l+2)}{2mr^2} - \hbar\omega(l + 1/2),$$
$$V_-(r) = \frac{m}{2}\,\omega^2 r^2 + \frac{\hbar^2 l(l+1)}{2mr^2} - \hbar\omega(l + 3/2), \tag{6.50}$$

where the latter is identical to the original potential V up to a constant negative shift by the ground-state energy $\tilde{E}_0 = \hbar\omega(l + 3/2)$. V_+ corresponds to the same potential but with l replaced by $l + 1$ and a different energy shift. Using the qc-SUSY formula (6.32)

$$\int_{x_L}^{x_R} dr \sqrt{2m\left(E - \frac{m}{2}\omega^2 r^2 - \frac{\hbar^2(l+1)^2}{2mr^2} + \hbar\omega(l+1)\right)} = \hbar\pi\left(n + \frac{1}{2} \pm \frac{1}{2}\right) \tag{6.51}$$

we immediately arrive at (see problem 6.2 below)

$$E_n^{\pm} = 2\hbar\omega\left(n + \frac{1}{2} \pm \frac{1}{2}\right), \tag{6.52}$$

which is the exact result as $\tilde{E}_n = E_n^- + \hbar\omega(l + 3/2)$.

In contrast, choosing the SUSY potential to

$$\Phi(r) := \sqrt{\frac{m}{2}}\,\omega r + \frac{\hbar l}{\sqrt{2m}\,r} \tag{6.53}$$

we realise a broken SUSY with partner potentials

$$V_+(r) = \frac{m}{2}\omega^2 r^2 + \frac{\hbar^2(l-1)l}{2mr^2} + \hbar\omega(l+1/2),$$

$$V_-(r) = \frac{m}{2}\omega^2 r^2 + \frac{\hbar^2 l(l+1)}{2mr^2} + \hbar\omega(l-1/2). \tag{6.54}$$

Here the latter is also identical to V with a positive shift by $\hbar\omega(l-1/2)$ and V_+ is similar to V_- with l replaced by $l-1$. In this case the qc-SUSY quantisation condition reads

$$\int_{x_L}^{x_R} dr \sqrt{2m\left(E - \frac{m}{2}\omega^2 r^2 - \frac{\hbar^2 l^2}{2mr^2} - \hbar\omega l\right)} = \hbar\pi\left(n + \frac{1}{2}\right) \tag{6.55}$$

and yields the energy eigenvalues

$$E_n^\pm = \hbar\omega(2n + 2l + 1), \tag{6.56}$$

which again are exact because of $\tilde{E}_n = E_n^- - \hbar\omega(l-1/2)$.

Similar to this example it is straightforward but tedious [24, 30] to show that all of the shape-invariant potentials listed in table 4.1 give rise to the exact energy eigenvalues if the qc-SUSY formula (6.32) is used. For those systems where SUSY is good this phenomenon can be explained according to our remark (3) in case 1 of section 6.2.

Note that it has been argued [31] that for the new shape-invariant potentials found by [32] the qc-SUSY does not yield the exact eigenvalues. In contrast to the arguments given in [31] this might happen because of the singular nature of these potentials. See the comments made in section 4.2. However, for those cases (marked with c in table 4.1) which also allow for a broken-SUSY potential, there is no explanation available for the exactness of formula (6.31). In fact, the question—why is the qc-SUSY approximation exact for shape-invariant potentials?—is still open. A possible explanation might be based on the discussion of the Nicolai map by Ezawa and Klauder [33], see also section 7.3 below. Ezawa and Klauder have shown that the (Euclidean-time) path integral for SUSY quantum mechanics can be transformed into a Gaussian path integral via this Nicolai map. Gaussian path integrals, in turn, are exactly evaluated by the method of stationary phase.

6.4.2 Numerical investigations

We have seen that the qc-SUSY approximation (6.32) does lead to exact bound-state spectra for (by other means) exactly solvable problems. Here, the following question immediately comes to mind: Does this approximation also yield better eigenvalues (than the WKB approximation (6.13) does) for those problems which are not exactly solvable? To answer this questions let us investigate a class of power SUSY potentials of the form [18]

$$\Phi(x) := \frac{\hbar a}{\sqrt{2m}} x^d \qquad \text{for } x \geqslant 0. \tag{6.57}$$

Here $a > 0$ and $d \geqslant 1$ are free parameters. Note that the above definition is only valid for the positive Euclidean half-line. For $x < 0$ we may define the SUSY potential either through an antisymmetric or symmetric continuation, which leads to a good and broken SUSY, respectively:

$$\Phi(x) = \frac{\hbar a}{\sqrt{2m}} |x|^d \, \text{sgn}(x) \qquad \text{for good SUSY,} \tag{6.58}$$

$$\Phi(x) = \frac{\hbar a}{\sqrt{2m}} |x|^d \qquad \text{for broken SUSY.} \tag{6.59}$$

The associated partner potentials read

$$V_\pm(x) = \frac{\hbar^2}{2m}(a^2 x^{2d} \pm ad \, |x|^{d-1}) = V_\pm(-x) \qquad \text{for good SUSY,} \tag{6.60}$$

$$V_\pm(x) = \frac{\hbar^2}{2m}(a^2 x^{2d} \pm ad \, |x|^{d-1}\text{sgn } x) = V_\mp(-x) \qquad \text{for broken SUSY} \tag{6.61}$$

and are shown in figures 6.3 and 6.4, respectively.

For the SUSY potentials (6.58) and (6.59) the qc-SUSY quantisation condition (6.32) can be evaluated analytically with the result (Γ denotes Euler's gamma function)

$$E_n^\pm = \frac{\hbar^2 a^2}{2m}\left[\frac{\Gamma\left(\frac{3d+1}{2d}\right)}{\Gamma\left(\frac{2d+1}{2d}\right)}\frac{\sqrt{\pi}}{a}\left(n + \frac{1}{2} \pm \frac{\Delta}{2}\right)\right]^{2d/(d+1)}, \tag{6.62}$$

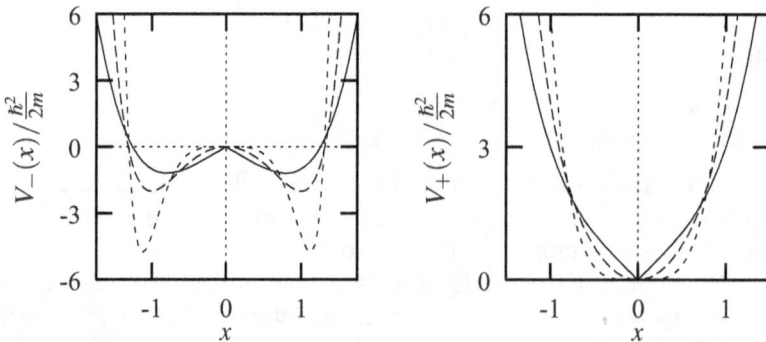

Figure 6.3. The partner potentials (6.60) for values $a = 1$ and $d = 2$ (solid line), 3 (long dashes), and 5 (short dashes).

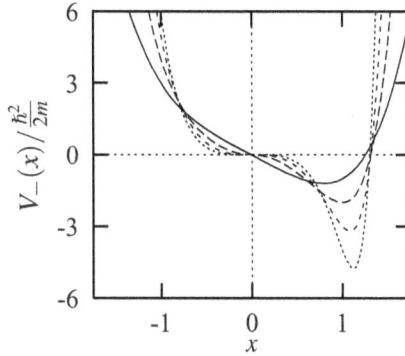

Figure 6.4. The potential V_- (6.61) for broken SUSY and $a = 1$, $d = 2$ (solid line), 3 (long dashes), 4 (short dashes), and 6 (dotted line).

where $\Delta = 1$ for the case of good SUSY and $\Delta = 0$ for broken SUSY, respectively. These energy eigenvalues are in general not exact. For arbitrary d and good SUSY only $E_0^- = 0$ will be exact. The associated normalised ground-state wave function reads

$$\phi_0^-(x) = \left[\frac{a}{\Gamma\left(\frac{1}{d+1}\right)} \left(\frac{d+1}{2a} \right)^{d/(d+1)} \right]^{1/2} \exp\left\{ -\frac{a}{d+1}|x|^{d+1} \right\}. \tag{6.63}$$

For $d = 1$ and good SUSY we arrive at the supersymmetric harmonic oscillator problem, where equation (6.62) provides the exact spectrum for all n. Formally, for good SUSY we may also include the case $d = 0$, if in this limit in the potential (6.60) the term $\pm ad|x|^{d-1}$ is interpreted as $\pm a\delta(x)$. Indeed, for $d = 0$ the SUSY potential (6.58) coincides with that for the δ-potential. See example 3.4 in section 3.6.1. In this case only $n = 0$ is allowed.

Another case, where equation (6.62) becomes exact for all n and for good as well as broken SUSY, is the limit $d \to \infty$:

$$\lim_{d\to\infty} E_n^\pm = \frac{\hbar^2\pi^2}{8m}\left(n + \frac{1}{2} \pm \frac{\Delta}{2} \right)^2. \tag{6.64}$$

That this is the exact spectrum can be seen by realising that both equations (6.60) and (6.61) become infinite square-well potentials in this limit:

$$\lim_{d\to\infty} V_\pm(x) = \begin{cases} 0 & \text{for } |x| < 1 \\ \infty & \text{for } |x| > 1 \end{cases}. \tag{6.65}$$

This limit has to be taken with some care. Actually, the Hilbert spaces $\mathcal{H}^\pm$ change in this limit, too: $L^2(\mathbb{R}) \to L^2([-1, 1])$. In addition one has to specify boundary conditions at $x = \pm 1$ in order to have a well-defined problem. The type of boundary conditions which has to be chosen is related to the requirement that the SUSY structure, that is, good or broken SUSY, is conserved in the limit $d \to \infty$. It is obvious that for this one has to impose, for good SUSY, Neumann conditions at $x = \pm 1$ for V_- and Dirichlet condition at $x = \pm 1$ for V_+. Thus parity as well as SUSY remain good symmetries. For broken SUSY (and, hence, broken parity) we have to choose for V_- Dirichlet conditions at $x = -1$ and Neumann conditions at $x = 1$, and vice versa for V_+. See also our discussion of example 3.10 in section 3.6.3. That these boundary conditions are indeed the ones which are simulated by the finite-d system in the limit $d \to \infty$ can also be seen by looking at the numerically calculated energy eigenfunction of the Schrödinger equation [18].

For finite d we have compared the eigenvalues (6.62) and those obtained via the WKB formula (6.13) with the numerical exact eigenvalues of the Schrödinger equation. Note that the potentials (6.60) and (6.61) obey the scaling property

$$x \to \lambda x, \qquad a \to a/\lambda^{d+1}, \qquad E \to E/\lambda^2. \qquad (6.66)$$

For the case of good SUSY we have considered the parameter values $d = 2, 3, 5$. The graphs for the corresponding partner potentials are shown in figure 6.3. Note that V_- is a double-well potential with a barrier at $x = 0$. Hence, the WKB formula cannot be applied for $n = 0$. These wells become smaller and deeper with increasing d and are located near $x = \pm 1$. In figures 6.5–6.7 we show the relative error

$$(E_{\text{exact}} - E_{\text{approx}})/E_{\text{exact}} \qquad (6.67)$$

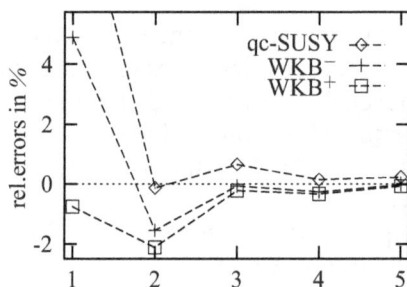

Figure 6.5. Relative errors for the qc-SUSY and the WKB approximations for $d = 2$ and good SUSY.

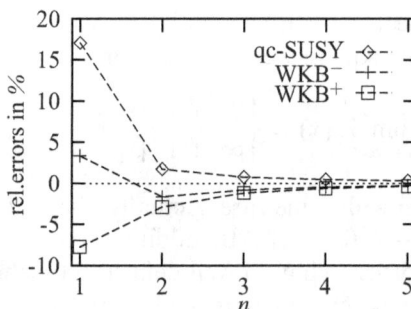

Figure 6.6. The same as figure 6.5 for $d = 3$ and good SUSY.

in percent. Because of the scaling property (6.66), the relative errors are independent of the coupling parameter a. The case of good SUSY with odd d has also been studied by Khare [34]. See, however, the corrections made in [19].

In the case of broken SUSY we have made explicit calculations for the parameter values $d = 2, 3, 4$, and 6. A graph of the corresponding potential V_- is given in figure 6.4. Note that $V_+(x) = V_-(-x)$ in this case. The relative errors are visualised in figures 6.8–6.11. Note that, because of the symmetry between V_+ and V_-, the WKB approximation also yields identical eigenvalues $E_n^- = E_n^+$ for both potentials.

By inspecting the numerical results we see that the qc-SUSY approximation is in general not better than the usual WKB approximation. However, the interesting observation we have made is that the qc-SUSY estimate is always above the WKB estimate:

$$E_{\text{qc–SUSY}} \geqslant E_{\text{WKB}^-} \geqslant E_{\text{WKB}^+}. \tag{6.68}$$

Furthermore, only for the case $d = 2$ and good SUSY did we find one value $E_{\text{qc–SUSY}}$ (for $n = 2$), which is slightly below the exact eigenvalue. Note that the corresponding partner potentials are not differentiable at $x = 0$. See the solid line in figure 6.3. For this example we also find an unusual behaviour of the relative error. That is, the absolute value of this error is not monotonically decreasing with increasing n. It

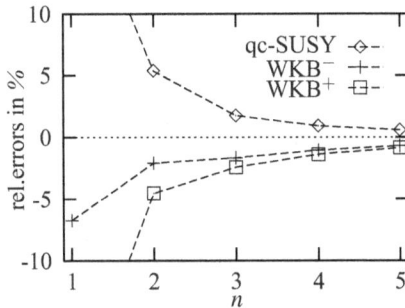

Figure 6.7. The same as figure 6.5 for $d = 5$ and good SUSY.

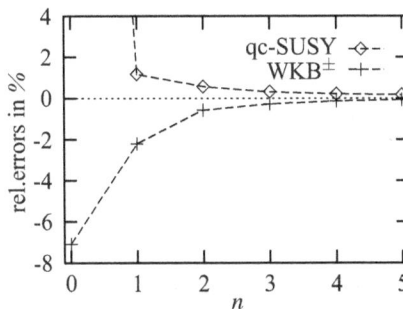

Figure 6.8. Relative errors of the qc-SUSY and the WKB approximation for $d = 2$ and broken SUSY.

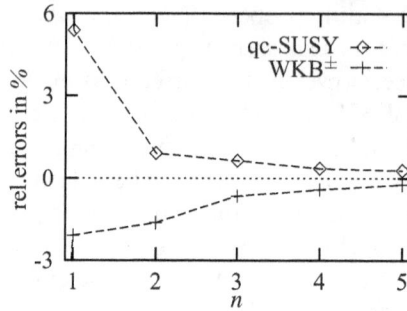

Figure 6.9. The same as figure 6.8 for $d = 3$ and broken SUSY.

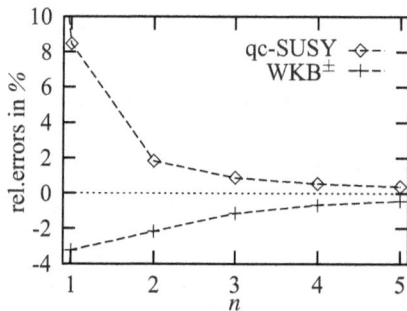

Figure 6.10. The same as figure 6.8 for $d = 4$ and broken SUSY.

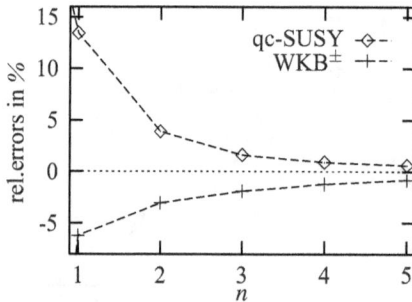

Figure 6.11. The same as figure 6.8 for $d = 6$ and broken SUSY.

rather shows an oscillatory behaviour (see figure 6.4). Excluding such cases we arrive at the more interesting observation

$$E_{\text{qc–SUSY}} \geqslant E_{\text{exact}} \geqslant E_{\text{WKB}^+} \tag{6.69}$$

for all continuously differentiable partner potentials $V_\pm$. In the case of broken SUSY we even found for such potentials that E_{WKB^-} is always below the exact value:

$$E_{\text{qc-SUSY}} \geqslant E_{\text{exact}} \geqslant E_{\text{WKB}^-}. \tag{6.70}$$

We again emphasise that these three relations are not strict inequalities based on some mathematical proof, but rather express the observation made in the numerical investigations.

To support these relations we have also considered other systems, which do not belong to the above class of power SUSY potentials. Here we present results for exponential-type SUSY potentials

$$\Phi(x) = \frac{\hbar}{\sqrt{2m}} \sinh(x) \qquad \text{good SUSY,}$$

$$\Phi(x) = \frac{\hbar}{\sqrt{2m}} \cosh(x) \qquad \text{broken SUSY,} \tag{6.71}$$

$$\Phi(x) = \frac{\hbar}{\sqrt{2m}} \exp\{x^2/2\} \qquad \text{broken SUSY.}$$

The numerical results displayed in figures 6.12–6.14 support the previously made observation. This, however, is no longer the case when considering spherically

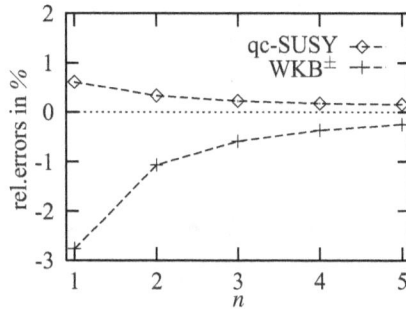

Figure 6.12. Relative errors for the good-SUSY potential $\Phi(x) = \sqrt{\hbar^2/2m} \sinh x$. Here the WKB approximation respects the exact relation $E_n^- = E_{n+1}^+$.

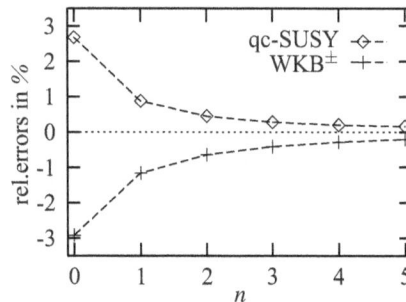

Figure 6.13. Relative errors for the broken-SUSY potential $\Phi(x) = \sqrt{\hbar^2/2m} \cosh x$.

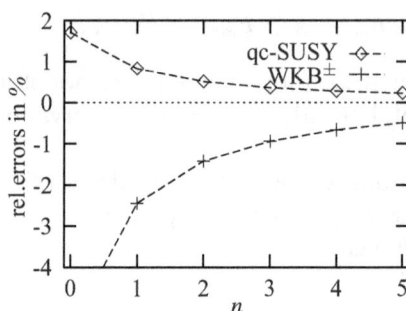

Figure 6.14. Relative errors for the broken-SUSY potential $\Phi(x) = \sqrt{\hbar^2/2m}\,\exp\{x^2/2\}$.

symmetric potentials. In reference [35] two classes of central potentials were considered with a broken SUSY and the relation

$$E_{\text{exact}} \geqslant E_{\text{qc-SUSY}} \geqslant E_{\text{WKB}^+} \geqslant E_{\text{WKB}^-} \qquad (6.72)$$

was observed. In conclusion, we can say that the qc-SUSY approximation is as good as and in particular cases even better than the WKB formula.

Finally, let us note that a similar phenomenon, namely the ordering of energy levels of spherical symmetric potentials, has been observed within a WKB approximation [36] and a rigorous proof based on the algebra of the generalised creation and annihilation operator (3.3) has been given by Grosse *et al* [37–39]. For a derivation of lower and upper bounds to the ground-state energy of a given Hamiltonian based on the factorisation method see [40].

6.5 Problems

Problem 6.1. WKB approximation for the radial harmonic oscillator

For the radial harmonic oscillator the effective potential reads

$$V(r) = \frac{m}{2}\omega^2 r^2 + \frac{\hbar^2\lambda^2}{2mr^2},$$

where $\lambda^2 := \ell(\ell + 1)$ with $\ell \in \mathbb{N}_0$ being the angular momentum quantum number.

(a) Show that the WKB approximation

$$\int_{r_L}^{r_R} dr\sqrt{2m(\tilde{E} - V(r))} = \hbar\pi\left(n + \frac{1}{2}\right), \quad n \in \mathbb{N}_0,$$

with $V(r_L) = \tilde{E} = V(r_R)$ results in energy eigenvalues given by $\tilde{E}_n = \hbar\omega(2n + \lambda + 1)$. *Hint:* Use the integral formula $\int_{x_L}^{x_R} dx\frac{1}{x}\sqrt{(x_R - x)(x - x_L)} = \frac{\pi}{2}(x_R + x_L) - \pi\sqrt{x_R x_L}$.

(b) Show that via the Langer modification $\lambda \to \ell + \frac{1}{2}$ the exact quantum mechanical eigenvalues are obtained. That is $\tilde{E}_n \to E_n = \hbar\omega(2n + \ell + 3/2)$.

Problem 6.2. qc-SUSY approximation for the radial harmonic oscillator

Consider the SUSY potential

$$\Phi(r) = \sqrt{\frac{m}{2}}\,\omega r - \frac{\hbar}{\sqrt{2m}}\frac{\eta}{r}.$$

(a) Show that for $\eta = \ell + 1 > 0$ SUSY is unbroken with $\Delta = +1$, and the SUSY partner potentials read

$$V_{\pm}(r) = \frac{m}{2}\omega^2 r^2 + \frac{\hbar^2(\ell + 1)(\ell + 1 \pm 1)}{2mr^2} - \hbar\omega\left(\ell + 1 \mp \frac{1}{2}\right).$$

(b) Show that for $\eta = -\ell < 0$ SUSY is broken, i.e. $\Delta = 0$, and the corresponding SUSY partner potentials are given by

$$V_{\pm}(r) = \frac{m}{2}\omega^2 r^2 + \frac{\hbar^2\ell(\ell \mp 1)}{2mr^2} + \hbar\omega\left(\ell \pm \frac{1}{2}\right).$$

(c) With the help of the hint in problem 6.1(a), evaluate the qc-SUSY approximation integral

$$\int_{r_L}^{r_R} dr\sqrt{2m(E^{\pm}{-}\Phi^2(r))} = \hbar\pi\left(n + \frac{1}{2} \pm \frac{\Delta}{2}\right), \quad n \in \mathbb{N}_0,$$

and show that this leads for both cases, i.e. broken and unbroken SUSY, to the exact quantum mechanical spectrum which may be put into the form

$$E_n^{\pm} = \hbar\omega[(2n + 2\ell + 1) - \Delta(2\ell \mp 1)].$$

References

[1] Comtet A, Bandrauk A D and Campbell D K 1985 Exactness of semiclassical bound state energies for supersymmetric quantum mechanics *Phys. Lett.* **B 150** 159–62

[2] Eckhardt B 1986 Maslov-WKB theory for supersymmetric Hamiltonians *Phys. Lett.* **B 168** 245–7

[3] Inomata A and Junker G 1993 Quasi-classical approach to path integrals in supersymmetric quantum mechanics *Lectures on Path Integration: Trieste 1991* ed H Cerdeira, S Lundqvist, D Mugnai, A Ranfagni, V Sa-yakanit and L S Schulman (Singapore: World Scientific) pp 460–82

[4] Feynman R P 1948 Space-time approach to non-relativistic quantum mechanics *Rev. Mod. Phys.* **20** 367–87

[5] Feynman R P and Hibbs A R 1965 *Quantum Mechanics and Path Integrals* (New York: McGraw-Hill)

[6] Gutzwiller M C 1967 Phase-integral approximation in momentum space and the bound states of an atom *J. Math. Phys.* **8** 1979–2000

[7] Schulman L S 1981 *Techniques and Applications of Path Integration* (New York: Wiley)

[8] Gutzwiller M C 1991 The semi-classical quantization of chaotic Hamiltonian systems *Chaos and Quantum Physics, Les Houches 1989* ed M-J Giannoni, A Voros and J Zinn-Justin (Amsterdam: North Holland) pp 201–49

[9] van Vleck J H 1928 The correspondence principle in the statistical interpretation of quantum mechanics *Proc. Natl Acad. Sci. USA* **14** 178–88

[10] Pauli W 1951 *Ausgewählte Kapitel aus der Feldquantisierung, Lecture Notes at the ETH Zürich 1950-1951* (Zürich: Haag)

[11] Morette C 1952 On the definition and approximation of Feynman's path integral *Phys. Rev.* **81** 848–52

[12] Felsager B 1987 *Geometry, Particles and Fields* 4th edn (Odense: University Press)

[13] Wenzel G 1926 Eine Verallgemeinerung der Quantenbedingungen für die Zwecke der Wellenmechanik *Z. Phys.* **38** 518–29

[14] Brillouin L 1926 La méchanique ondulatoire de Schrödinger; une méthode générale de résolution par approximations successives *C. R. Acad. Sci.* **183** 24–6

[15] Kramers M A 1926 Wellenmechanik und halbzahlige Quantisierung *Z. Phys.* **39** 828–40

[16] Dunham J L 1931 The Wentzel–Brillouin–Kramers method of solving the wave equation *Phys. Rev.* **41** 713–20

[17] Inomata A and Junker G 1993 Quasi-classical approach in supersymmetric quantum mechanics *Proc. International Symposium on Advanced Topics of Quantum Physics* ed J Q Liang, M L Wang, S N Qiao and D C Su (Beijing: Science Press) pp 61–74

[18] Inomata A and Junker G 1994 Quasiclassical path-integral approach to supersymmetric quantum mechanics *Phys. Rev.* A **50** 3638–49

[19] Dutt R, Khare A and Sukhatme U P 1986 Exactness of supersymmetric WKB spectra for shape-invariant potentials *Phys. Lett.* B **181** 295–8

[20] Rosenzweig C and Krieger J B 1968 Exact quantization conditions *J. Math. Phys.* **9** 849–60

[21] Raghunathan K, Seetharaman M and Vasan S S 1987 On the exactness of the SUSY semiclassical quantization rule *Phys. Lett.* B **188** 351–2

[22] Barclay D T and Maxwell C J 1991 Shape invariance and the SWKB series *Phys. Lett.* A **157** 357–60

[23] Inomata A, Junker G and Suparmi A 1993 Remarks on semiclassical quantization rule for broken SUSY *J. Phys. A: Math. Gen.* **26** 2261–64

[24] Koleci C K, Suparmi A and Inomata A 1994 Semiclassical quantization of the Gendenshtein systems *Path Integrals in Physics* ed V Sa-yakanit, J-O Berananda and W Sritrakool (Singapore: World Scientific) pp 127–36

[25] Murali N R, Govindarajan T R and Khare A 1995 Exactness of the broken supersymmetric, semiclassical quantization rule *Phys. Rev.* A **52** 4259–61

[26] Junker G, Matthiesen S and Inomata A 1996 Classical and quasi-classical aspects of supersymmetric quantum mechanics *Symmetry Methods in Physics* vol 1 ed A N Sissakian and G S Pogosyan (Dubna: Joint Institute for Nuclear Research) pp 290–7

[27] Fricke S H, Balantekin A B, Hatchell P J and Uzer T 1988 Uniform semiclassical approximation to supersymmetric quantum mechanics *Phys. Rev.* A **37** 2797–804

[28] Murayama Y 1989 Supersymmetric Wentzel–Kramers–Brillouin (SWKB) method *Phys. Lett.* A **136** 455–7

[29] Pagnamenta A and Sukhatme U 1990 Non-divergent semiclassical wave functions in supersymmetric quantum mechanics *Phys. Lett.* A **151** 7–11

[30] Suparmi A 1992 Semi-classical quantization rules in supersymmetric quantum mechanics *PhD Thesis* State University of New York, Albany

[31] Barclay D T, Khare A and Sukhatme U 1993 Is the lowest order supersymmetric WKB approximation exact for all shape invariant potentials? *Phys. Lett.* A **183** 263-6

[32] Khare A and Sukhatme U P 1993 New shape-invariant potentials in supersymmetric quantum mechanics *J. Phys. A: Math. Gen.* **26** L901-4

[33] Ezawa H and Klauder J R 1985 Fermions without fermions: the Nicolai map revisited *Prog. Theor. Phys.* **74** 904-15

[34] Khare A 1985 How good is the supersymmetry-inspired WKB quantization condition? *Phys. Lett.* B **161** 131-5

[35] Junker G, Roy P and Varshni Y P 1997 Quasi-classical investigation of nonpolynomial central potentials with broken supersymmetry *Can. J. Phys.* **75** 695-703

[36] Feldman G, Fulton T and Devoto A 1979 Energy levels and level ordering in the WKB approximation *Nucl. Phys.* B **154** 441-62

[37] Grosse H and Martin A 1984 Two theorems on the level order in potential models *Phys. Lett.* B **134** 368-72

[38] Baumgartner B, Grosse H and Martin A 1984 The Laplacian of the potential and the order of energy levels *Phys. Lett.* B **146** 363-6

[39] Grosse H 1991 Supersymmetric quantum mechanics *Recent Developments in Quantum Mechanics, Mathematical Physics Studies Nr. 12* ed A Boutet de Monvel, P Dita, G Nenciu and R Purice (Dordrecht: Kluwer) pp 299-327

[40] Schmutz M 1985 The factorization method and ground state bounds *Phys. Lett.* A **108** 195-6

IOP Publishing

Supersymmetric Methods in Quantum, Statistical and Solid State Physics

Enlarged and revised edition
Georg Junker

Chapter 7

Supersymmetry in classical stochastic dynamics

In 1979 Parisi and Sourlas [1] pointed out that there is a hidden supersymmetry in systems characterised via classical stochastic differential equations. In fact, it is possible to reformulate some supersymmetric models of field theory in terms of classical stochastic equations [2]. The existence or non-existence of a stationary solution of a classical stochastic system is related to good and broken SUSY in the corresponding field theory [3], respectively. In the case of one Cartesian degree of freedom, whose averaged dynamics is characterised by a one-dimensional Fokker–Planck equation, it can be shown [2, 4–6] that this equation may be put into the form of a supersymmetric Schrödinger equation in imaginary time. The corresponding Hamiltonian can be identified with that of Witten's model. This analogy explains the fact that the Fokker–Planck operator for a given drift potential and diffusion constant is essential iso-spectral to the Fokker–Planck operator with the inverted drift potential and the same diffusion constant [6, 7]. The breaking of SUSY in the Langevin dynamics of non-potential systems is also related to the occurrence of corrections to the linear fluctuation–dissipation theorem [8].

In this chapter, we will review the relations between the classical stochastic dynamics of one Cartesian degree of freedom and supersymmetric quantum mechanics. We will show that SUSY may be utilised to derive decay rates for bistable and metastable Fokker–Planck equations. The methods developed in chapter 4 on constructing models with exact solutions in quantum mechanics can similarly be used for the diffusion problem. In particular, we discuss the construction of conditionally exactly solvable drift potentials. We will also consider stationary stochastic processes. For applications to field theories of statistical and condensed matter physics see, for example, [9, 10].

7.1 Langevin and Fokker–Planck equation

The dynamics of many complex systems in physics, chemistry, and biology can be described phenomenologically by the so-called *Langevin equation* [4, 11–13]:

$$\dot{\eta} = F(\eta) + \xi(t). \tag{7.1}$$

Here $\eta \in \mathbb{R}$ denotes a macroscopic degree of freedom, the time-evolution of which we are interested in. The quantity $\xi \in \mathbb{R}$ is assumed to be a random function of time. The stochastic differential equation (7.1) first appeared in the work of Langevin [11] who studied a simplified approach of Einstein's [14, 15] and Smoluchowski's [16] description of Brownian motion. In this model η denotes the momentum (or for highly overdamped motion the position) of the Brownian particle, F stands for an external deterministic force, and ξ is a random function characterising the fluctuations of the medium in which the Brownian particle is immersed. Properties of this random function, which is usually called *noise*, are characterised by expectation values. Without loss of generality, one may assume

$$\langle \xi(t) \rangle = 0, \qquad t \geqslant 0, \tag{7.2}$$

because a non-vanishing expectation value can be absorbed in the deterministic force F. The Langevin equation (7.1), as it stands, may characterise physical systems with noise ξ correlated on a given timescale $\tau_c > 0$ like

$$\langle \xi(t)\xi(t') \rangle = \frac{D}{2\tau_c} \exp\{-|t - t'|/\tau_c\}, \qquad D > 0. \tag{7.3}$$

See for example [13]. Such a noise is called *coloured noise* and is a realistic description of various physical systems. For many systems, however, it is justified to consider the idealisation of so-called δ-correlated or *white noise* by taking the limit $\tau_c \searrow 0$:

$$\langle \xi(t)\xi(t') \rangle = D\delta(t - t'). \tag{7.4}$$

To make the problem definite, in addition to equation (7.4), we will assume that this white noise has a Gaussian distribution, which is uniquely characterised by equations (7.2) and (7.4):

$$\langle \cdot \rangle := \int \mathcal{D}\xi \exp\left\{ -\frac{1}{2D} \int_0^\infty \mathrm{d}\tau \, \xi^2(\tau) \right\}(\cdot). \tag{7.5}$$

Usually, one is not interested in a particular solution of equation (7.1) for a given realisation ξ. Of interest, however, are properties of typical realisations. These properties may be deduced from the probability density for arriving at $x \in \mathbb{R}$ in a given time t if the particle initially started in $x_0 \in \mathbb{R}$:

$$m_t(x, x_0) := \langle \delta(\eta(t) - x) \rangle, \qquad x_0 := \eta(0). \tag{7.6}$$

This transition-probability density can be shown [4, 17] to be determined by the so-called *Fokker–Planck equation* [18, 19]:

$$\frac{\partial}{\partial t}\, m_t(x,\, x_0) = \frac{D}{2}\, \frac{\partial^2}{\partial x^2} m_t(x,\, x_0) - \frac{\partial}{\partial x}\, F(x) m_t(x,\, x_0),$$

$$m_0(x,\, x_0) = \delta(x - x_0).$$

(7.7)

This is a diffusion equation with diffusion constant $D/2$ and an additional drift coefficient F. For the case of a test particle moving in an external potential U being coupled to a heat bath of temperature T these are given by $D = 2k_B T$, with k_B denoting Boltzmann's constant, and $F(x) = -U'(x)$ being the force acting on it.

Let us assume that this system has a non-trivial stationary distribution

$$P(x) := \lim_{t \to \infty} m_t(x,\, x_0) \quad \text{with} \quad \int_{-\infty}^{+\infty} \mathrm{d}x\, P(x) = 1.$$

(7.8)

It is straightforward to show that this is then explicitly given by

$$P(x) = N \exp\{-2U(x)/D\} = N \mathrm{e}^{-U(x)/k_B T},$$

(7.9)

with N being a proper constant of normalisation. Such systems are called stable and it is obvious that normalisation implies $U(x) \to \infty$ for $x \to \pm\infty$.

In the following we will show that the Fokker–Planck equation can be put into the form of an imaginary-time Schrödinger equation with supersymmetry of the type of Witten's model. The drift coefficient F will turn out to play the role of the SUSY potential Φ, whereas the drift potential U in essence is being identified with the superpotential (3.12). As a consequence the diffusion problem with drift $F = \Phi$ and that with inverted drift $F = -\Phi$ will turn out to be SUSY partners and unbroken SUSY guarantees the existence of a stationary distribution (7.9).

7.2 Supersymmetry of the Fokker–Planck equation

In this section we will study the Fokker–Planck equation for two drift coefficients which differ by an overall sign, $F_\pm := \pm\Phi$, being associated with a pair of drift potentials $U_\pm' = -F_\pm$ differing only by an overall sign

$$U_\pm(x) := \mp \int^x \mathrm{d}z\, \Phi(z) = -U_\mp(x).$$

(7.10)

Typical drift potentials $U_\pm$ and their related drift coefficients are shown in figure 7.1. Here we adopt the convention that in the case a stable drift potential exists it is given by U_-. This convention will turn out to be equivalent to our convention (3.20) made for the Witten model. The corresponding Fokker–Planck equation reads

$$\frac{\partial}{\partial t}\, m_t^\pm(x,\, x_0) = \frac{D}{2}\, \frac{\partial^2}{\partial x^2} m_t^\pm(x,\, x_0) \mp \frac{\partial}{\partial x}\, \Phi(x) m_t^\pm(x,\, x_0),$$

(7.11)

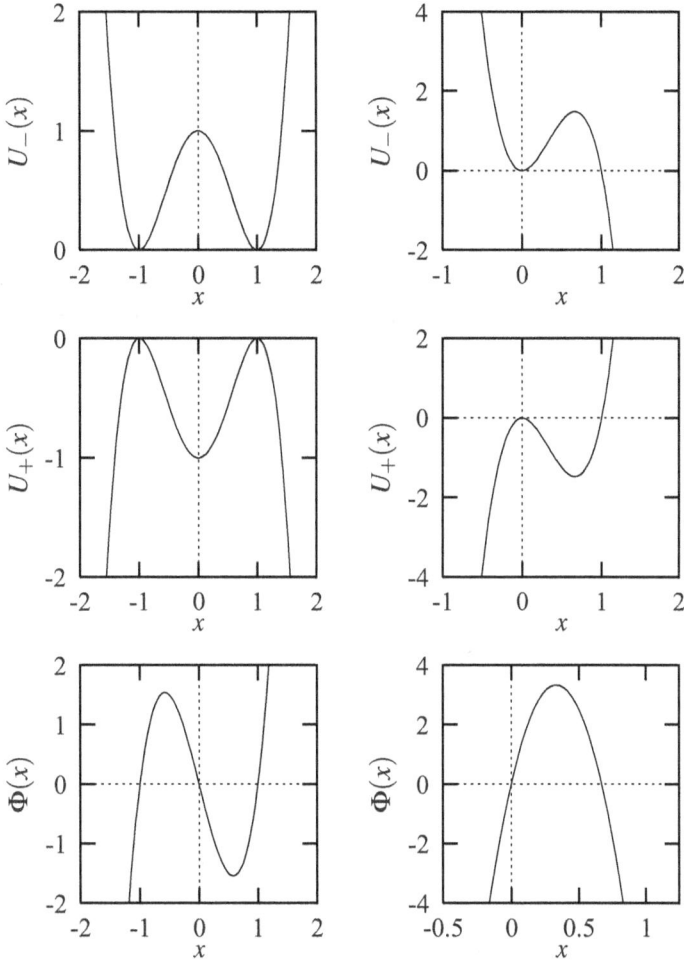

Figure 7.1. Typical drift potentials U_-, inverted drift potentials $U_+ = -U_-$ and drift coefficients $\Phi = \mp U'_\pm$ for good SUSY (left column) and broken SUSY (right column).

where we have introduced an additional superscript in the transition-probability density in order to discriminate the solution for the drift potential U_- from the one for drift potential U_+. Making the ansatz [4]

$$m_t^\pm(x, x_0) =: \exp\left\{-\frac{1}{D}[U_\pm(x) - U_\pm(x_0)]\right\} K_t^\pm(x, x_0),$$

$$K_0^\pm(x, x_0) = \delta(x - x_0),$$

$$\tag{7.12}$$

one arrives at the imaginary-time Schrödinger equation

$$-D\frac{\partial}{\partial t}K_t^\pm(x, x_0) = H_\pm K_t^\pm(x, x_0), \tag{7.13}$$

where

$$H_\pm := -\frac{D^2}{2}\frac{\partial^2}{\partial x^2} + \frac{1}{2}\Phi^2(x) \pm \frac{D}{2}\Phi'(x) \tag{7.14}$$

appear to be the partner Hamiltonians of Witten's model upon the substitution

$$\Phi \to \sqrt{2}\,\Phi, \qquad D \to \hbar/\sqrt{m}. \tag{7.15}$$

Clearly, the drift coefficient Φ plays the role of the SUSY potential and, hence, the drift potential U_- may be identified with the superpotential (3.12). With the initial condition (7.12) the solution of equation (7.13) is given by the heat kernel

$$K_t^\pm(x, x_0) = \langle x|\exp\{-tH_\pm/D\}|x_0\rangle. \tag{7.16}$$

Denoting the eigenvalues of $H_\pm$ by $\lambda_n^\pm$ (we will assume a purely discrete spectrum for simplicity) and the associated eigenfunctions by $\phi_n^\pm$, that is,

$$H_\pm\phi_n^\pm = \lambda_n^\pm\phi_n^\pm, \qquad n \in \mathbb{N}_0, \tag{7.17}$$

we arrive at the spectral representation [4]

$$\begin{aligned}
m_t^\pm(x, x_0) &= \exp\left\{-\frac{1}{D}[U_\pm(x) - U_\pm(x_0)]\right\} \\
&\times \sum_{n=0}^{\infty} \exp\left\{-\frac{1}{D}\,t\lambda_n^\pm\right\} \phi_n^\pm(x)\phi_n^{\pm*}(x_0).
\end{aligned} \tag{7.18}$$

Because of SUSY we know that H_- and H_+ are essential iso-spectral. That is, the strictly positive decay rates for the drift potential U_- and for the inverted one $U_+ = -U_-$ are identical [7]. This fact is even true for coloured noise, where it is also related in some sense to an underlying SUSY [20].

In addition, in the case of a good SUSY, we have a vanishing eigenvalue $\lambda_0^- = 0$ which, due to our convention, belongs to U_-. The corresponding ground-state wave function is given by, cf. equation (3.11),

$$\phi_0^-(x) = C\exp\left\{-\frac{1}{D}\int^x dz\,\Phi(z)\right\} = C\exp\left\{-\frac{1}{D}U_-(x)\right\}, \tag{7.19}$$

As a consequence, good SUSY implies the existence of a stationary non-trivial distribution:

$$P(x) = \lim_{t\to\infty} m_t^-(x, x_0) = \left|\phi_0^-(x)\right|^2. \tag{7.20}$$

Obviously the normalisation constant N of the stationary distribution (7.9) and the above normalisation for the SUSY ground state are related by $N = |C|^2$.

7.3 Supersymmetry of the Langevin equation

The Langevin equation associated with the Fokker–Planck equation (7.11) of the previous section reads

$$\dot{\eta} = \pm\Phi(\eta) + \xi^{\pm}(t), \tag{7.21}$$

where, as in section 7.1, ξ^+ and ξ^- denote independent Gaussian white-noise variables with the same diffusion constant, which for simplicity we set to unity in the discussion below, i.e. $D = 1$,

$$\langle\xi^+(t)\xi^+(t')\rangle_+ = \langle\xi^-(t)\xi^-(t')\rangle_- = \delta(t - t'),$$

$$\langle\cdot\rangle_{\pm} := \int \mathcal{D}\xi^{\pm}\exp\left\{-\frac{1}{2}\int_0^{\infty} d\tau[\xi^{\pm}(\tau)]^2\right\}(\cdot). \tag{7.22}$$

It is interesting to note that $\xi^{\pm} = \dot{\eta} \mp \Phi(\eta)$ are in essence the Euclidean versions of the classical canonical momenta introduced in equation (5.26). In other words, they may be considered as the canonical momenta of the Euclidean version of the quasi-classical Lagrangian (5.25):

$$\tilde{L}_{qc,eucl.}^{\pm}(\dot{\eta}, \eta) := \frac{1}{2}(\dot{\eta} \mp \Phi(\eta))^2. \tag{7.23}$$

Let us now consider the transition-probability density associated with the Langevin equation (7.21):

$$m_t^{\pm}(x, x_0) = \langle\delta(\eta(t) - x)\rangle_{\pm}, \qquad x_0 = \eta(0). \tag{7.24}$$

Using the definition (7.22) we find the following path-integral representation:

$$m_t^{\pm}(x, x_0) = \int_{\eta(0)=x_0} \mathcal{D}\xi^{\pm}\exp\left\{-\frac{1}{2}\int_0^{\infty} d\tau[\xi^{\pm}(\tau)]^2\right\}\delta(\eta(t) - x). \tag{7.25}$$

This Gaussian path integral may be transformed into a Wiener-type path integral upon the transformation $\xi^{\pm} \to \eta$, which is called Nicolai map [21–25]:

$$\xi^{\pm}(t) = \dot{\eta}(t) \mp \Phi(\eta(t)). \tag{7.26}$$

Noting that

$$\frac{\delta\xi^{\pm}(t)}{\delta\eta(t')} = \left(\frac{\partial}{\partial t} \mp \Phi'(\eta(t))\right)\delta(t - t') \tag{7.27}$$

we find

$$m_t^{\pm}(x, x_0) = \int_{\eta(0)=x_0} \mathcal{D}\eta \exp\left\{-\frac{1}{2}\int_0^{\infty} d\tau[\dot{\eta}(\tau) \mp \Phi(\eta(\tau))]^2\right\}$$

$$\times \delta(\eta(t) - x)\det\left[\left(\frac{\partial}{\partial t} \mp \Phi'(\eta(t))\right)\delta(t - t')\right]$$

$$= \int_{\eta(0)=x_0}^{\eta(t)=x} \mathcal{D}\eta \exp\left\{-\int_0^t d\tau\tilde{L}_{qc,eucl.}^{\pm}(\dot{\eta}(\tau), \eta(\tau))\right\}$$

$$\times \det\left[\left(\frac{\partial}{\partial t} \mp \Phi'(\eta(t))\right)\delta(t - t')\right]. \tag{7.28}$$

Here several remarks are in order. The functional determinant appearing in equation (7.28) can be given a definite meaning if the associated Nicolai map (7.26) is properly interpreted as a change from the stochastic Gaussian process in $\xi^{\pm}$ to a Wiener process in η. Ezawa and Klauder [23] considered a Stratonovich-(Str) and an Itô-related interpretation of equation (7.26) leading to

$$
\det\left[\frac{\delta\xi^{\pm}(t)}{\delta\eta(t')}\right]_{\text{Str}} = \text{const.} \times \exp\left\{\mp\frac{1}{2}\int_0^t d\tau\,\Phi'(\eta(\tau))\right\},
$$

$$
\det\left[\frac{\delta\xi^{\pm}(t)}{\delta\eta(t')}\right]_{\text{Itô}} = \text{const.} \times 1,
$$

(7.29)

where the (infinite) constant may be absorbed by a normalisation factor. Similar, the Euclidean action appearing in the exponent of equation (7.28) has to be interpreted in a consistent way:

$$
\int_0^t d\tau\,\tilde{L}^{\pm}_{\text{qc,eucl.}}(\dot{\eta}(\tau),\,\eta(\tau)) = \frac{1}{2}\int_0^t d\tau[\dot{\eta}^2(\tau) + \Phi^2(\eta(\tau))]
$$

$$
\mp \int_0^t d\tau\,\dot{\eta}(\tau)\Phi(\eta(\tau)).
$$

(7.30)

It is the last integral of this relation which requires a proper interpretation, because of the stochastic nature of $\eta(\tau)$. For the Stratonovich and the Itô interpretations we have [23], respectively,

$$
\pm\int_0^t d\tau\,\dot{\eta}(\tau)\,\Phi(\eta(\tau))\Big|_{\text{Str}} = \pm\int_{x_0}^x dz\,\Phi(z) = -[U_{\pm}(x) - U_{\pm}(x_0)],
$$

$$
\pm\int_0^t d\tau\,\dot{\eta}(\tau)\,\Phi(\eta(\tau))\Big|_{\text{Itô}} = -[U_{\pm}(x) - U_{\pm}(x_0)] \mp \frac{1}{2}\int_0^t d\tau\,\Phi'(\eta(\tau)),
$$

(7.31)

where $U_{\pm}$ is defined in equation (7.10). Both interpretations lead, of course, to the same Wiener-type path-integral expression

$$
m_t^{\pm}(x, x_0) = \exp\{-[U_{\pm}(x) - U_{\pm}(x_0)]\}
$$

$$
\times \int_{\eta(0)=x_0}^{\eta(\tau)=x} \mathcal{D}\eta\,\exp\left\{-\frac{1}{2}\int_0^t d\tau[\dot{\eta}^2(\tau) + \Phi^2(\eta(\tau)) \pm \Phi'(\eta(\tau))]\right\} \quad (7.32)
$$

$$
= \exp\{-[U_{\pm}(x) - U_{\pm}(x_0)]\}\langle x|e^{-tH_{\pm}}|x_0\rangle,
$$

where $H_{\pm}$ is given in equation (7.14). This result coincides, as expected, with equations (7.12)–(7.16).

Let us note, that the functional determinant in equation (7.28) may also be expressed in terms of a fermionic path integral

$$
\det\left[\left(\frac{\partial}{\partial t} \mp \Phi'(\eta(t))\right)\delta(t - t')\right]_{\text{Str}}
$$

$$
= \int \mathcal{D}\psi \int \mathcal{D}\bar{\psi}\,\exp\left\{\int_0^t d\tau\,\bar{\psi}(\tau)\left(\frac{\partial}{\partial\tau} \mp \Phi'(\eta(\tau))\right)\psi(\tau)\right\}.
$$

(7.33)

Here we have taken the Stratonovich-related interpretation for the determinant. Because of equation (7.29) the Itô-related interpretation will only lead to a trivial fermion path integral. There would be no coupling between the bosonic variable η and the fermionic variables ψ, $\bar{\psi}$. The fermionic path integral, as it stands, is not well defined. To give it a meaning Ezawa and Klauder [23] used a time-lattice definition $\epsilon := t/N$, $\psi_j := \psi(j\epsilon)$, and showed that the right-hand side of equation (7.33) may be given by

$$\lim_{N\to\infty} \int \prod_{j=0}^{N} (\mathrm{d}\psi_j \mathrm{d}\bar{\psi}_j) \exp\left\{ \sum_{j=0}^{N}\left[\bar{\psi}_j(\psi_j - \psi_{j-1}) \mp \frac{\epsilon}{2}\Phi'(\eta(\epsilon j))\bar{\psi}_j(\psi_j + \psi_{j-1}) \right] \right\}, \quad (7.34)$$

where $\psi_{-1} := 0$, which they called a half-Dirichlet boundary condition. See equation (3.19) in [23]. Note, however, that our convention (5.63) for the Grassmann integration rule differs from that of Ezawa and Klauder. Having this in mind we may represent the transition-probability density as follows:

$$m_t^{\pm}(x, x_0) = \int_{\eta(0)=x_0}^{\eta(t)=x} \mathcal{D}\eta \int \mathcal{D}\psi \int \mathcal{D}\bar{\psi}\, \exp\left\{ -\int_0^t \mathrm{d}\tau\, L_0^{\pm} \right\}, \quad (7.35)$$

where

$$L_0^{\pm} := \frac{1}{2}\dot{\eta}^2 + \frac{1}{2}\Phi^2(\eta) - \bar{\psi}\dot{\psi} \pm \Phi'(\eta)\bar{\psi}\psi. \quad (7.36)$$

The Lagrangian $L_0^{\pm}$ is, up to a total time derivative, invariant under the following SUSY transformations

$$\delta\eta = \bar{\epsilon}\psi - \bar{\psi}\epsilon, \qquad \delta\psi = \epsilon(\dot{\eta} \pm \Phi(\eta)), \qquad \delta\bar{\psi} = (\dot{\eta} \mp \Phi(\eta))\bar{\epsilon}. \quad (7.37)$$

Compare this also with the SUSY transformations (5.11) discussed in section 5.2.

There exists another time-lattice definition for the right-hand side of equation (7.33) which reads (cf. equation (A.7) of [23])

$$\lim_{N\to\infty} \int \prod_{j=0}^{N} (\mathrm{d}\psi_j \mathrm{d}\bar{\psi}_j)\psi_N \exp\left\{ \sum_{j=0}^{N}\left[\bar{\psi}_j(\psi_j - \psi_{j-1}) \pm \frac{\epsilon}{2}\Phi'(\eta(\epsilon j))\bar{\psi}_j(\psi_j + \psi_{j-1}) \right] \right\}\bar{\psi}_0, \quad (7.38)$$

where again the half-Dirichlet boundary condition $\psi_{-1} := 0$ is used. Here, in essence, the overall sign of the SUSY potential Φ has been changed and the exponential is now sandwiched between $\psi_N = \psi(t)$ and $\bar{\psi}_0 = \bar{\psi}(0)$. With this definition we can put the transition-probability density into the form

$$m_t^{\pm}(x, x_0) = \int_{\eta(0)=x_0}^{\eta(t)=x} \mathcal{D}\eta \int \mathcal{D}\psi \int \mathcal{D}\bar{\psi}\, \psi(t)\exp\left\{ -\int_0^t \mathrm{d}\tau\, L_1^{\pm} \right\}\bar{\psi}(0), \quad (7.39)$$

where

$$L_1^{\pm} := \frac{1}{2}\dot{\eta}^2 + \frac{1}{2}\Phi^2(\eta) - \bar{\psi}\dot{\psi} \mp \Phi'(\eta)\bar{\psi}\psi. \quad (7.40)$$

Again, because of $L_1^{\pm} = L_0^{\mp}$, this Lagrangian is invariant under the SUSY transformation (7.37).

It should be noted that these two representations of the formal fermionic path integral (7.33) are related to the invariance of Witten's model under a change of an overall sign of the SUSY potential. See our discussion in section 3.3.2. This two-fold SUSY has been interpreted by Gozzi [26] as a macroscopic manifestation of Onsagers principle of microscopic reversibility. Indeed, the backward Fokker–Planck equation can be obtained from equation (7.11) by taking its adjoint on the right-hand side. This, however, is equivalent to a change of the overall sign of the SUSY potential. In the following we will study some implications of this symmetry between the forward and the backward process.

7.4 Implications of supersymmetry

7.4.1 Unbroken SUSY

First we will consider the case of unbroken SUSY. Typical drift and SUSY potentials have already been shown in the left column of figure 7.1. From our discussion of the Witten model we know that the decay rates $\lambda_n^{\pm}$ in the potentials $U_{\pm}$ are related by

$$\lambda_0 := \lambda_0^- = 0, \qquad \lambda_n := \lambda_n^- = \lambda_{n-1}^+ > 0, \qquad n = 1, 2, 3, \dots, \qquad (7.41)$$

and the ground-state wave function of H_- is given by

$$\phi_0^-(x) = C \exp\left\{-\frac{1}{D}\int^x \mathrm{d}z\,\Phi(z)\right\} = C \exp\left\{-\frac{1}{D}U_-(x)\right\}. \qquad (7.42)$$

Assuming, without loss of generality $C \in \mathbb{R}$, the transition-probability densities read

$$
\begin{aligned}
m_t^-(x, x_0) &= |\phi_0^-(x)|^2 + \frac{\phi_0^-(x)}{\phi_0^-(x_0)}\sum_{n=1}^{\infty}\mathrm{e}^{-\lambda_n t/D}\phi_n^-(x)\phi_n^{-*}(x_0), \\
m_t^+(x, x_0) &= \frac{\phi_0^-(x_0)}{\phi_0^-(x)}\sum_{n=1}^{\infty}\mathrm{e}^{-\lambda_n t/D}\phi_{n-1}^+(x)\phi_{n-1}^{+*}(x_0).
\end{aligned}
\qquad (7.43)
$$

As already mentioned, because of good SUSY, there exists a stationary distribution for U_-, which is given by the SUSY ground state

$$P(x) = |\phi_0^-(x)|^2 = |C|^2 \exp\left\{-\frac{2}{D}U_-(x)\right\}. \qquad (7.44)$$

Due to the SUSY transformation (3.9) we can also relate the normalised decay modes $\phi_n^{\pm}$ of U_- with that of U_+ and vice versa:

$$
\begin{aligned}
\phi_{n-1}^+(x) &= \frac{1}{\sqrt{2\lambda_n}}\left(D\frac{\partial}{\partial x} + \Phi(x)\right)\phi_n^-(x), \\
\phi_n^-(x) &= \frac{1}{\sqrt{2\lambda_n}}\left(D\frac{\partial}{\partial x} - \Phi(x)\right)\phi_{n-1}^+(x),
\end{aligned}
\qquad n = 1, 2, 3, \dots. \qquad (7.45)
$$

SUSY is not only useful for obtaining the above connections between the diffusion in a given drift potential and the diffusion in the inverted potential. It also proves to be useful for explicit calculations. For example, for all shape-invariant SUSY potentials listed in table 4.1, one can easily find exact solutions of the associated Fokker–Planck equation [27–29]. Approximate results can also be obtained, for example, via a variational approach. Here, in essence, one obtains from a variational ansatz the ground-state properties of H_+, that is, λ_1 and ϕ_0^+. Then via equations (7.41) and (7.45) one finds the timescale $1/\lambda_1$ and mode ϕ_1^- with which the system U_- relaxes into the stationary distribution (7.44). This process can in principle be continued to obtain a hierarchy of diffusion potentials in analogy to our discussion at the end of section 4.1.

Here we limit ourselves to an approximate derivation of the smallest non-vanishing eigenvalue λ_1 based on a SUSY method originally devised for tunnelling problems [30, 31]. We want to find the ground-state properties of the Hamiltonian H_+ of the pair of partner Hamiltonians

$$H_\pm = -\frac{D^2}{2} \frac{\partial^2}{\partial x^2} + V_\pm(x) \tag{7.46}$$

with partner potentials

$$V_\pm(x) := \frac{1}{2} \Phi^2(x) \pm \frac{D}{2} \Phi'(x). \tag{7.47}$$

See the left column of figure 7.2 for a plot of V_-, V_+, and Φ^2, which are associated with the drift potential given by the left column of figure 7.1. For simplicity, we will assume that Φ is an odd function, which implies that $U_\pm$ and $V_\pm$ are even functions. We further assume that U_- is a bistable potential as shown in the left column of figure 7.1 (for the corresponding partner potentials see the left column of figure 7.2) with a high barrier between the two minima. The last assumption implies that the eigenvalue $\lambda_1 > 0$ will be very close to zero. Here we note that a non-normalisable eigenfunction of H_+ with zero eigenvalue is given by (cf. equation (3.14))

$$\frac{1}{\phi_0^-(x)} \propto \exp\{U_-(x)/D\} = \exp\{-U_+(x)/D\}. \tag{7.48}$$

Obviously, for $x \to \pm\infty$ this function diverges. However, for small values of x it may be a good approximation to the exact ground state ϕ_0^+ of H_+. In fact, regularising equation (7.48) for large x values one can use it as a trial function for a variational approach [6, 32].

Here, however, we choose a perturbative approach. Let us consider the following nodeless and normalisable wave function [30, 31]:

$$\phi(x) := \frac{C}{\phi_0^-(|x|)} \int_{|x|}^{+\infty} dz \left(\phi_0^-(z)\right)^2. \tag{7.49}$$

For small x values this wave function is similar to equation (7.48) and therefore is a good approximation to ϕ_0^+. It is, however, not differentiable at $x = 0$. In fact, it is

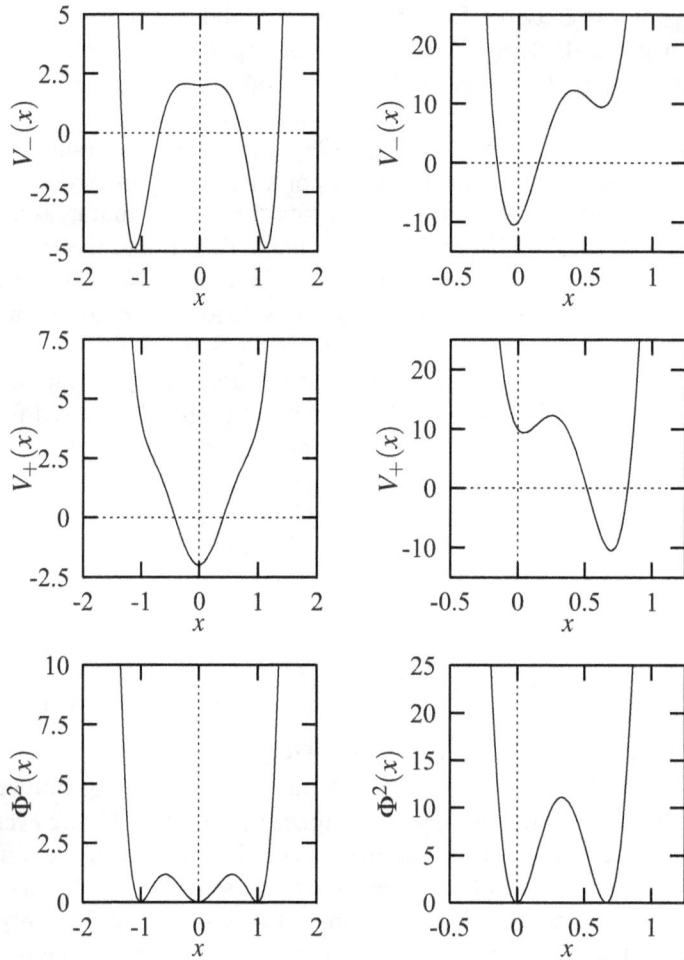

Figure 7.2. The partner potentials $V_{\pm}$ and Φ^2 for the drift potentials shown in figure 7.1. Again the left column corresponds to the good-SUSY and the right to the broken-SUSY case. The diffusion constant D has been set to unity.

easily verified that ϕ is the exact zero-energy ground-state wave function for the Hamiltonian

$$H := H_+ - 2D^2\big(\phi_0^-(0)\big)^2\delta(x). \qquad (7.50)$$

The Hamiltonian H_+ may be viewed as Hamiltonian H with additional δ-like perturbation. Knowing the ground-state properties of H one may obtain those of H_+ via a perturbation expansion. In a first-order perturbation expansion the ground-state eigenvalue of H_+ is given as

$$\lambda_1 \simeq \int_{\mathbb{R}} dx \, (H_+ - H)\phi^2(x) = 2D^2\big(\phi_0^-(0)\big)^2\phi^2(0) = D^2C^2. \qquad (7.51)$$

Hence, the eigenvalue λ_1 is essentially given by the normalisation constant C introduced in equation (7.42). Using the Laplace method for small D in the calculation of C one arrives [31] at the result of Kramers [33]

$$\lambda_1 \simeq \frac{D}{\pi}\sqrt{U_-''(x_{\min})|U_-''(x_{\max})|}\exp\left\{-\frac{2}{D}[U_-(x_{\max}) - U_-(x_{\min})]\right\}. \quad (7.52)$$

Here $x_{\min}$ is the position of the (right) minimum of U_- ($x_{\min} = 1$ in figure 7.1, left column) and $x_{\max}$ is the position of the barrier of U_- ($x_{\max} = 0$ in figure 7.1 left column). Despite the fact that this derivation has been given for a symmetric drift potential it can be used for asymmetric potentials, too [7].

Higher eigenvalues $\lambda_n^\pm$ may be estimated via the WKB approximation (6.13), which reads in the current context

$$\int_{q_L}^{q_R} dx\sqrt{2(\lambda^\pm - V_\pm(x))} = D\pi\left(n + \frac{1}{2}\right), \qquad V_\pm(q_L) = \lambda^\pm = V_\pm(q_R), \quad (7.53)$$

or better via the CBC formula (6.30)

$$\int_{x_L}^{x_R} dx\sqrt{2\lambda - \Phi^2(x)} = D\pi n, \qquad \Phi^2(x_L) = 2\lambda = \Phi^2(x_R), \quad (7.54)$$

which respects the spectral symmetry (7.41). Remember that in equation (6.30) we need to replace $\hbar/\sqrt{m} \to D$, $\Phi \to \Phi/\sqrt{2}$, and $E \to \lambda$. The CBC formula may even be used to approximate λ_1 for the case $D \sim U(x_{\max}) - U(x_{\min})$, where Kramer's result is not applicable.

7.4.2 Broken SUSY

The discussion for the broken-SUSY case is similar to the previous one. Here the decay rates for $U_\pm$ (see figure 7.1 right column) are identical, that is,

$$\lambda_n := \lambda_n^- = \lambda_n^+ > 0, \qquad n = 0, 1, 2, \dots. \quad (7.55)$$

The corresponding transition-probability densities read

$$m_t^\pm(x, x_0) = \exp\left\{\pm\frac{1}{D}[U_-(x) - U_-(x_0)]\right\} \times \sum_{n=0}^\infty e^{-\lambda_n t/D}\phi_n^\pm(x)\phi_n^{\pm*}(x_0), \quad (7.56)$$

where the decay modes are related by

$$\phi_n^\pm(x) = \frac{1}{\sqrt{2\lambda_n}}\left(D\frac{\partial}{\partial x} \pm \Phi(x)\right)\phi_n^\mp(x). \quad (7.57)$$

Note that because of broken SUSY ($\lambda_0 > 0$) there exists no stationary distribution.

As for the good-SUSY case there are shape-invariant SUSY potentials with broken SUSY. Hence, it is also possible to study the unstable or metastable Fokker–Planck equation exactly. As an interesting example we mention

$$\Phi(x) := \Phi_0 \tanh x + a, \qquad \Phi_0 > 0, \quad a > 0,$$
$$U_-(x) = \Phi_0 \ln(\cosh x) + ax. \tag{7.58}$$

For $a < \Phi_0$ SUSY is good (see table 4.1). However, for $a \geqslant \Phi_0$ SUSY will be broken. The corresponding drift potential U_- for various values of a is shown in figure 7.3. Of particular interest is the broken case $a = \Phi_0$, which describes the diffusion of a free particle near a soft wall. The corresponding partner potentials read

$$V_{\pm}(x) = \frac{1}{2}\left[\Phi_0^2 + a^2 - \frac{\Phi_0(\Phi_0 \mp D)}{\cosh^2 x} + 2a\Phi_0 \tanh x\right], \tag{7.59}$$

which is the so-called non-symmetric Rosen–Morse potential [34]. The eigenvalue problem of the associated Schrödinger equation has been studied in various ways [34–37].

For obtaining approximate decay rates of not exactly solvable potentials one can choose the same techniques as in the good-SUSY case. For example, for a high barrier of a metastable potential U_- the smallest decay rate $\lambda_0 > 0$ will be very close to zero. Hence, for ϕ_0^- we may take the same ansatz as before,

$$\phi(x) = C e^{U_-(|x|)/D} \int_{|x|}^{\infty} dz\, e^{-2U_-(z)/D}, \tag{7.60}$$

and arrive at

$$\lambda_0 \simeq \frac{1}{2} C^2 D^2. \tag{7.61}$$

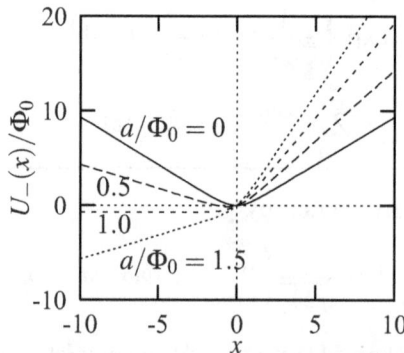

Figure 7.3. The drift potential (7.58) for various values of a/Φ_0.

Again the normalisation constant C may be evaluated via the Laplace method which yields the result

$$\lambda_0 \simeq \frac{D}{2\pi}\sqrt{U_-''(x_{\min})||U_-''(x_{\max})|}\exp\left\{-\frac{2}{D}[U_-(x_{\max}) - U_-(x_{\min})]\right\}. \qquad (7.62)$$

As before $x_{\min}$ and $x_{\max}$ denote the positions of the local minimum and maximum of U_-, respectively. The above result differs from that of the bistable potential by a factor two which is due to the fact that the metastable potential has only one instead of two local minima.

For lower potential barriers or for the other decay rates one may also apply the WKB approximation (7.53) or better the EIJ formula (6.31) respecting the iso-spectral property (7.55)

$$\int_{x_L}^{x_R} dx\sqrt{2\lambda - \Phi^2(x)} = D\pi\left(n + \frac{1}{2}\right). \qquad (7.63)$$

Finally, let us point out that in the Fokker–Planck equation one may allow for a singular drift coefficient, which produces a cusp-like barrier in the drift potential. As an example, let us consider the following drift:

$$\Phi(x) := -\frac{\operatorname{sgn} x}{\sqrt{|x|}} - x^2, \qquad U_\pm(x) = \pm\left(\frac{1}{3}x^3 + \sqrt{|x|}\right). \qquad (7.64)$$

For the shape of the drift potentials see figure 7.4. Obviously, these metastable potentials do not have a stationary distribution and, hence, SUSY is broken. However, they do not necessarily have identical decay rates. Mathematically, this could happen because then the superalgebra holds only formally. The corresponding partner Hamiltonians are only formal operators and one has to define their proper domains carefully. See also our remark in point (c) of section 5.1. Physically, the system with U_- may decay more rapidly than that with U_+. This is because, once the particle has reached the top of the cusp-like barrier of U_- with a finite velocity it will never return into the well. In the case of U_+, however, because of the flatness of the

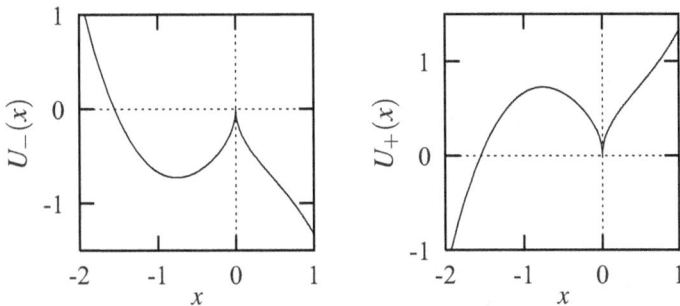

Figure 7.4. The metastable drift potentials (7.64) showing a cusp-like barrier and well due to a singularity in the SUSY potential.

barrier, the particle has a good change to move back into the well. Hence, we expect U_+ to have a smaller decay rate λ_0^+ then U_-. For an overview of the results see [38].

7.4.3 Conditionally exactly solvable Fokker–Planck potentials

In section 7.2 we have seen that the solution of the Fokker–Planck equation with drift coefficient $F_\pm = \pm \Phi$ can be reduced to the eigenvalue problem for the associated partner Hamiltonians (again for simplicity we set in this section $D = 1$)

$$H_\pm^0 := -\frac{1}{2} \frac{\partial^2}{\partial x^2} + \frac{1}{2} \Phi^2(x) \pm \frac{1}{2} \Phi'(x). \tag{7.65}$$

With the SUSY potential Φ being one of the shape-invariant ones as listed in table 4.1, this eigenvalue problem is exactly solvable and we can obtain from the eigenvalues λ_n^0 and the associated eigenfunctions $\phi_n^\pm$ the spectral representation for the transition-probability density according to equation (7.43) in the case of good SUSY and equation (7.56) in case of broken SUSY, respectively.

Let us now consider a perturbation of the SUSY potential in the form

$$W(x) = \Phi(x) + g(x), \tag{7.66}$$

where Φ is as above, that is, the spectral properties of $H_\pm^0$ are explicitly known. If we now restrict the perturbation g to obey the Riccati equation

$$g^2(x) + 2\Phi(x)g(x) + g'(x) = b, \qquad b = \text{const.} \tag{7.67}$$

the perturbed Hamiltonians explicitly read

$$\begin{aligned}
H_+ &= -\frac{1}{2} \frac{\partial^2}{\partial x^2} + \frac{1}{2} \Phi^2(x) + \frac{1}{2} \Phi'(x) + \frac{b}{2}, \\
H_- &= -\frac{1}{2} \frac{\partial^2}{\partial x^2} + \frac{1}{2} \Phi^2(x) - \frac{1}{2} \Phi'(x) - g'(x) + \frac{b}{2}.
\end{aligned} \tag{7.68}$$

Hence the eigenvalues λ_n of H_+ are up to the constant shift by $b/2$ identical to those of H_+^0. The corresponding eigenfunctions coincide. The spectral properties of H_-, which are *a priori* not known, can be obtained from those of H_+ via the SUSY transformations (7.45) in the case of unbroken SUSY and (7.57) in the case of broken SUSY, respectively.

In the following we will only consider the case of unbroken SUSY that is the SUSY potential W, such that H_- in addition to the positive eigenvalue being identical to those of H_+, that is, $\lambda_n = \lambda_n^0 + b/2$, $n = 1, 2, 3, \ldots$, also has a zero-energy eigenvalue $\lambda_0 = 0$. The corresponding eigenfunction is given by

$$\phi_0^-(x) = C \exp\left\{ -\int^x dz[\Phi(x) + g(x)] \right\} = Ce^{-U_-(x)}. \tag{7.69}$$

The above Riccati equation (7.67) can be linearised with the ansatz $g(x) = \frac{v'(x)}{v(x)}$ to

$$v''(x) + 2\Phi(x)v'(x) = bv(x). \tag{7.70}$$

This in turn can be put into a Schrödinger-like equation by making a further substitution as $u(x) = v(x)\exp\left\{\int^x dz\ \Phi(z)\right\}$:

$$\left(-\frac{1}{2}\frac{\partial^2}{\partial x^2} + \frac{1}{2}\Phi^2(x) + \frac{1}{2}\Phi'(x)\right)u(x) = -\frac{b}{2}u(x). \tag{7.71}$$

This may be compared with the discussion in section 4.3.1, where we have constructed conditionally exactly solvable quantum mechanical potentials. In fact, we may identify the parameter ε there with the current arbitrary constant b as follows, $\varepsilon = -b/2$. In order to have a nodeless solution u this parameter is restricted to $\varepsilon = -b/2 < \lambda_1$, that is, $b > -2\lambda_1$. This is required to have a non-singular g and hence a non-singular SUSY potential W. Compare this also with problem 4.3. Any nodeless solution of equation (7.71) actually leads to a new exactly solvable drift potential. Note that these drift potentials are directly expressible in terms of the solution of equation (7.71) as follows;

$$U_\pm(x) = \mp\int^x dz\ W(z) = \mp\int^x dz\left(\Phi(z) + \frac{v'(x)}{v(x)}\right) = \mp\ln u(x). \tag{7.72}$$

Let us consider the special case of the harmonic SUSY potential $\Phi(x) = \frac{1}{2}x^2$. Obviously, the nodeless solution of equation (7.71) is given by the solution (4.60) related to the family of SUSY partners associated with the harmonic oscillator, with additional parameter β obeying condition (4.61). Hence, we have found a two-parameter family of conditionally exactly solvable drift potentials whose associated Fokker–Planck equation is exactly solvable,

$$U_\pm(x) = \pm\frac{x^2}{2} \mp \ln\left[{}_1F_1\left(\frac{1 - 2\varepsilon}{4}, \frac{1}{2}; x^2\right) + \beta\, x\, {}_1F_1\left(\frac{3 - 2\varepsilon}{4}, \frac{3}{2}; x^2\right)\right]. \tag{7.73}$$

For the special case $\beta = 0$ this reduces to a one-parameter family of symmetric mono- or bistable potentials depending on the value of ε. This family is actually identical to that found by Hongler and Zheng [27]. Figure 7.5 shows this family of drift potentials for a particular range of allowed values in the parameter ε. For ε close to the limit $-1/2$ the drift potential is bistable whereas for large negative ε it becomes a mono-stable linear confining potential. The case of non-vanishing β has been discussed by Hongler and Zheng in [39]. Let us note that in principle we could also allow for complex values in β which would lead to a stable but complex-valued drift potential.

7.5 Continuous stationary Markov processes

In this section we will briefly analyse the implications of SUSY, and in particular the shape-invariance condition we have met in section 4.2, on continuous stationary

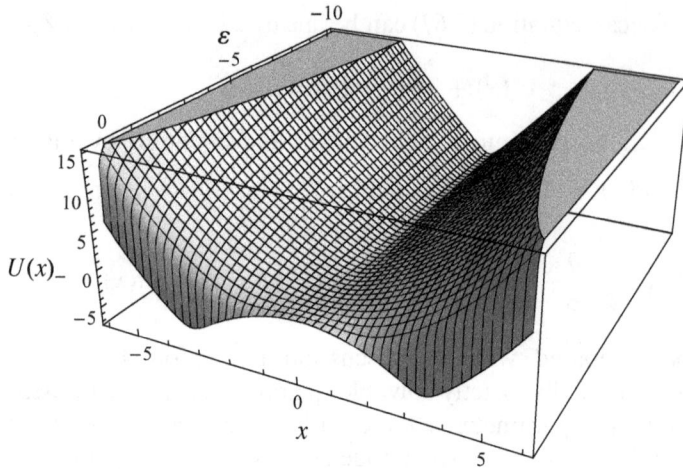

Figure 7.5. The conditionally exactly solvable drift potential (7.72) associated with the harmonic oscillator.

Markov processes. First we need to recall some basic aspects of stationary stochastic processes.

Let us begin with considering a well-defined transition-probability density on some configuration space Q, that is, $m_t(x, x_0)$, $t \geqslant 0$, x, $x_0 \in Q \subseteq \mathbb{R}$, and assume that it obeys the following conditions:

Positivity: $m_t(x, x_0) > 0$.

Normalisation: $\int_Q \mathrm{d}x \, m_t(x, x_0) = 1$.

Initial condition: $\lim_{t \to 0} m_t(x, x_0) = \delta(x - x_0)$.

Chapman–Kolmogorov condition: $\int_Q \mathrm{d}x' \, m_t(x, x')m_{t'}(x', x_0) = m_{t+t'}(x, x_0)$.

Lindeberg condition: $\lim_{t \to 0} \frac{1}{t} \int_{|x-x_0|>\epsilon} \mathrm{d}x \, m_t(x, x_0) = 0$, $\forall \epsilon > 0$.

The first three conditions are basic properties of a transition-probability density. The fourth condition is also known as the Markov property, thus the name Markov process. Finally, the last condition assures the continuity of the process where no jumps may occur. It follows from Kolmogorov's fundamental theorem [40] that under these conditions the transition-probability density uniquely defines a continuous stationary stochastic process on the configuration space Q. To be a bit more precise, let $X_t(\omega) \in Q$ be a family of random variables, with $t \geqslant 0$ and $\omega \in \Omega$ being an element in a probability space Ω equipped with a probability measure P. Then the stochastic process denoted by X_t is uniquely defined via the relation

$$P\left(X_{t_n}(\omega) = x_n, X_{t_{n-1}}(\omega) = x_{n-1}, \cdots, X_{t_1}(\omega) = x_1\right) = $$
$$m_{t_n - t_{n-1}}(x_n, x_{n-1})m_{t_{n-1} - t_{n-2}}(x_{n-1}, x_{n-2}) \cdots m_{t_1 - t_0}(x_1, x_0) \tag{7.74}$$

for all $n \in \mathbb{N}$ with $0 =: t_0 < t_1 < \cdots < t_n$ and initial condition $P(X_0(\omega) = x_0) = 1$. Note that for a fixed $\omega \in \Omega$ the realisations $X_t(\omega)$ are called paths. These paths are with probability one continuous. The set of all these paths starting at $t_0 = 0$ in x_0 is denoted by $C(Q, x_0)$. The probability measure P induces a probability measure $\mathcal{M}$ on $C(Q, x_0)$ with the help of so-called cylinder functions $f_n(x_1, \ldots, x_n)$ via the definition

$$
\int_{C(Q,x_0)} \mathrm{d}\mathcal{M}[x] f_n(x(t_1), \ldots, x(t_n)) :=
$$

$$
\int_Q \mathrm{d}x_n \cdots \int_Q \mathrm{d}x_1 m_{t_n-t_{n-1}}(x_n, x_{n-1}) m_{t_{n-1}-t_{n-2}}(x_{n-1}, x_{n-2}) \cdots m_{t_1-t_0}(x_1, x_0) \tag{7.75}
$$

$$
\times f_n(x_1, \ldots, x_n).
$$

Hence, in addition to the transition-probability density the stochastic process is alternatively characterisable via the path measure $\mathcal{M}$.

A third way of characterising such a process is via its generator. For this let us consider the Fokker–Planck equation (7.7) with drift coefficient F and $D = 1$ in the form

$$
-\partial_t \mathrm{m}_t = \mathrm{M}\,\mathrm{m}_t, \tag{7.76}
$$

with $\mathrm{m}_0 = 1$ being the unit operator acting on $\mathcal{H} = L^2(Q)$ and the generator M is defined by

$$
\mathrm{M} := \frac{1}{2}P^2 + \mathrm{i}P\,F(Q), \tag{7.77}
$$

with $P = -\mathrm{i}\partial_x$ and Q being the momentum and position operator, respectively, acting on $\mathcal{H} := L^2(Q)$. The above equation is easily integrated, yielding

$$
\mathrm{m}_t = \exp\{-t\mathrm{M}\}. \tag{7.78}
$$

Obviously the transition-probability density is the matrix element in the position representation of the above Euclidean propagator generated by M. Hence, we have three ways to characterise a continuous stationary stochastic process. They are related by

$$
m_t(x, x_0) = \langle x|\mathrm{e}^{-t\mathrm{M}}|x_0\rangle = \int_{C(Q,x_0)} \mathrm{d}\mathcal{M}[x]\,\delta(x(t) - x). \tag{7.79}
$$

Finally let us consider a generalised Hamiltonian $H := \mathrm{M} + V(Q)$, then the associated Euclidean propagator may be expressed in terms of an Euclidean path integral via the well-known Feynman–Kac formula [41]:

$$
\langle x|\mathrm{e}^{-tH}|x_0\rangle = \int_{C(Q,x_0)} \mathrm{d}\mathcal{M}[x]\,\delta(x(t) - x)\exp\left\{-\int_0^t \mathrm{d}\tau\,V(x(\tau))\right\}. \tag{7.80}
$$

For the case where $Q = \mathbb{R}$ and vanishing drift $F = 0$ the above stochastic process is the well-studied Wiener process describing the Brownian motion on the real line.

In the following let us consider a family of drifts F_s where s is an index enumerating the associated process generated by

$$\mathrm{M}_s := \frac{1}{2} P^2 + iP \, F_s(Q). \tag{7.81}$$

With the help of the auxiliary function

$$g_{rs}(x) := \exp \left\{ \int^x dz [F_r(z) - F_s(z)] \right\} \tag{7.82}$$

it is straightforward to show the below relation between the generators of the process indexed by s and that indexed by r

$$g_{rs}(Q) \mathrm{M}_s g_{rs}^{-1}(Q) = \mathrm{M}_r - \frac{1}{2} \Big[F_r^2(Q) - F_s^2(Q) + F_s'(Q) - F_r'(Q) \Big]. \tag{7.83}$$

This immediately leads to a relation between the corresponding Euclidean propagators

$$\langle x | \exp \left\{ -t[\mathrm{M}_s + V(Q)] \right\} | x_0 \rangle = \frac{g_{rs}(x_0)}{g_{rs}(x)} \langle x | \exp \left\{ -t[\mathrm{M}_r + \tilde{V}(Q)] \right\} | x_0 \rangle, \tag{7.84}$$

with

$$\tilde{V}(x) := V(x) - \frac{1}{2} \Big[F_r^2(x) - F_s^2(x) + F_r'(x) - F_s'(x) \Big]. \tag{7.85}$$

Note that

$$\frac{g_{rs}(x_0)}{g_{rs}(x)} = \exp \left\{ -\int_{x_0}^x dz [F_r(z) - F_s(z)] \right\}. \tag{7.86}$$

The above relation explicates that a change in drift also leads to a change in the potential. Applying the Feynman–Kac relation to expression (7.83) leads to the so-called Cameron–Martin–Girsanov theorem. This has attracted much attention in financial stochastics, where one tries to eliminate drift-terms by changing the underlying process. See, for example, [42]. In particular, one may eliminate potential terms by a suitable change of the drift thus leading to exactly solvable path integrals [43].

Let us discuss a few special cases for the above relation. As a first example we consider the case $V(x) = V_r(x) := \frac{1}{2} F_r^2(x) + \frac{1}{2} F_r'(x)$. Hence it follows that $\tilde{V}(x) = V_s(x) := \frac{1}{2} F_s^2(x) + \frac{1}{2} F_s'(x)$ and we arrive at the relation

$$\langle x | \exp \left\{ -t[\mathrm{M}_s + V_r(Q)] \right\} | x_0 \rangle = \frac{g_{rs}(x_0)}{g_{rs}(x)} \langle x | \exp \left\{ -t[\mathrm{M}_r + V_s(Q)] \right\} | x_0 \rangle. \tag{7.87}$$

If we further impose $Q = \mathbb{R}$ and $F_s(x) = 0$, that is, the process with index s is the usual Wiener process, we explicitly see how a quantum mechanical potential term may be changed into a pure drift:

$$\langle x|\exp\left\{-t[P^2/2 + V_r(Q)]\right\}|x_0\rangle = \frac{g_{rs}(x_0)}{g_{rs}(x)}\langle x|\exp\left\{-t\mathrm{M}_r\right\}|x_0\rangle. \tag{7.88}$$

Specifying further to $F_r(x) = \pm\Phi(x)$ with $V_r(x) = V_\pm(x) = \frac{1}{2}\Phi^2(x) \pm \frac{1}{2}\Phi'(x)$ we obtain

$$\langle x|\exp\left\{-t[P^2/2 + V_\pm(Q)]\right\}|x_0\rangle = \frac{g_{rs}(x_0)}{g_{rs}(x)}\langle x|\exp\left\{-t[P^2/2 \pm iP\Phi(Q)]\right\}|x_0\rangle. \tag{7.89}$$

With the obvious relations $g_{rs}(x) = g_{rs}(x_0)\exp\{-[U_\pm(x) - U_\pm(x_0)]\}$, where $U_\pm$ is given by equation (7.10), and $m_t^\pm(x, x_0) = \langle x|\exp\left\{-t\mathrm{M}_r\right\}|x_0\rangle$ we obtain the already know result (7.12) with equation (7.16),

$$m_t^\pm(x, x_0) = \exp\{-[U_\pm(x) - U_\pm(x_0)]\}\langle x|\exp\left\{-tH_\pm\right\}|x_0\rangle. \tag{7.90}$$

Finally let us now consider the special case $F_r(x) = \tanh x$ leading to a constant quantum potential $V_+(x) = \frac{1}{2}$. The corresponding drift potential and auxiliary functions are given by $U_+(x) = -\ln\cosh x$ and $g_{rs}(x) = \cosh x$, respectively. Hence, the transition-probability density for the diffusion process in the potential U_+, see figure 7.3 case $a = 0$ with $U_+ = -U_-$, can directly be related to the well-known transition-probability density of the free Brownian motion,

$$m_t^+(x, x_0) = \langle x|\exp\left\{-t[P^2/2 + iP\tanh Q]\right\}|x_0\rangle$$
$$= \sqrt{\frac{1}{2\pi t}}\frac{\cosh x}{\cosh x_0}\exp\left\{-\frac{(x - x_0)^2}{2t} - \frac{t}{2}\right\}. \tag{7.91}$$

The above example can be generalised to the case of shape-invariant drift coefficients. Let us go back to relation (7.84) consider the case $V(x) = 0$ and specialise the drift to a shape-invariant SUSY potential, that is, $F_s(x) := \Phi(a_s, x)$, cf. section 4.2. Furthermore, we choose $F_r(x) = -F_{s+1}(x) = -\Phi(a_{s+1}, x)$. Using the shape-invariance condition in the form (4.33), note that we need to make the replacements $\hbar/\sqrt{m} \to D = 1$ and $\Phi \to \Phi/\sqrt{2}$, we find that

$$\tilde{V}(x) = -\frac{1}{2}\Phi^2(a_{s+1}, x) + \frac{1}{2}\Phi^2(a_s, x) + \frac{1}{2}\Phi'(a_{s+1}, x) + \frac{1}{2}\Phi'(a_s, x) \tag{7.92}$$
$$= R(a_{s+1})$$

is constant as well. As a consequence we find the following recursion relation between the transition-probability densities for systems with shape-invariant drifts and alternating sign:

$$\langle x|\exp\left\{-t[P^2/2 + iP\Phi(a_s, Q)]\right\}|x_0\rangle = \exp\left\{\int_{x_0}^x dz[\Phi(a_{s+1}, z) + \Phi(a_s, z)]\right\}$$
$$\times e^{-tR(a_{s+1})}\langle x|\exp\left\{-t[P^2/2 - iP\Phi(a_{s+1}, Q)]\right\}|x_0\rangle. \tag{7.93}$$

Assuming that we have a family of shape-invariant SUSY potentials given by $\{\Phi(a_s, x)\}$ with $s = 0, 1, \ldots, n$ the above recursion relation results in the identity

$$\langle x|\exp\left\{-t[P^2/2 + iP\Phi(a_0, Q)]\right\}|x_0\rangle = \exp\left\{\int_{x_0}^{x} dz[\Phi(a_0, z) - (-1)^n\Phi(a_n, z)]\right\}$$
(7.94)

$$\times e^{-tE_n}\langle x|\exp\left\{-t[P^2/2 + i(-1)^nP\Phi(a_n, Q)]\right\}|x_0\rangle,$$

where $E_n = \sum_{s=1}^{n} R(a_s)$ is identical to the first relation in equation (4.23). For the special case $\Phi(a_s, x) := a_s \tanh x$ with $a_s = n - s$ this results in, see problem 7.4 below,

$$\langle x|\exp\left\{-t[P^2/2 + inP \tanh Q]\right\}|x_0\rangle$$

$$= \sqrt{\frac{1}{2\pi t}}\left(\frac{\cosh x}{\cosh x_0}\right)^n \exp\left\{-\frac{(x - x_0)^2}{2t} - \frac{n^2}{2}t\right\}.$$
(7.95)

Note that this family of drift potentials is identical to the shape-invariant SUSY potential discussed in section 4.2. For an integer $a_0 = n \in \mathbb{N}$ the relation (7.94) provides a recursion relation for transition-probability densities and relates them to that of the free Brownian motion as shown in equation (7.95).

7.6 Problems

Problem 7.1. Derivation of the stationary distribution for unbroken SUSY

Derive expression (7.9) from the Fokker–Planck (7.7) by assuming that the limit (7.8) exists.

Problem 7.2. Derivation of the imaginary-time Schrödinger equation

Derive the imaginary-time Schrödinger equation (7.13) and show that the corresponding Fokker–Planck Hamiltonians (7.14) are identical to those of the quantum mechanical Witten model.

Problem 7.3. On the Cameron–Martin–Girsanov formula

Prove the relations (7.83) and (7.84) and derive via the Feynman–Kac formula (7.80) the path-integral version of the Cameron–Martin–Girsanov formula [43]

$$\int_{C(Q,x_0)} d\mathcal{M}_s[x]\,\delta(x(t) - x)\exp\left\{-\int_0^t d\tau\, V(x(\tau))\right\}$$

$$= \exp\left\{-\int_{x_0}^{x} dz[F_r(z) - F_s(z)]\right\}$$

$$\times \int_{C(Q,x_0)} d\mathcal{M}_r[x]\,\delta(x(t) - x)\exp\left\{-\int_0^t d\tau\, \tilde{V}(x(\tau))\right\}.$$

Here $\mathcal{M}_s$ is the path measure associated with the generator $\mathrm{M}_s := \frac{1}{2} P^2 + iP\, F_s(Q)$ and similar for index r. The potential $\tilde{V}$ is given by relation (7.85).

Problem 7.4. An explicit example for a family of shape-invariant drifts

Consider the relation (7.93) for a family of drifts given by $\Phi(a_s, x) := a_s \tanh x$ with $a_s = a_0 - s$, $s = 0, 1, 2, \ldots \leqslant a_0$. Show that this relation explicitly reads

$$\langle x|\exp\left\{-t[P^2/2 + \mathrm{i}(a_0 - s)P \tanh Q]\right\}|x_0\rangle = \left(\frac{\cosh x}{\cosh x_0}\right)^{2a_0 - 2s - 1} \mathrm{e}^{-t(a_0 - s - 1/2)}$$

$$\times \langle x|\exp\left\{-t[P^2/2 - \mathrm{i}(a_0 - s - 1)P \tanh Q]\right\}|x_0\rangle.$$

Show that for $s = a_0$ this relation reduces to that given in equation (7.90).

References

[1] Parisi G and Sourlas N 1979 Random magnetic fields, supersymmetry, and negative dimensions *Phys. Rev. Lett.* **43** 744–5
[2] Parisi G and Sourlas N 1982 Supersymmetric field theories and stochastic differential equations *Nucl. Phys.* B **206** 321–32
[3] Feĭgel'man M V and Tsvelik A M 1982 Hidden supersymmetry of stochastic dissipative dynamics *Sov. Phys. JETP* **56** 823–30
[4] van Kampen N G 2007 *Stochastic Processes in Physics and Chemistry* 3rd edn (Amsterdam: North-Holland)
[5] Cooper F and Freedman B 1983 Aspects of supersymmetric quantum mechanics *Ann. Phys.* **146** 262–88
[6] Bernstein M and Brown L S 1984 Supersymmetry and the bistable Fokker–Planck equation *Phys. Rev. Lett.* **52** 1933–5
[7] Risken H 1989 *The Fokker–Planck Equation Methods of Solution and Applications* 2nd edn (Berlin: Springer)
[8] Trimper S 1990 Supersymmetry breaking for dynamical systems *J. Phys. A: Math. Gen.* **23** L169–74
[9] Sourlas N 1985 Introduction to supersymmetry in condensed matter physics *Physica* D **15** 115–22
[10] Zinn-Justin J 2002 *Quantum Field Theory and Critical Phenomena* 4th edn (Oxford: Clarendon)
[11] Langevin P 1908 Sur la théorie du mouvement Brownien *C. R. Acad. Sci.* **146** 530–3
[12] Arnold L 1973 *Stochastische Differentialgleichungen* (München: Oldenbourg)
[13] Gardiner C W 1990 *Handbook of Stochastic Methods for Physics Chemistry and the Natural Sciences Springer Series in Synergetics* vol 13 2nd edn (Berlin: Springer)
[14] Einstein A 1905 Über die von der molekularkinetischen Theorie der Wärme geforderte Bewegung von in ruhenden Flüssigkeiten suspendierten Teilchen *Ann. Physik* **17** 549–60
[15] Einstein A 1906 Zur Theorie der Brownschen Bewegung *Ann. Phys.* **19** 371–81
[16] von Smoluchowski M 1906 Zur kinetischen Theorie der Brownschen Molekularbewegung und der Suspensionen *Ann. Phys.* **21** 756–80
[17] Leschke H 1981 Path integral approach to fluctuations in dynamic processes *Chaos and Order in NatureSpringer Series in Synergetics* vol 11 ed H Haken (Berlin: Springer) pp 157–63
[18] Fokker A D 1914 Die mittlere Energie rotierender elektrischer Dipole im Strahlungsfeld *Ann. Phys.* **43** 810–20

[19] Planck M 1917 Über einen Satz der statistischen Dynamik und seine Erweiterung in der Quantentheorie *Sitzungsber. Preuss. Akad. Wiss.* **24** 324–41

[20] Leiber T, Marchesoni F and Risken H 1987 Colored noise and bistable Fokker–Planck equations *Phys. Rev. Lett.* **59** 1381–4
Leiber T, Marchesoni F and Risken H 1988 Colored noise and bistable Fokker–Planck equations *Phys. Rev. Lett.* **60** 659 (erratum)

[21] Cecotti S and Girardello L 1983 Stochastic and parastochastic aspects of supersymmetric functional measures: a new non-perturbative approach to supersymmetry *Ann. Phys.* **145** 81–99

[22] Kihlberg A, Salomonson P and Skagerstam B S 1985 Witten's index and supersymmetric quantum mechanics *Z. Phys.* C **28** 203–9

[23] Ezawa H and Klauder J R 1985 Fermions without fermions: the Nicolai map revisited *Prog. Theor. Phys.* **74** 904–15

[24] Graham R and Roekaerts D 1985 Supersymmetric quantum mechanics and stochastic processes in curved configuration space *Phys. Lett.* A **109** 436–40

[25] Singh L P and Steiner F 1986 Fermionic path integrals, the Nicolai map and the Witten index *Phys. Lett.* B **166** 155–9

[26] Gozzi E 1984 Onsager principle of microscopic reversibility and supersymmetry *Phys. Rev.* D **30** 1218–27

[27] Hongler M O and Zheng W M 1982 Exact solution for the diffusion in bistable potentials *J. Stat. Phys.* **29** 317–27

[28] Englefield M J 1988 Exact solutions of a Fokker–Planck equation *J. Stat. Phys.* **52** 369–81

[29] Jauslin H R 1988 Exact propagator and eigenfunction for multistable models with arbitrary prescribed N lowest eigenvalues *J. Phys. A: Math. Gen.* **21** 2337–50

[30] Keung W-Y, Kovacs E and Sukhatme U 1988 Supersymmetry and double-well potentials *Phys. Rev. Lett.* **60** 41–4

[31] Wittmer S 1992 Supersymmetrische Quantenmechanik und semiklassische Näherung *Diploma thesis* University of Stuttgart

[32] Marchesoni F, Sodano P and Zannetti M 1988 Supersymmetry and bistable soft potentials *Phys. Rev. Lett.* **61** 1143–6

[33] Kramers H A 1940 Brownian motion in a field of force and the diffusion model of chemical reactions *Physica* **7** 284–304

[34] Rosen N and Morse P M 1932 On the vibrations of polyatomic molecules *Phys. Rev.* **42** 210–7

[35] Nieto M M 1978 Exact wave-function normalization for the $B_0 \tanh z - U_0 \cosh^{-2} z$ and Pöschl–Teller potentials *Phys. Rev.* A **17** 1273–83

[36] Junker G and Inomata A 1986 Path integral on S^3 and its application to the Rosen–Morse oscillator *Path Integrals from meV to MeV, Bielefeld Encounters in Physics and Mathematics* vol 7 ed M C Gutzwiller, A Inomata, J R Klauder and L Streit (Singapore: World Scientific) pp 315–34

[37] Barut A O, Inomata A and Wilson R 1987 Algebraic treatment of second Pöschl–Teller, Morse-Rosen and Eckart equations *J. Phys. A: Math. Gen.* **20** 4083–96

[38] Hänggi P, Talkner P and Borkovec M 1990 Reaction-rate theory: fifty years after Kramers *Rev. Mod. Phys.* **62** 251–341

[39] Hongler M-O and Zheng W M 1983 Exact results for the diffusion in a class of asymmetric bistable potentials *J. Math. Phys.* **24** 336–40

[40] Kolmogorow A N 1933 *Grundbegriffe der Wahrscheinlichkeitstheorie* (Berlin: Springer)

[41] Roepstorff G 1994 *Path Integral Approach to Quantum Physics* (Berlin: Springer)

[42] van der Vaart A W 2005 *Financial Stochastics, Lecture Notes* (Amsterdam: Vrije Universiteit)

[43] Fischer W, Leschke H and Müller P 1993 Path integration in quantum mechanics by changing drift and time of the underlying diffusion process *Path Integrals from meV to MeV: Tutzing '92* ed H Grabert, A Inomata, L Schulman and U Weiss (Singapore: World Scientific) pp 259–67

IOP Publishing

Supersymmetric Methods in Quantum, Statistical and Solid State Physics

Enlarged and revised edition

Georg Junker

Chapter 8

Supersymmetric Pauli Hamiltonians

For the Pauli Hamiltonian [1], characterising the dynamics of a non-relativistic spin-$\frac{1}{2}$ particle, the Z_2-grading of the Hilbert space is rather inherent, and SUSY is naturally expected to be its symmetry. For the example of an electron in an external magnetic field presented in section 2.1.1 the Pauli Hamiltonian has been identified as a square of the self-adjoint supercharge.

Currently the idea of SUSY has been employed in a variety of areas. For example, to study systems of strongly correlated electrons, a supersymmetric extension of the Hubbard model is proposed [2, 3], by which an explicit expression for the ground-state wave function is obtained. Other supersymmetric extensions of integrable quantum models can be found in [4, 5], and a SUSY application to fractional statistics in [6]. Another area where supersymmetric methods are extensively used is that of disordered systems [7–9], for which SUSY provides an alternative to the replica trick.

In the present chapter we discuss implications of supersymmetry for systems described by the Pauli Hamiltonian. First, we make a few remarks on the Pauli Hamiltonian for electrons in a stationary magnetic field. Then, in section 8.2, we consider the Pauli paramagnetism of two- and three-dimensional non-interacting electron gases, and show that the Witten index (IDOS regularised) is related to the zero-temperature magnetisation of such systems. We also show that breaking of SUSY gives rise to counterexamples of the so-called paramagnetic conjecture. In section 8.3, we deal with the Pauli Hamiltonian for an electron in a spherical scalar potential, including a vanishing potential and a Coulomb potential. We also briefly discuss an electron in a spherical tensor potential and the Pauli oscillator.

8.1 Electrons in an external magnetic field

The standard Schrödinger Hamiltonian for an electron with mass m and electric charge e ($e < 0$ for electrons) in an external electromagnetic field

doi:10.1088/2053-2563/aae6d5ch8

described by a vector potential $\vec{A}: \mathbb{R}^3 \to \mathbb{R}^3, \vec{r} \mapsto \vec{A}(\vec{r})$ and a scalar potential $\phi: \mathbb{R}^3 \to \mathbb{R}, \vec{r} \mapsto \phi(\vec{r})$,

$$H_S := \frac{1}{2m}\left(\vec{p} - \frac{e}{c}\vec{A}(\vec{r})\right)^2 + e\phi(\vec{r}), \tag{8.1}$$

being the same in form as the classical Hamiltonian, acts on state functions in Hilbert space $L^2(\mathbb{R}^3)$. Here $\vec{p}$ is the linear momentum operator, and $\vec{r}$ is the position operator. It shows that the electron interacts with the electric field $\vec{E}(\vec{r}) := -\vec{\nabla}\phi(\vec{r})$ and the magnetic field $\vec{B}(\vec{r}) := \vec{\nabla} \times \vec{A}(\vec{r})$ only through the coupling introduced in the minimal manner $\vec{p} \to \vec{p} - (e/c)\vec{A}$ and $H_S \to H_S - e\phi$ under the requirement of $U(1)$ gauge invariance [10].

In the case of a constant magnetic field $\vec{B}$, where $\vec{A} = \frac{1}{2}\vec{B} \times \vec{r}$, the term $(e/2mc)\vec{L} \cdot \vec{B}$ contained in H_S coincides with the classical coupling between the external magnetic field $\vec{B}$ and the magnetic moment $\vec{\mu}_L = (e/2mc)\vec{L}$ induced by an electron circulating in a classical orbit with the angular momentum $\vec{L}$. Note that the gyromagnetic ratio $\gamma = e/(2mc)$ is in common. Historically, the predictions based on the Schrödinger Hamiltonian did not agree with experiments in the following two points. The observed number of energy states is double the predicted number, and the gyromagnetic ratio varies and differs from $e/(2mc)$. In order to resolve these discrepancies, Goudsmit and Uhlenbeck [11] put forth a postulate that each electron has an intrinsic angular momentum $\vec{S}$, i.e. spin, of magnitude $\frac{1}{2}\hbar$. The spin quantum numbers $\pm\frac{1}{2}$ explain the duplexity of states and the magnetic moment of the electron $\vec{\mu}_S = g(e/2mc)\vec{S}$ written with the Landé g-factor gives, if $g \approx 2$, a value of the gyromagnetic ratio in good agreement with experiments. The predicted magnitude of $\vec{\mu}_S$ is one Bohr magneton, that is, $\mu_B := |e|\hbar/(2mc)$. Pauli [1], even having rejected the classical view that every electron spins, formulated a way to integrate the idea of spin into the Schrödinger scheme by extending Hilbert space from $L^2(\mathbb{R}^3)$ to $L^2(\mathbb{R}^3) \otimes \mathbb{C}^2$ of two-component functions and by treating the vector components of spin $\vec{S}$ as elements of the rotational algebra of $su(2) \approx so(3)$ acting on $\mathbb{C}^2$. The spin is expressed in terms of the Pauli matrices as $\vec{S} = \frac{\hbar}{2}\vec{\sigma}$, and the spin coupling term with a chosen value $g = 2$ is added to the Schrödinger Hamiltonian as

$$H_P := H_S - \frac{e\hbar}{2mc}\vec{B} \cdot \vec{\sigma}, \tag{8.2}$$

where $\vec{\sigma}$ is a vector whose components are the Pauli matrices. This is indeed the Pauli Hamiltonian discussed in section 2.1.1 with $\phi = 0$. The last term in equation (8.2) appears to be a non-minimal coupling. However, insofar as $g = 2$ is chosen and going back to the general case of an arbitrary magnetic field, this Hamiltonian can be put into the form

$$H_P = \frac{1}{2m}\left[\vec{\sigma} \cdot \left(\vec{p} - \frac{e}{c}\vec{A}(\vec{r})\right)\right]^2 + e\phi(\vec{r}), \tag{8.3}$$

in which the electron couples minimally with the electric and magnetic field. Note that the minimal coupling of equation (8.3) contains the coupling term $(e/2mc)\vec{B} \cdot (\vec{L} + 2\vec{S})$.

Although the external fields can be time-dependent, we limit our discussion, for simplicity, to the stationary case only. The Pauli Hamiltonian in above form can be reduced to the standard form,

$$H_P = \frac{1}{2m}\left(\vec{p} - \frac{e}{c}\vec{A}(\vec{r})\right)^2 - \frac{e\hbar}{2mc}\vec{B}(\vec{r}) \cdot \vec{\sigma} + e\phi(\vec{r}). \tag{8.4}$$

Immediately after Pauli's formulation, Dirac [12] developed his relativistic theory of the electron by expanding further the Hilbert space from $L^2(\mathbb{R}^3) \otimes \mathbb{C}^2$ to $L^2(\mathbb{R}^3) \otimes \mathbb{C}^4$ of four-component state functions. It was shown that the Pauli Hamiltonian can be derived from the Dirac Hamiltonian as a non-relativistic approximation. While Pauli choses $g = 2$ to meet the experiment, Dirac obtained the same g as a theoretical consequence. In quantum electrodynamics (QED), Dirac's theoretical value $g = 2$ corresponds to the contribution from the basic tree level of the Feynman diagram (i.e. the structureless electron coupling with a single photon). Taking radiative corrections into account, QED predicts $g = 2 + \mathcal{O}(\alpha)$ where $\alpha = e^2/\hbar c$ is Sommerfeld's fine structure constant. The deviation of the magnetic moment from the one in Dirac's theory, $\mu - \mu_B = a\mu_B$ where $a = (g - 2)/2$, is called the anomalous magnetic moment. Schwinger [13] calculated the anomalous magnetic moment to the first order correction, which is the one-loop correction to the tree diagram, and found $a = \alpha/2\pi \approx 0.001162$. Higher order corrections have also been computed by a number of theorists. The last computations including the tenth order corrections [14] gave $a = 0.001\ 159\ 652\ 181\ 78(77)$, which may be compared with the experimental results $a = 0.001\ 159\ 652\ 188\ 4(43)$ in [15] and $a = 0.001159\ 652\ 180\ 73(28)$ in [16]. For a review on the history of the electron spin see [17, 18].

According to our discussion in section 2.1.1 the Pauli Hamiltonian in three space dimension possesses an $N = 1$ SUSY if the scalar potential vanishes, $\phi = 0$, and the g-factor is set to $g = 2$. Thus, in the absence of any external scalar potential, the self-adjoint supercharge and the Pauli Hamiltonian for an electron read ($e < 0$)

$$Q_1 := \frac{1}{\sqrt{4m}}\left(\vec{p} - \frac{e}{c}\vec{A}\right) \cdot \vec{\sigma},$$

$$H_P^{(3)} := 2Q_1^2 = \frac{1}{2m}\left(\vec{p} - \frac{e}{c}\vec{A}\right)^2 - \frac{e\hbar}{2mc}\vec{B} \cdot \vec{\sigma}, \tag{8.5}$$

where the superscript indicates the space dimension. Here we note that because of $N = 1$ we cannot construct a Witten operator analogous to equation (2.32). Indeed, in the general case none of the Pauli matrices σ_i commutes with $H_P^{(3)}$ because of the Zeeman term. Due to the property $W^2 = 1$ the Witten operator has to be represented by a linear combination of these σ_i's. Consequently, the result derived for $N = 2$ SUSY cannot be applied to the general three-dimensional Pauli Hamiltonian.

However, it is still possible to introduce a grading of the Hilbert space. In fact, with the help of the gauge-invariant velocity operator

$$\vec{v} := \dot{\vec{r}} = \frac{i}{\hbar}[H_P^{(3)}, \vec{r}] = \frac{1}{m}\left(\vec{p} - \frac{e}{c}\vec{A}\right) \qquad (8.6)$$

and its property

$$(\vec{v} \cdot \vec{\sigma})^2 = 4Q_1^2/m = 2H_P^{(3)}/m \qquad (8.7)$$

one can introduce the helicity operator

$$\Lambda := \frac{m\vec{v} \cdot \vec{\sigma}}{\sqrt{2mH_P^{(3)}}} = \operatorname{sgn} Q_1 \qquad (8.8)$$

as an alternative to the missing Witten parity. Note that

$$\Lambda^\dagger = \Lambda, \quad \Lambda^2 = 1, \quad [\Lambda, H_P^{(3)}] = 0 \qquad (8.9)$$

and, therefore, one can grade the Hilbert space into subspaces of positive and negative helicity. However, Λ commutes with the supercharge $Q_1 = \Lambda\sqrt{H_P^{(3)}/2}$ and, therefore, leaves the helicity eigenspaces invariant. It does not generate transformations between these spaces. However, see our discussion below in section 8.3 where we show that a supercharge similar to the helicity operator can be defined which generates SUSY transformations between eigenspaces of the spin–orbit operator in the absence of a magnetic field and the presence of a spherically symmetric scalar potential.

If the magnetic field is chosen such that $\vec{B}(\vec{r}) = B(x_1, x_2)\vec{e}_3$ we may, however, introduce a Witten parity[1] by setting $W := \sigma_3$. This operator now commutes with $H_P^{(3)}$. Note that such a magnetic field, being perpendicular to the (x_1, x_2)-plane, is generated by a vector potential of the form

$$\vec{A}(\vec{r}) := \begin{pmatrix} a_1(x_1, x_2) \\ a_2(x_1, x_2) \\ 0 \end{pmatrix}, \quad B(x_1, x_2) := \frac{\partial a_2}{\partial x_1} - \frac{\partial a_1}{\partial x_2}, \qquad (8.10)$$

where $a_i: \mathbb{R}^2 \to \mathbb{R}$, $i = 1, 2$. The three-dimensional Pauli Hamiltonian can be expressed in terms of the two-dimensional one,

$$H_P^{(3)} = H_P^{(2)} + \frac{p_3^2}{2m}, \qquad (8.11)$$

---

[1] Another class of magnetic fields which gives rise to a Witten parity consists of those with a definite space parity $\vec{B}(-\vec{r}) = \pm\vec{B}(\vec{r})$ [19]. Here, in essence, the Witten operator is given by the space-parity operator Π (cf. equation (3.31)). Also for the case of a magnetic monopole field it has been shown that the Pauli Hamiltonian possesses a dynamical $OSp(1, 1)$ supersymmetry [20].

where

$$H_P^{(2)} := \frac{1}{2m} \sum_{i=1}^{2} \left(p_i - \frac{e}{c} a_i \right)^2 - \frac{e\hbar}{2mc} B\sigma_3. \tag{8.12}$$

It is this two-dimensional Pauli Hamiltonian in a perpendicular magnetic field which possesses an $N = 2$ SUSY. The associated complex supercharge may be defined by

$$Q := \frac{1}{\sqrt{2m}} \left[\left(p_1 - \frac{e}{c} a_1 \right) - \mathrm{i} \left(p_2 - \frac{e}{c} a_2 \right) \right] \otimes \sigma^+, \quad \sigma^+ := \frac{1}{2}(\sigma_1 + \mathrm{i}\sigma_2) \tag{8.13}$$

and obeys the superalgebra

$$Q^2 = 0, \quad \{Q, Q^\dagger\} = H_P^{(2)}. \tag{8.14}$$

Because of $W = \sigma_3$ the eigenstates of the Witten operator with positive (negative) Witten parity are spin-up (spin-down) eigenstates. Hence, as a result of SUSY all positive eigenvalues of $H_P^{(2)}$ are spin degenerate. These degenerate eigenstates are related via the SUSY transformation (2.55) or (2.57). Note that the generalised annihilation operator in this model reads

$$A := \frac{1}{\sqrt{2m}} \left[\left(p_1 - \frac{e}{c} a_1 \right) - \mathrm{i} \left(p_2 - \frac{e}{c} a_2 \right) \right] \tag{8.15}$$

and the Hamiltonians restricted to the spin-up and spin-down subspace, respectively, are given by

$$H_P^{(2)} \lceil \mathcal{H}^+ = AA^\dagger, \quad H_P^{(2)} \lceil \mathcal{H}^- = A^\dagger A. \tag{8.16}$$

Under some mild conditions [21] (B is assumed to be bounded with compact support) it has been shown by Aharonov and Casher [22] that for the two-dimensional Pauli Hamiltonian SUSY is always good if the magnetic flux through the (x_1, x_2)-plane,

$$F := \int_{\mathbf{R}^2} \mathrm{d}x_1 \mathrm{d}x_2 \, B(x_1, x_2), \tag{8.17}$$

is sufficiently large. To be more precise, the degeneracy of the zero-energy eigenvalue of $H_P^{(2)}$ is given by

$$d := \left[\left[\frac{|eF|}{2\pi\hbar c} \right] \right], \tag{8.18}$$

where $[[z]]$ denotes the largest integer which is strictly less than z and $[[0]] = 0$. In other words, the magnitude $|F|$ of the flux (8.17) has to be larger than one flux-quantum $2\pi\hbar c/|e|$ for the existence of a zero-energy eigenstate. The d degenerate zero-energy eigenstates are all spin-up or spin-down states for $\mathrm{sgn}(eF) = 1$ or

$\mathrm{sgn}(eF) = -1$, respectively. Let us note that the above degeneracy is related to the Witten index by

$$\Delta = -d\ \mathrm{sgn}(eF). \qquad (8.19)$$

In concluding this section, we remark that one may also consider the supercharge

$$\tilde{Q} := \frac{1}{\sqrt{2m}}\left[\left(p_1 - \frac{e}{c}\,a_1\right) + \mathrm{i}\left(p_2 - \frac{e}{c}\,a_2\right)\right] \otimes \sigma^+, \qquad (8.20)$$

which differs from that in equation (8.13) by the sign in front of the imaginary unit. As a consequence, one arrives at a Pauli Hamiltonian:

$$\tilde{H}_\mathrm{P}^{(2)} := \frac{1}{2m}\sum_{i=1}^{2}\left(p_i - \frac{e}{c}\,a_i\right)^2 + \frac{e\hbar}{2mc}\,B\sigma_3, \qquad (8.21)$$

which differs from that in equation (8.12) by the sign in front of the Zeeman term. Note that $H_\mathrm{P}^{(2)}$ and $\tilde{H}_\mathrm{P}^{(2)}$ are not related by a charge conjugation ($e \to -e$). They are related by a reflection at the (x_1, x_2)-plane ($x_3 \to -x_3$). As it stands $\tilde{H}_\mathrm{P}^{(2)}$ characterises the same particle as $H_\mathrm{P}^{(2)}$ but with a g-factor $g = -2$, that is, $g \to -g$, which is approximately that of a positron but still having a negative charge $e < 0$. Let us also remark that the case of a charged spin-$\frac{1}{2}$ particle in two dimensions with $g = 2n$, $n \in \mathbb{N}$, and orthogonal magnetic field can be characterised by a nonlinear supersymmetry of order n [23].

Finally, it is worth mentioning that for constant magnetic field $B > 0$ the eigenvalues and eigenfunctions of the two-dimensional Pauli Hamiltonian are well-known, see [24]. Without the Zeeman (spin) term the Pauli Hamiltonian reduces to the so-called Landau Hamiltonian whose eigenvalue problem is basically that of a harmonic oscillator with cyclotron frequency $\omega_c := |eB|/mc$. These *Landau levels* are given by $\hbar\omega_c(n + 1/2)$ with $n \in \mathbb{N}_0$ and are d-fold degenerate. Here we need to limit our constant magnetic field to a large but finite area ℓ^2 to result in a finite flux $F = B\ell^2$ and a finite degeneracy. If we now add the constant Zeeman term, which actually reads $\frac{1}{2}\hbar\omega_c\sigma_3$ and thus results in a splitting of the energy levels being precisely the distance between two Landau levels, we obtain the eigenvalues of the Pauli Hamiltonian $H_\mathrm{P}^{(2)}$ for a constant magnetic field,

$$E_n = \hbar\omega_c n, \quad n \in \mathbb{N}_0, \qquad (8.22)$$

where $n = 0$ is only allowed for the spin-down states as $e < 0$.

8.2 Pauli paramagnetism of non-interacting electrons

Here we will investigate the implications of the SUSY structure in Pauli's Hamiltonian for the paramagnetic magnetisation of a system of non-interacting electrons in two and three dimensions. This magnetisation arises from the magnetic moment associated with the spin of the electrons. Moreover, we will confine ourselves to zero temperature, that is, we are looking at the ground state of non-interacting fermions obeying Pauli's exclusion principle. Without interaction the ground-state properties of the electron system are characterised by the single-electron Hamiltonian $H_P^{(2)}$ and $H_P^{(3)}$, respectively. For both values of the space dimension we assume that the magnetic field is perpendicular to the (x_1, x_2)-plane and does not depend on time for simplicity. Since we are not interested in diamagnetic effects due to the orbital motion, the magnetisation M of $\mathcal{N} = \mathcal{N}_+ + \mathcal{N}_-$ electrons can be written as

$$M := \mu_B(\mathcal{N}_+ - \mathcal{N}_-), \tag{8.23}$$

where $\mathcal{N}_+$ and $\mathcal{N}_-$ is the number of electrons with spin up and spin down, respectively, in the ground state of the electron system.

8.2.1 Two-dimensional electron gas

According to the Pauli principle, the ground state of $\mathcal{N}$ non-interacting electrons with single-electron Hamiltonian $H_P^{(2)}$ is characterised by the reduced single-electron density operator $\Theta(\varepsilon_F - H_P^{(2)})$, where the *Fermi energy* ε_F can be determined from the 'normalisation' condition

$$\mathrm{Tr}\left(\Theta\big(\varepsilon_F - H_P^{(2)}\big)\right) = \mathcal{N}. \tag{8.24}$$

The magnetisation is then given by

$$M = \mu_B\,\mathrm{Tr}\left(\sigma_3\Theta\big(\varepsilon_F - H_P^{(2)}\big)\right). \tag{8.25}$$

Let us note that because of $W = \sigma_3$ the right-hand side of equation (8.25) can be interpreted as a regularised Witten index. In other words, the magnetisation is given by the IDOS regularised index (2.64)

$$M = -\mu_B\tilde{\Delta}(\varepsilon_F), \tag{8.26}$$

where the SUSY Hamiltonian in equation (2.64) is given by equation (8.21).

For a purely discrete spectrum and a finite degeneracy of each eigenvalue the operator $H_P^{(2)}$ is Fredholm and the regularised index $\tilde{\Delta}(\varepsilon)$ becomes identical to Witten's index:

$$M = -\mu_B\Delta = \mu_B d\,\mathrm{sgn}(eF). \tag{8.27}$$

This is the expected result. Because of SUSY we know that all positive eigenvalues are spin degenerate. Hence, the contribution to the trace in equation (8.25) of the corresponding eigenstates cancel each other and only the degeneracy of the ground-state energy of $H_P^{(2)}$ contributes. In other words, the Pauli paramagnetism stems from these unpaired zero-energy states only. Because of the SUSY pairing of the excited states only a fraction of all of the electrons contributes to the magnetisation. We have illustrated these facts in figure 8.1.

As an example, let us consider the case of a constant magnetic field $B(x_1, x_2) = B > 0$. In order to have well-defined quantities we restrict the

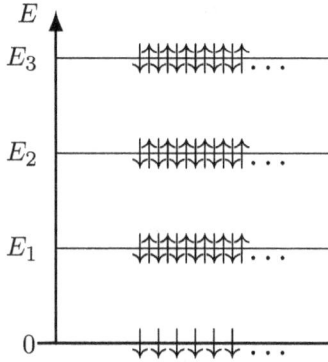

Figure 8.1. Typical ground state for the two-dimensional electron system. Only the zero-energy states of the single-electron Hamiltonian $H_{\mathrm{P}}^{(2)}$ are unpaired and contribute to the paramagnetic magnetisation

configuration space (sample) to a large but finite region with area ℓ^2 such that boundary effects, nevertheless, may be neglected. The flux (8.17) is then given by

$$F = B\ell^2. \tag{8.28}$$

Note that in this case the spectrum of $H_{\mathrm{P}}^{(2)}$ is discrete and consists of (approximately) the well-known equidistant Landau levels $E_n = n\frac{\hbar\,|eB|}{mc} = n\hbar\omega_c$. The magnetisation per area reads

$$\frac{M}{\ell^2} = \mu_{\mathrm{B}}\frac{|e|B}{2\pi\hbar c}, \tag{8.29}$$

which is independent of the electron density. Of course, we have assumed that there are sufficiently many electrons in the sample to fill up the zero-energy eigenstates. The paramagnetic susceptibility per unit area then reads

$$\chi := \frac{1}{\ell^2}\frac{\partial M}{\partial B} = \mu_{\mathrm{B}}\frac{|e|}{2\pi\hbar c} = \frac{e^2}{4\pi mc^2}, \tag{8.30}$$

which is the expected result [25, 26].

We note that in addition to the regularisation by a finite region ℓ^2 we have also neglected possible unpaired spins at the Fermi energy. These unpaired spins can only occur when the Fermi energy coincides with an eigenvalue of $H_{\mathrm{P}}^{(2)}$. That is, if we have not completely filled Landau levels.

Despite the fact that we have confined ourselves to the discussion of zero-temperature effects, let us briefly indicate the extension of the above results to finite inverse temperature $\beta < \infty$. As they are non-interacting fermions, the magnetisation of the electrons at finite temperature reads

$$M(\beta) = \mu_{\mathrm{B}}\,\mathrm{Tr}\left(\frac{\sigma_3}{1 + \exp\left\{\beta\left(H_{\mathrm{P}}^{(2)} - \mu\right)\right\}}\right), \tag{8.31}$$

where the chemical potential $\mu = \mu(\beta, \mathcal{N})$ is determined by the 'normalisation' condition

$$\mathrm{Tr}\left(\frac{1}{1 + \exp\{\beta(H_\mathrm{P}^{(2)} - \mu)\}}\right) = \mathcal{N}. \tag{8.32}$$

For a vanishing magnetic field the chemical potential can be expressed in terms of the Fermi energy $\varepsilon_\mathrm{F} = \mu(\infty, \mathcal{N}) \geqslant 0$ [27, 28]:

$$\exp\{\beta\mu\} = \exp\{\beta\varepsilon_\mathrm{F}\} - 1. \tag{8.33}$$

Assuming, as before, that $H_\mathrm{P}^{(2)}$ is the Fredholm contributions of the positive energy eigenstates to the magnetisation cancel each other due to SUSY. Hence, we arrive at

$$M(\beta) = \mu_\mathrm{B} d \, \mathrm{sgn}(eF)(1 + e^{-\beta\mu})^{-1}. \tag{8.34}$$

Assuming further that for an infinitesimal constant magnetic field $B > 0$ the chemical potential is approximately given by the zero-field expression (8.33) the zero-field susceptibility at finite temperature reads

$$\chi_0(\beta) := \lim_{B \to 0} \frac{1}{\ell^2} \frac{\partial M(\beta)}{\partial B} = \frac{e^2}{4\pi m c^2}(1 - e^{-\beta\varepsilon_\mathrm{F}}). \tag{8.35}$$

This result may be interpreted as follows. The fraction of zero-energy states which is not occupied by electrons due to thermal excitation (into available states at the Fermi energy ε_F) is given by $e^{-\beta\varepsilon_\mathrm{F}}$. Hence, for the magnetisation and the susceptibility these electrons do not contribute.

8.2.2 Three-dimensional electron gas

In this section we will utilise our previous results by considering only the case $\vec{B}(\vec{r}) = B(x_1, x_2)\vec{e}_3$ for the three-dimensional problem. Again we will neglect the interaction of the electrons and, hence, have to consider the Hamiltonian $H_\mathrm{P}^{(3)}$ as given in equation (8.11). Obviously, in each subspace of the Hilbert space with a fixed eigenvalue $\hbar k_3$ of p_3 we have the SUSY structure of the two-dimensional Pauli Hamiltonian. Hence, from each of these subspaces we have a contribution to the zero-temperature magnetisation as given in the above section. In order to find the total magnetisation we simply have to count the occupied eigenvalues of k_3. Assuming a finite range ℓ_3 for the x_3 coordinate the possible eigenvalues are given by

$$k_3 = \frac{2\pi}{\ell_3} n, \quad n \in \mathbb{Z}. \tag{8.36}$$

However, only those states are occupied with $|k_3|$ less than the *Fermi momentum*

$$k_\mathrm{F} := \frac{\sqrt{2m\varepsilon_\mathrm{F}}}{\hbar}. \tag{8.37}$$

For large ℓ_3 the number of these occupied states is approximately given by $\ell_3 k_F/\pi$. Multiplying the result (8.26) by this factor we arrive at the three-dimensional magnetisation at zero temperature:

$$M = -\mu_B \tilde{\Delta}(\varepsilon_F) \frac{k_F \ell_3}{\pi}. \tag{8.38}$$

Note that here ε_F is the Fermi energy of the three-dimensional system, that is, in equation (8.24) $H_P^{(2)}$ has to be replaced by $H_P^{(3)}$. The IDOS regularised index, however, is to be calculated with $H_P^{(2)}$.

For the particular case of a constant magnetic field the magnetisation and paramagnetic susceptibility per unit volume at zero temperature read

$$\frac{M}{\ell^2 \ell_3} = \mu_B \frac{eB}{2\pi^2 \hbar c} k_F = \mu_B^2 B \frac{m k_F}{\hbar^2 \pi^2}, \qquad \chi = \mu_B^2 \frac{m k_F}{\hbar^2 \pi^2} = \frac{e^2}{4\pi^2 mc^2} k_F. \tag{8.39}$$

Note that in contrast to the two-dimensional case, the three-dimensional zero-temperature paramagnetic susceptibility depends on the electron density and the magnetic field B via the Fermi momentum and has no physical dimension.

In the three-dimensional case there practically always exist two eigenvalues of p_3 for which the corresponding eigenstates of $H_P^{(3)}$ coincide with the Fermi energy. Hence, unpaired spins of such states near ε_F are certainly neglected. The above result (8.39) coincides in the limit of vanishing magnetic field with that given in textbooks [25, 29, 30].

8.2.3 The paramagnetic conjecture and SUSY

Another consequence of SUSY is that it may provide a counterexample to the paramagnetic conjecture by Hogreve, Schrader, and Seiler [31]. This conjecture states that the ground-state energy of the general Pauli Hamiltonian with an additional scalar potential $V := e\phi$,

$$H_P^{(3)}(\vec{A}, V) := \frac{1}{2m}\left(\vec{p} - \frac{e}{c}\vec{A}\right)^2 - \frac{e\hbar}{2mc}\vec{B}\cdot\vec{\sigma} + V, \tag{8.40}$$

is always less or equal to that with zero magnetic field:

$$\inf \operatorname{spec}\left(H_P^{(3)}(\vec{A}, V)\right) \leqslant \inf \operatorname{spec}\left(H_P^{(3)}(0, V)\right). \tag{8.41}$$

A proof of this inequality exist for arbitrary scalar potential V and a magnetic field of essentially the form $\vec{B} = B(x_1^2 + x_2^2)\vec{e}_3$ [32]. However, in the general case Avron and Simon [33] found one counterexample. Nevertheless, it is believed, see [21, p 131], that the above inequality (8.41) '... still holds for general $\vec{A}$ and selected sets of V...'. Here we note that for the particular case $V = 0$ the factorisability of $H_P^{(3)}(\vec{A}, 0) = 2Q_1^2 \geqslant 0$ implies the inequality

$$\inf \operatorname{spec}\left(H_P^{(3)}(\vec{A}, 0)\right) \geqslant \inf \operatorname{spec}\left(H_P^{(3)}(0, 0)\right) = 0. \tag{8.42}$$

This inequality is in the opposite direction of the paramagnetic conjecture (8.41). For arbitrary magnetic fields such that SUSY is a good symmetry the inequality (8.42) can be replaced by an equality. However, for any magnetic field which does break SUSY we have a strict inequality in equation (8.42) and thereby a counter-example to the conjecture (8.41). Or vice versa, for any magnetic field, for which the paramagnetic conjecture with $V = 0$ can be proven, SUSY will be a good symmetry and hence the equality holds in equation (8.42). To conclude let us mention that the magnetic field generated by a straight current has been shown to exhibit an $N = 3$ SUSY structure with broken SUSY [34] and hence being another counterexample to the paramagnetic conjecture.

8.3 Electrons in a spherically symmetric scalar field

In this section we will consider the case of a vanishing magnetic field but non-vanishing rotationally invariant scalar field. That is, we assume a vanishing vector potential and a spherically symmetric scalar potential $V(r) := e\phi(r)$ with $r = |\vec{r}|$. In other words the corresponding Pauli Hamiltonian is of the form

$$H_P = \frac{\vec{p}^2}{2m} + V(r) \tag{8.43}$$

acting on $\mathcal{H} = L^2(\mathbb{R}^3) \otimes \mathbb{C}^2$. Throughout this section we will set $\hbar = 1$ to keep formulas simple.

It is obvious that the above Hamiltonian is invariant under rotations. In other words the orbital angular momentum $\vec{L} := \vec{r} \times \vec{p}$ acting on $L^2(\mathbb{R}^3)$, the spin $\vec{S} := \frac{1}{2}\vec{\sigma}$ acting on $\mathbb{C}^2$, and the total angular momentum $\vec{J} := \vec{L} + \vec{S}$ acting on $\mathcal{H}$ are conserved observables. Let us recall that the eigenvalues of $\vec{L}^2$ are given by $\ell(\ell + 1)$, $\ell = 0, 1, 2, 3, \ldots$, that of $\vec{S}^2$ is $\frac{1}{2}(1 + \frac{1}{2}) = \frac{3}{4}$ and those of $\vec{J}^2$ are given by $j(j + 1)$ with $j = \frac{1}{2}, \frac{3}{2}, \frac{5}{2}, \ldots$.

Now we introduce the non-relativistic spin–orbit operator

$$K_{nr} := 2\vec{S} \cdot \vec{L} + 1 = \vec{\sigma} \cdot \vec{L} + 1 \tag{8.44}$$

and leave it as an exercise to the reader to verify the following relations:

$$K_{nr} = \vec{J}^2 - \vec{L}^2 + \frac{1}{4}, \quad \vec{L}^2 = K_{nr}(K_{nr} - 1), \quad K_{nr}^2 = \vec{J}^2 + \frac{1}{4}. \tag{8.45}$$

Let us denote the eigenvalue of K_{nr} by $-\kappa$, to be consistent with the notation used in Dirac's theory [35, 36], then the last of above relations implies that $\kappa^2 = j(j + 1) + \frac{1}{4} = (j + \frac{1}{2})^2$. In other words κ takes the values

$$\kappa = \pm\left(j + \frac{1}{2}\right) = \pm 1, \pm 2 \pm 3, \ldots \in \mathbb{Z}/\{0\}. \tag{8.46}$$

The second relation in equation (8.45) also implies

$$\kappa = \ell = j + \frac{1}{2} > 0, \quad \kappa = -\ell - 1 = -j - \frac{1}{2} < 0. \tag{8.47}$$

The joint eigenfunctions of the operators $\vec{J}^2$, J_z, and K_{nr} form a complete set in the space $L^2(S^2) \otimes \mathbb{C}^2$, where S^2 denotes the unit sphere embedded in $\mathbb{R}^3$. Let us denote these by $|j, m_j, s\rangle$, where we have defined the reduced quantum number $s := -\mathrm{sgn}(\kappa)$, then we arrive at the decomposition

$$L^2(S^2) \otimes \mathbb{C}^2 = \bigoplus_{j=\frac{1}{2},\frac{3}{2},\dots}^{\infty} \bigoplus_{m_j=-j}^{j} \bigoplus_{s=\pm 1} |j, m_j, s\rangle\langle j, m_j, s|, \tag{8.48}$$

with

$$\vec{J}^2 |j, m_j, s\rangle = j(j+1)|j, m_j, s\rangle, \quad j = \frac{1}{2}, \frac{3}{2}, \frac{5}{2}, \dots,$$

$$J_z |j, m_j, s\rangle = m_j |j, m_j, s\rangle, \quad m_j = -j, -j+1, \dots, j-1, j, \tag{8.49}$$

$$K_{nr} |j, m_j, s\rangle = s(j+\frac{1}{2})|j, m_j, s\rangle, \quad s = -1, +1.$$

These states are the well-known Pauli spinors and in coordinate representation they can be expressed in terms of the usual spherical harmonics [35, 36]

$$\varphi_{jm_j}^{(s)}(\theta, \varphi) := \langle \theta, \varphi | j, m_j, s \rangle = \begin{pmatrix} \sqrt{\dfrac{\ell + sm_j + \frac{1}{2}}{2\ell+1}}\, Y_\ell^{m_j-\frac{1}{2}}(\theta, \varphi) \\[2em] s\sqrt{\dfrac{\ell - sm_j + \frac{1}{2}}{2\ell+1}}\, Y_\ell^{m_j+\frac{1}{2}}(\theta, \varphi) \end{pmatrix}. \tag{8.50}$$

Note that in our notation we have the relation $j = \ell + s/2$ which follows from equation (8.47). Hence $s = +1$ can be associated with the situation where spin $\vec{S}$ and orbital angular momentum $\vec{L}$ are parallel and for $s = -1$ they are antiparallel. It was already observed by Bjorken and Drell [35] that the Pauli spinors obey the relation

$$(\vec{\sigma} \cdot \vec{e}_r)\varphi_{jm_j}^{(s)}(\theta, \varphi) = \varphi_{jm_j}^{(-s)}(\theta, \varphi), \tag{8.51}$$

which reflects the fact that K_{nr} and $(\vec{\sigma} \cdot \vec{e}_r)$ anticommute. More generally it is straightforward to show that any vector operator[2] $\vec{v}$ being orthogonal to the orbital angular momentum, i.e. $\vec{L} \cdot \vec{v} = 0 = \vec{v} \cdot \vec{L}$, obeys the relation

$$\{K_{nr}, (\vec{\sigma} \cdot \vec{v})\} = 0. \tag{8.52}$$

---

[2] The components of a vector operator $\vec{v} = (v_1, v_2, v_3)^T$ obey the algebra $[L_i, v_j] = i\varepsilon_{ijk}v_k$, where we use Einstein's summation convention, that is, a sum is implicitly taken for repeated indices.

Typical examples of such a vector $\vec{v}$ would be the position, velocity, and momentum operators or the Laplace–Runge–Lenz–Pauli vector, which plays an important role in the Coulomb problem.

Finally let us conclude that for spherically symmetric Pauli Hamiltonians there exists a natural Witten parity $W := K_{nr}/|K_{nr}|$ which commutes with the Hamiltonian. Restricting that to a subspace with fixed j and m_j, where also $|K_{nr}|$ is fixed, the subspace with Witten parity $W = \mp 1$ belongs to $s = \pm 1$. Furthermore, in may be possible to find a proper vector operator $\vec{v}$ defining a supercharge $Q = (\vec{\sigma} \cdot \vec{v})$ such that $[H, Q] = 0 = [H, Q^\dagger]$ and $H = \{Q, Q^\dagger\}$, which then naturally anti-commutes with W, $\{W, Q\} = 0$, thus forming a $N = 2$ SUSY system. This will be our programme for the next section.

8.3.1 Partial wave subspaces and their grading

The decomposition (8.48) implies that the Hilbert space $\mathcal{H} = L^2(\mathbb{R}^3) \otimes \mathbb{C}^2$ may be decomposed into partial wave subspaces with fixed angular momentum quantum numbers j and m_j,

$$\mathcal{H} = \bigoplus_{j=\frac{1}{2},\frac{3}{2},\dots}^{\infty} \bigoplus_{m_j=-j}^{j} \mathcal{H}_{j,m_j}. \tag{8.53}$$

Indeed, this leads to a decomposition of the wave functions into partial waves of the form

$$\Psi(\vec{r}) = \sum_{j,m_j,s} R_j^s(r)\varphi_{jm_j}^{(s)}(\theta, \varphi), \tag{8.54}$$

with the radial wave function R_j^s being an element of $\mathcal{H}_{j,m_j} = L^2(\mathbb{R}^+, r^2 dr) \otimes \mathbb{C}^2$. The notation $L^2(\mathbb{R}^+, r^2 dr)$ stands for the Hilbert space of square-integrable functions defined on the positive half-line equipped with the scalar product $\langle \psi_1 | \psi_2 \rangle = \int_0^\infty dr\, r^2 \psi_1^*(r)\psi_2(r)$.

It is obvious that the operator $(\vec{\sigma} \cdot \vec{e}_r)$ acts within such a subspace by transforming states with $s = +1$ into those with $s = -1$ and vice versa. Thus this implies a natural grading of $\mathcal{H}_{j,m_j}$ into subspaces with a positive and negative s,

$$\mathcal{H}_{j,m_j} = \mathcal{H}_{j,m_j}^+ \oplus \mathcal{H}_{j,m_j}^-, \quad \mathcal{H}_{j,m_j}^\pm = L^2(\mathbb{R}^+, r^2 dr). \tag{8.55}$$

Note also that within this partial wave subspace the total angular momentum j is fixed. Hence, as $j = \ell(s) + s/2$ the orbital angular momentum quantum number depends on s and it differs by one between these two s-subspaces, that is, we have the relation $\ell(+1) = \ell(-1) - 1$. Let us fix $\ell(-1) \equiv \ell = 1, 2, 3, \dots$ and $\ell(+1) = \ell - 1$. As a consequence, within $\mathcal{H}_{j,m_j}^+$ the angular momentum quantum number is fixed to $(\ell - 1)$ whereas in $\mathcal{H}_{j,m_j}^-$ it takes the value ℓ. Hence, a general element of $\mathcal{H}_{j,m_j}$ may be put into the form

$$\psi(r) = R_{\ell-1}(r)\chi^+ + R_\ell(r)\chi^-, \quad \chi^+ := \begin{pmatrix} 1 \\ 0 \end{pmatrix}, \quad \chi^- := \begin{pmatrix} 0 \\ 1 \end{pmatrix} \tag{8.56}$$

and $K_{nr}\chi^\pm = \pm\ell\chi^\pm$ following from the above definition. The Pauli Hamiltonian (8.43) reads in polar coordinates as follows:

$$H_P = \frac{1}{2m}\left[p_r^2 + \frac{\vec{L}^2}{r^2}\right] + V(r) = \frac{1}{2m}p_r^2 + \frac{K_{nr}(K_{nr}-1)}{2mr^2} + V(r). \tag{8.57}$$

Here $p_r := -i(\partial_r + 1/r) = (-i/r)\partial_r r$ is the radial momentum operator being self-adjoint on $L^2(\mathbb{R}^+, r^2 dr)$. Restricting this Pauli operator onto the partial wave subspace $\mathcal{H}_{j,m_j}$ and using the notation (8.56) we arrive at the matrix representation

$$H_P\Big|\mathcal{H}_{jm_j} = \begin{pmatrix} H^+ & 0 \\ 0 & H^- \end{pmatrix} \quad \text{with} \quad H^\pm := \frac{1}{2m}p_r^2 + V_{\text{eff}}^\pm(r) \tag{8.58}$$

and effective potential

$$V_{\text{eff}}^\pm(r) := \frac{\ell(\ell \mp 1)}{2mr^2} + V(r). \tag{8.59}$$

The above result indicates that the operator

$$\vec{\sigma} \cdot \vec{e}_r = \begin{pmatrix} 0 & 1 \\ 1 & 0 \end{pmatrix} \quad \text{indeed acts as} \quad (\vec{\sigma} \cdot \vec{e}_r)\chi^\pm = \chi^\mp \tag{8.60}$$

and thus transforms between subspaces where the orbital angular momentum quantum number differs by one. Here we recall the examples of Witten models on the half-line discussed in section 3.6.2. There we presented two examples, the radial harmonic oscillator and the radial Coulomb problem, as a pair of supersymmetric systems relating the effective radial quantum systems of one with angular momentum ℓ to one with angular momentum $\ell - 1$. However, only for the Coulomb problem (3.56) do we obtain the structure (8.59) by identifying $V(r) = -\frac{\alpha}{r} + \frac{m\alpha^2}{2\ell^2}$. This includes also the special case of the free particle $\alpha = 0$. The radial harmonic oscillator does not directly allow such an identification as the example (3.54) would require a different additive constant in the two s-subspaces $V(r) = \frac{m}{2}\omega^2 r^2 - \omega(\ell \pm \frac{1}{2})$. Note that we do not have a complete SUSY structure as we have not yet found an appropriate supercharge. In fact the special examples of section 3.6.2 mentioned above indicate that this may only be possible for these two particular potentials V.

8.3.2 The free particle $V(r) = 0$

In order to find a proper supercharge transforming between the subspaces $\mathcal{H}_{jm_j}^\pm$ we note that it should be proportional to $(\vec{\sigma} \cdot \vec{e}_r)$ and thus anticommute with K_{nr}. The simplest and only candidate at hand is

$$\begin{aligned} Q_1 &:= \frac{1}{\sqrt{4m}}(\vec{\sigma} \cdot \vec{p}) = \frac{-i}{\sqrt{4m}}(\vec{\sigma} \cdot \vec{e}_r)\left[\partial_r - \frac{K_{nr}-1}{r}\right] \\ &= \frac{1}{\sqrt{4m}}(\vec{\sigma} \cdot \vec{e}_r)\left[p_r + i\frac{K_{nr}}{r}\right], \end{aligned} \tag{8.61}$$

which obeys the relation $H_P = 2Q_1^2 = \vec{p}^2/2m$ as required. Note that Q_1 in essence represents the helicity operator. In matrix notation restricted to the partial wave subspace $\mathcal{H}_{j,m_j}$ this supercharge explicitly reads

$$Q_1 = -\frac{i}{\sqrt{4m}}\begin{pmatrix} 0 & \partial_r + \dfrac{\ell+1}{r} \\ \partial_r - \dfrac{\ell-1}{r} & 0 \end{pmatrix} = \frac{1}{\sqrt{4m}}\begin{pmatrix} 0 & p_r - i\dfrac{\ell}{r} \\ p_r + i\dfrac{\ell}{r} & 0 \end{pmatrix}. \quad (8.62)$$

The corresponding complex supercharges are then given by

$$Q := \begin{pmatrix} 0 & A \\ 0 & 0 \end{pmatrix}, \quad A := \frac{1}{\sqrt{2m}}\left(p_r - i\frac{\ell}{r}\right)$$
$$Q^\dagger = \begin{pmatrix} 0 & 0 \\ A^\dagger & 0 \end{pmatrix}, \quad A^\dagger = \frac{1}{\sqrt{2m}}\left(p_r + i\frac{\ell}{r}\right) \quad (8.63)$$

and the matrix representation of the SUSY Hamiltonian in this subspace is

$$H = 2Q_1^2 = \{Q, Q^\dagger\} = \begin{pmatrix} AA^\dagger & 0 \\ 0 & A^\dagger A \end{pmatrix} = \frac{1}{2m}\begin{pmatrix} p_r^2 + \dfrac{\ell(\ell-1)}{r^2} & 0 \\ 0 & p_r^2 + \dfrac{\ell(\ell+1)}{r^2} \end{pmatrix}. (8.64)$$

The eigenfunctions of the free particle in three-dimensional space are given by plane waves, i.e. $\Psi(\vec{r}) = (2\pi)^{-3/2}e^{i\vec{k}\cdot\vec{r}}$ with $\vec{k} \in \mathbb{R}^3$ being the wave vector. The associated energy eigenvalue is given by $E = \vec{k}^2/2m$. The radial component of the plane wave associate with a fixed orbital angular momentum ℓ is given by the spherical Bessel function of the first kind, $R_\ell(r) = j_\ell(kr)$, where $k = |\vec{k}|$. Hence, the SUSY transformations induced by Q_1 explicitly read

$$Q_1 j_{\ell-1}(kr)\chi^+ = -\frac{i}{\sqrt{4m}}\chi^-\left(\partial_r - \frac{\ell-1}{r}\right)j_{\ell-1}(kr) = i\sqrt{E/2}\,j_\ell(kr)\chi^-,$$
$$Q_1 j_\ell(kr)\chi^- = -\frac{i}{\sqrt{4m}}\chi^+\left(\partial_r + \frac{\ell+1}{r}\right)j_\ell(kr) = -i\sqrt{E/2}j_{\ell-1}(kr)\chi^+. \quad (8.65)$$

As expected we have Q_1: $\mathcal{H}_{j,m_j}^\pm \to \mathcal{H}_{j,m_j}^\mp$. Note that the SUSY transformations reflect the know recurrence relations of the spherical Bessel functions

$$\left(\partial_z + \frac{\ell+1}{z}\right)j_\ell(z) = j_{\ell-1}(z), \quad \left(-\partial_z + \frac{\ell}{z}\right)j_\ell(z) = j_{\ell+1}(z). \quad (8.66)$$

Here we explicitly have a non-vanishing phase factor $e^{\pm i\pi/2}$ in the SUSY transformation relations.

8.3.3 The hydrogen atom $V(r) = -\alpha/r$

The Pauli Hamiltonian for the hydrogen atom is given by

$$H_{\mathrm{P}} = \frac{\vec{p}^2}{2m} - \frac{\alpha}{r} \tag{8.67}$$

with α denoting Sommerfeld's fine structure constant. This system is known to exhibit an accidental degeneracy which can be attributed to the existence of an additional constant of motion, the Laplace–Runge–Lenz–Pauli vector [37–40]. The dimensionless version of this vector is defined by

$$\vec{R} := \frac{1}{2\alpha m}\left(\vec{p} \times \vec{L} - \vec{L} \times \vec{p}\right) - \vec{e}_r \tag{8.68}$$

and obeys the following relations:

$$[H_{\mathrm{P}}, \vec{R}] = 0, \quad \vec{R} \cdot \vec{L} = 0 = \vec{L} \cdot \vec{R}, \quad |\vec{R}|^2 = \frac{2H_{\mathrm{P}}}{m\alpha^2}\left(\vec{L}^2 + 1\right) + 1$$

$$\left[L_i, R_j\right] = \mathrm{i}\varepsilon_{ijk}R_k, \quad \left[R_i, R_j\right] = -\frac{2\mathrm{i}H_{\mathrm{P}}}{m\alpha^2}\varepsilon_{ijk}L_k, \tag{8.69}$$

$$\left(\vec{\sigma} \cdot \vec{R}\right) = \left(\vec{\sigma} \cdot \vec{e}_r\right)\left[\frac{-K_{nr}}{m\alpha}\left(\partial_r - \frac{K_{nr}-1}{r}\right) - 1\right], \quad \left(\vec{\sigma} \cdot \vec{R}\right)^2 = 1 + \frac{2H_{\mathrm{P}}K_{nr}^2}{m\alpha^2}.$$

For proofs see, for example, the book by Hirshfeld [41]. Note, however, that Hirshfeld's definition of the non-relativistic spin–orbit operator differs by an overall minus sign from the usual one. The last relation above suggests defining the supercharge as follows:

$$Q_1 := \sqrt{\frac{m\alpha^2}{4K_{nr}^2}}\left(\vec{\sigma} \cdot \vec{R}\right). \tag{8.70}$$

This immediately leads to the SUSY Hamiltonian

$$H = 2Q_1^2 = H_{\mathrm{P}} + \frac{m\alpha^2}{2K_{nr}^2} = \frac{p_r^2}{2m} + \frac{K_{nr}(K_{nr}-1)}{2mr^2} - \frac{\alpha}{r} + \frac{m\alpha^2}{2K_{nr}^2}. \tag{8.71}$$

That is, within the partial wave subspace $\mathcal{H}_{j,\,m_j}$, where the operator K_{nr}^2 is constant and takes the value ℓ^2, we have

$$H = \begin{pmatrix} H^+ & 0 \\ 0 & H^- \end{pmatrix}, \tag{8.72}$$

with

$$H^{\pm} = \frac{p_r^2}{2m} + \frac{\ell(\ell \mp 1)}{2mr^2} - \frac{\alpha}{r} + \frac{m\alpha^2}{2\ell^2}. \tag{8.73}$$

The eigenfunctions and eigenvalues of the Coulomb Hamiltonian (8.67) are well-know and read

$$R_{n_r,\ell}(r) = \frac{2(ma)^{3/2}}{(n_r + \ell + 1)^2}\left[\frac{n_r!}{(n_r + 2\ell + 1)!}\right]^{1/2}\rho^\ell e^{-\rho/2}L_{n_r}^{2\ell+1}(\rho),$$

$$\tilde{E}_{n_r,\ell} = -\frac{ma^2}{2(n_r + \ell + 1)^2},$$

(8.74)

with $\rho := 2mar/(n_r + \ell + 1)$, radial quantum number $n_r = 0, 1, 2, \ldots$, and $L_{n_r}^k$ stands for the generalised Laguerre polynomial of order n_r. Hence the eigenfunctions and eigenvalues of $H^\pm$, that is, $H^\pm\psi_{n_r,\ell}^\pm(r) = E_{n_r,\ell}\psi_{n_r,\ell}^\pm(r)$, are given by

$$\psi_{n_r,\ell}^+(r) = R_{n_r,\ell-1}(r), \quad \psi_{n_r,\ell}^-(r) = R_{n_r-1,\ell}(r),$$

$$E_{n_r,\ell} = \frac{ma^2}{2}\left[\frac{1}{\ell^2} - \frac{1}{(\ell + n_r)^2}\right].$$

(8.75)

The corresponding SUSY transformations which are given by the usual relation $Q_1\psi_{n_r,\ell}^\pm(r) = \sqrt{E_{n_r,\ell}/2}\,\psi_{n_r,\ell}^\mp(r)$ lead to the explicit transformations

$$\left[-\frac{\ell}{ma}\left(\partial_r - \frac{\ell - 1}{r}\right) - 1\right]R_{n_r,\ell-1}(r) = \sqrt{1 - \frac{\ell^2}{(n_r + \ell)^2}}\,R_{n_r-1,\ell}(r),$$

$$\left[\frac{\ell}{ma}\left(\partial_r + \frac{\ell + 1}{r}\right) - 1\right]R_{n_r-1,\ell}(r) = \sqrt{1 - \frac{\ell^2}{(n_r + \ell)^2}}\,R_{n_r,\ell-1}(r).$$

(8.76)

For such recurrence relations as well as more general relations see also [42]. Note that for the second relation $n_r > 0$ whereas the first relation for $n_r = 0$ explicitly shows that the SUSY ground state $\psi_{0,\ell}^+$ is annihilated by the supercharge,

$$\left(-\partial_r + \frac{\ell - 1}{r}\right)R_{0,\ell-1}(r) = \frac{ma}{\ell}R_{0,\ell-1}(r).$$

(8.77)

This is easily integrated and results in the solution $R_{0,\ell-1}(r) = Cr^{\ell-1}\exp\{-mar/\ell\}$. With the help of the SUSY transformations (8.76) one can then reconstruct the radial wave functions (8.74). For more details we refer to the book by Hirshfeld [41], where also the eccentricity operator

$$E := \sqrt{(\vec{\sigma} \cdot \vec{R})^2} = \sqrt{1 + \frac{2H_P K_{nr}^2}{ma^2}}$$

(8.78)

is explicitly discussed. Note that the coefficients on the right-hand side of the SUSY transformations (8.76) are the eigenvalues of this operator as $E = \sqrt{\frac{4K_{nr}^2}{ma^2}}|Q_1|$ with corresponding eigenstates $\psi_{n_r,\ell}^\pm(r)$. Finally let us note that the SUSY transformations also apply for the continuous eigenstates of the Coulomb problem, that is, for the so-called Coulomb wave functions. We leave it as an exercise for the reader to verify that the transformation (8.76) applied to the continuous states resemble the recurrence relations of the Coulomb wave functions [43] which is closely related to the factorisation method [44] and hence to SUSY.

8.3.4 Spherical tensor potentials and the Pauli oscillator

In this subsection we will now consider a vanishing scalar potential but introduce a so-called spherically symmetric tensor potential $U(r)$ via the minimal coupling principle $\vec{p} \rightarrow \vec{p} - (e/c)\vec{A}(\vec{r})$ with $(e/c)\vec{A}(\vec{r}) = -i\nabla U(r) = -iU'(r)\vec{e}_r$. This is in analogy to the Dirac oscillator introduced by Moshinsky and Szczepaniak [45] in 1989. Hence we consider a supercharge of the form

$$Q := \begin{pmatrix} 0 & A \\ 0 & 0 \end{pmatrix}, \quad A := \frac{1}{\sqrt{2m}}\vec{\sigma} \cdot (\vec{p} + iU'(r)\vec{e}_r) = \frac{(\vec{\sigma} \cdot \vec{e}_r)}{\sqrt{2m}}\left[p_r + i\frac{K_{nr}}{r} - U'(r)\right]. \quad (8.79)$$

Obviously, $Q^2 = 0 = Q^{\dagger 2}$ and the SUSY algebra is closed with a SUSY Hamiltonian defined by

$$H := \{Q, Q^\dagger\} = \begin{pmatrix} AA^\dagger & 0 \\ 0 & A^\dagger A \end{pmatrix} = \begin{pmatrix} H_+ & 0 \\ 0 & H_- \end{pmatrix}. \quad (8.80)$$

After some calculations one actually arrives at [46]

$$H_\pm = \frac{\vec{p}^2}{2m} + \frac{1}{2m}U'^2(r) \mp \frac{1}{2m}U''(r) \mp \frac{K_{nr}}{mr}U'(r). \quad (8.81)$$

Restricting these Hamiltonians onto the associated subspaces $\mathcal{H}_{j,m_j}^\pm$ the effective partner potentials read

$$V_{\text{eff}}^\pm(r) := \frac{\ell(\ell \mp 1)}{2mr^2} + \frac{1}{2m}U'^2(r) \mp \frac{1}{2m}U''(r) - \frac{\ell}{mr}U'(r). \quad (8.82)$$

The matrix representation of the corresponding real supercharge $Q_1 = \frac{1}{\sqrt{2}}(Q + Q^\dagger)$ then reads

$$Q_1 = -\frac{i}{\sqrt{4m}}\begin{pmatrix} 0 & \partial_r + \frac{\ell + 1}{r} - U'(r) \\ \partial_r - \frac{\ell - 1}{r} + U'(r) & 0 \end{pmatrix} \quad (8.83)$$

and generalises that of the free particle (8.62). SUSY is unbroken if the ground state is annihilated by this supercharge. This leads to the conditions

$$\left(\partial_r - \frac{\ell - 1}{r} + U'(r)\right)R_{\ell-1}^{0+}(r) \quad \Rightarrow \quad R_{\ell-1}^{0+}(r) = Cr^{\ell-1}\exp\{-U(r)\},$$

$$\left(\partial_r + \frac{\ell + 1}{r} - U'(r)\right)R_{\ell}^{0-}(r) \quad \Rightarrow \quad R_{\ell}^{0-}(r) = Cr^{-\ell-1}\exp\{U(r)\}. \tag{8.84}$$

Obviously, for a well-behaved superpotential U bounded from below and diverging fast enough for $r \to \infty$, SUSY will be unbroken and the ground state belongs to $\mathcal{H}_{j,m_j}^{+}$ and is given by the first relation above. The general SUSY transformations are then expected to read

$$\frac{-i}{\sqrt{4m}}\left(\partial_r - \frac{\ell - 1}{r} + U'(r)\right)R_{\ell-1}^{+}(r) = \sqrt{E/2}\,R_{\ell}^{-}(r)$$

$$\frac{-i}{\sqrt{4m}}\left(\partial_r + \frac{\ell + 1}{r} - U'(r)\right)R_{\ell}^{-}(r) = \sqrt{E/2}\,R_{\ell-1}^{+}(r), \tag{8.85}$$

with E being the unknown energy eigenvalue associated with the subspace $\mathcal{H}_{j,m_j}$. In general the eigenvalue problem for $H_\pm$ cannot be solved analytically. However, there are several particular choices of the tensor potential U for which this is actually the case.

As a first example, let us consider a harmonic tensor potential $U(r) = \frac{1}{2}m\omega r^2$. The quantum system characterised by this potential will be called the Pauli oscillator in analogy to the relativistic version where this represents the Dirac oscillator. For this particular choice the effective partner potentials are explicitly given by

$$V_{\text{eff}}^{\pm}(r) = \frac{\ell(\ell \mp 1)}{2mr^2} + \frac{m}{2}\,\omega^2 r^2 - \omega\left(\ell \pm \frac{1}{2}\right). \tag{8.86}$$

Hence we have recovered the SUSY example of the radial harmonic oscillator, see example 3.5 of section 3.6.2, where $V_\pm(r) = V_{\text{eff}}^{\mp}(r)$. Here we only obtain the unbroken SUSY case as $\eta = \ell = 1, 2, 3, \ldots > 0$. This example also shows that the tensor potential actually plays the role of a superpotential. We leave it as an exercise to the reader to verify that the SUSY transformations (8.85) lead to the known recurrence relation [42] of the radial harmonic oscillator wave functions where $E = \omega n_r$, $n_r = 0, 1, 2, \ldots$.

A second example which leads to an exactly solvable SUSY Hamiltonian is a linear superpotential $U(r) = \gamma r$, $\gamma > 0$. Here the effective partner potentials read

$$V_{\text{eff}}^{\pm}(r) = \frac{\ell(\ell \mp 1)}{2mr^2} + \frac{\gamma^2}{2m} - \frac{\ell\gamma}{mr}, \tag{8.87}$$

which basically is that of the hydrogen atom with an angular momentum dependent coupling constant.

As a final example let us mention a logarithmic superpotential $U(r) = \gamma \ln r$, $\gamma > 0$. Here the effective partner potentials turn out to be in essence corrections to the centrifugal term of the free particle

$$V_{\text{eff}}^{\pm}(r) = \frac{1}{2mr^2}(\ell - \gamma)(\ell - \gamma \mp 1).$$ (8.88)

8.4 Problems

Problem 8.1. On the $N = 2$ SUSY of the two-dimensional Pauli operator

Consider the generalised annihilation operator $A := \frac{1}{\sqrt{2m}}[(p_1 - \frac{e}{c} a_1) \mp i(p_2 - \frac{e}{c} a_2)]$, where a_i and B are defined in equation (8.10). Show that

$$AA^{\dagger} = \frac{1}{2m}\left(\vec{p} - \frac{e}{c}\vec{A}\right)^2 \mp \frac{e\hbar}{2mc} B(x_1, x_2), \qquad A^{\dagger}A = \frac{1}{2m}\left(\vec{p} - \frac{e}{c}\vec{A}\right)^2 \pm \frac{e\hbar}{2mc} B(x_1, x_2).$$

Problem 8.2. The Aharonov–Casher theorem [22]

Consider a magnetic field being perpendicular to the x_1–x_2-plane, that is, $\vec{B}(\vec{r}) = B(\vec{r})\vec{e}_3$, $\vec{r} = (x_1, x_2)^T$, and the associated magnetic scalar field defined by

$$S(\vec{r}) := \int_{\mathbb{R}^2} d^2\vec{r}' \; B(\vec{r}') \ln|\vec{r} - \vec{r}'|.$$

(a) Show with the help of the two-dimensional Green's function relation

$$\Delta \ln|\vec{r} - \vec{r}'| = (\partial_{x_1}^2 + \partial_{x_2}^2)\ln|\vec{r} - \vec{r}'| = 2\pi\delta(\vec{r} - \vec{r}')$$

that $\Delta S(\vec{r}) = 2\pi B(\vec{r})$ and the vector potential defined by

$$\vec{A}(\vec{r}) = \begin{pmatrix} a_1(\vec{r}) \\ a_2(\vec{r}) \end{pmatrix} := \frac{1}{2\pi}\begin{pmatrix} -\partial_{x_2}S(\vec{r}) \\ \partial_{x_1}S(\vec{r}) \end{pmatrix}$$

represents the magnetic field via $B(\vec{r}) = \partial_{x_1}a_2(\vec{r}) - \partial_{x_2}a_1(\vec{r}) = \Delta S(\vec{r})/2\pi$.

(b) Show that the generalised annihilation operator defined in problem 7.1 can be put into the form (here consider only the upper sign and $\Phi_0 := 2\pi\hbar c/e$)

$$A = \frac{\hbar}{\sqrt{2m}}e^{S(\vec{r})/\Phi_0}[-\partial_{x_2} - i\partial_{x_1}]e^{-S(\vec{r})/\Phi_0}.$$

(c) Assume now that SUSY is unbroken, i.e. the ground state is either annihilated by A and/or by $A^{\dagger}$. Hence, $A\psi_0^{-}(\vec{r}) = 0$ and/or $A^{\dagger}\psi_0^{+}(\vec{r}) = 0$. Show that this results in the conditions

$$(\partial_{x_1} - i\partial_{x_2})e^{-S/\Phi_0}\psi_0^{-}(\vec{r}) = 0 \quad \text{and/or} \quad (\partial_{x_1} + i\partial_{x_2})e^{+S/\Phi_0}\psi_0^{+}(\vec{r}) = 0.$$

In other words $f_{+}(z) := e^{+S/\Phi_0}\psi_0^{+}$ is an analytic function in $z := x_1 + ix_2 \in \mathbb{C}$ and $f_{-}(z^*) := e^{-S/\Phi_0}\psi_0^{-}$ is analytic in z^*.

(d) For a finite magnetic field with compact support the flux $F := \int_{\mathbb{R}^2} d^2\vec{r}\; B(\vec{r})$ is finite. Show that the corresponding magnetic scalar field then behaves like $S(\vec{r}) = F \ln r$ for large $r = |\vec{r}|$.

(e) Assume $eF > 0$ and conclude from the fact that $\psi_0^-(\vec{r}) = e^{S/\Phi_0} f_-(z^*) \in L^2(\mathbb{R}^2)$ and $\psi_0^-(\vec{r}) = |z|^{F/\Phi_0} f_-(z^*)[1 + O(1/|z|)]$ that the analytic function $f_-(z^*)$ has to vanish for large z and hence $f_-(z^*) \equiv 0$. In summary, for $eF > 0$ there are no spin-down states for the two-dimensional Pauli Hamiltonian with vanishing eigenvalue.

(f) Again assume $eF > 0$ and conclude from $\psi_0^+(\vec{r}) = e^{-S/\Phi_0} f_+(z) \in L^2(\mathbb{R}^2)$ that $\psi_0^+(\vec{r}) = |z|^{-F/\Phi_0} f_+(z)[1 + O(1/|z|)]$. Hence the analytic function $f_+(z)$ has to be polynomially bound and therefore $f_+(z) = z^k$ with $k = 0, 1, 2, \ldots < (F/\Phi_0 - 1)$. In summary, there exist exact $d := [F/\Phi_0]$ linearly independent spin-up states for the two-dimensional Pauli Hamiltonian with vanishing eigenvalue. These states are given by

$$\psi_{0,k}^+(\vec{r}) = C \exp\{-S(\vec{r})/\Phi_0\}(x_1 + ix_2)^k, \quad k = 0, 1, 2, d - 1.$$

Problem 8.3. Some relations of Pauli matrices

The Pauli matrices σ_i, $i = 1, 2, 3$, obey the relation

$$\sigma_i\sigma_j = \delta_{ij} + i\varepsilon_{ijk}\sigma_k,$$

where ε_{ijk} is the Levi-Civita symbol, also called the epsilon tensor. Here summation over the repeated index k is implied.

(a) Prove for arbitrary vector-valued operators $\vec{A}$ and $\vec{B}$ the following formula:

$$\left(\vec{\sigma} \cdot \vec{A}\right)\left(\vec{\sigma} \cdot \vec{B}\right) = \vec{A} \cdot \vec{B} + i\vec{\sigma} \cdot \left(\vec{A} \times \vec{B}\right).$$

(b) Prove $(\vec{\sigma} \cdot \vec{e}_r)^2 = 1$ and $(\vec{\sigma} \cdot \vec{p})^2 = \vec{p}^2$.

(c) Show that $(\vec{\sigma} \cdot \vec{p}) = -i(\vec{\sigma} \cdot \vec{e}_r)(\partial_r - \frac{K_{nr}-1}{r}) = (\vec{\sigma} \cdot \vec{e}_r)(p_r + i\frac{K_{nr}}{r})$.
    Hint: $\vec{\sigma} \cdot \vec{p} = (\vec{\sigma} \cdot \vec{e}_r)^2 \vec{\sigma} \cdot \vec{p}$ and recall that we have set $\hbar = 1$.

(d) Show that $[\vec{\sigma} \cdot \vec{e}_r, \vec{\sigma} \cdot \vec{p}] = \frac{2i}{r} K_{nr}$.

Problem 8.4. Properties of the non-relativistic spin–orbit operator

(a) Prove the relations (8.45) between the non-relativistic spin–orbit operator K_{nr}, the orbital angular momentum $\vec{L}$, and the total angular momentum $\vec{J}$.

(b) Show that for an arbitrary vector operator $\vec{v} = (v_1, v_2, v_3)^T$ which is orthogonal to the orbital angular momentum $\vec{L} = (L_1, L_2, L_3^T)$, i.e. $\vec{L} \cdot \vec{v} = 0 = \vec{L} \cdot \vec{v}$, the pseudo-scalar $(\vec{v} \cdot \vec{\sigma})$ anti-commutes with the spin–orbit operator K_{nr}, that is, $\{K_{nr}, (\vec{v} \cdot \vec{\sigma})\} = 0$.

Problem 8.5. SUSY Hamiltonians for spherically symmetric tensor potentials

(a) Derive the relation (8.81), that is, prove

$$H_\pm := \frac{\vec{p}^2}{2m} + \frac{1}{2m}[U'(r)]^2 \mp \frac{1}{2m} U''(r) \mp \frac{U'(r)}{mr} K_{nr},$$

where $H_+ = AA^\dagger$, $H_- = A^\dagger A$ and $A := \frac{1}{\sqrt{2m}}[\vec{\sigma} \cdot (\vec{p} + i\vec{e}_r U'(r))]$.

Hint: Show for an arbitrary differentiable function $f(r) = U'(r)/r$ the relations

$$f(r)(\vec{\sigma} \cdot \vec{r})(\vec{\sigma} \cdot \vec{p}) = f(r)(\vec{r} \cdot \vec{p}) + if(r)(\vec{\sigma} \cdot \vec{L})$$
$$(\vec{\sigma} \cdot \vec{p})(\vec{\sigma} \cdot \vec{r})f(r) = f(r)(\vec{r} \cdot \vec{p}) - if(r)(\vec{\sigma} \cdot \vec{L}) - irf'(r) - 3if(r).$$

(b) Derive the SUSY groundstates (8.84) for the harmonic and the linear superpotential, $U(r) = \frac{1}{2}\omega r^2$ and $U(r) = \gamma r$, and show that they belong to $\mathcal{H}^+_{jm_j}$.

References

[1] Pauli W 1927 Zur Quantenmechanik des magnetischen Elektrons *Z. Phys.* **43** 601–23
[2] Essler F H L, Korepin V E and Schoutens K 1990 New exactly solvable model of strongly correlated electrons motivated by high-T_c superconductivity *Phys. Rev. Lett.* **68** 2960–3
[3] Essler F H L, Korepin V E and Schoutens K 1993 Electronic model for superconductivity *Phys. Rev. Lett.* **70** 73–6
[4] Baker T H and Jarvis P D 1994 Quantum superspin chains *Int. J. Mod. Phys.* B **8** 3623–35
[5] Bracken A J, Gould M D, Links J R and Zhang Y-Z 1995 A new supersymmetric and exactly solvable model of correlated electrons *Phys. Rev. Lett.* **74** 2768–71
[6] Girvin S M, MacDonald A H, Fisher M P A, Rey S-J and Sethna J P 1990 Exactly soluble models of fractional statistics *Phys. Rev. Lett.* **65** 1671–4
[7] Efetov K B 1982 Supersymmetry method in localized theory *Sov. Phys. JETP* **55** 514–21
[8] Bohr T and Efetov K B 1982 Derivation of Green function for disordered chain by integrating over commuting and anticommuting variables *J. Phys. C: Solid State Phys.* **15** L249–54
[9] Efetov K 1983 Supersymmetry and theory of disordered metals *Adv. Phys.* **32** 53–127
[10] Galindo A and Pascual P 1991 *Quantum Mechanics II* (Berlin: Springer) section 12.5
[11] Uhlenbeck G and Goudsmit S 1925 Ersetzung der Hypothese vom unmechanischen Zwang durch eine Forderung bezüglich des inneren Verhaltens jedes einzelnen Elektrons *Naturwissenschaften* **13** 953–4
[12] Dirac P A M 1928 The quantum theory of the electron *Proc. R. Soc. Lond.* A **117** 610–24
[13] Schwinger J 1948 On quantum-electrodynamics and the magnetic moment of the electron *Phys. Rev.* **73** 416
[14] Aoyama T, Hayakawa M, Kinoshita T and Nio M 2012 Tenth-order QED contribution to the electron $g - 2$ and an improved value of the fine structure constant *Phys. Rev. Lett.* **109** 111807

[15] Van Dyke R S Jr, Schwinberg P B and Dehmelt H G 1987 New high-precision comparison of electron and positron g factors *Phys. Rev. Lett.* **59** 26–9

[16] Hanneke D, Fogwell Hoogerheide S and Gabrielse G 2011 Cavity control of a single-electron quantum cyclotron: measuring the electron magnetic moment *Phys. Rev.* A **83** 052122

[17] Tomonaga S-i 1997 *The Story of Spin* (Chicago: University of Chicago Press)

[18] Commins E D 2012 Electron spin and its history *Annu. Rev. Nucl. Part. Sci.* **62** 133–57

[19] Gendenshteîn L É and Krive I V 1985 Supersymmetry in quantum mechanics *Sov. Phys. Usp.* **28** 645–66

[20] D'Hoker E and Vinet L 1984 Supersymmetry of the Pauli equation in the presence of a magnetic monopole *Phys. Lett.* B **137** 72–6

[21] Cycon H L, Froese R G, Kirsch W and Simon B 1987 *Schrödinger Operators with Application to Quantum Mechanics and Global Geometry* (Berlin: Springer)

[22] Aharonov Y and Casher A 1979 Ground state of a spin-1/2 charged particle in a two-dimensional magnetic field *Phys. Rev.* A **19** 2461–2

[23] Klishevich S M and Plyushchayab M S 2001 Nonlinear supersymmetry on the plane in magnetic field and quasi-exactly solvable systems *Nucl. Phys.* B **616** 403–18

[24] Landau L 1930 Diamagnetismus der Metalle *Z. Phys.* **64** 629–37

[25] Isihara A 1991 *Condensed Matter Physics* (New York: Oxford University Press)

[26] Isihara A 1993 *Electron Liquids Springer Series in Solid-State Sciences* vol 96 (Berlin: Springer)

[27] Lee M H 1989 Chemical potential of a D-dimensional free Fermi gas at finite temperature *J. Math. Phys.* **30** 1837–9

[28] Lee M H 1995 Polylogarithmic analysis of chemical potential and fluctuations in a D-dimensional free Fermi gas at low temperatures *J. Math. Phys.* **36** 1217–31

[29] Ashcroft N W and Mermin N D 1976 *Solid State Physics* (New York: Holt, Reinhart and Holt)

[30] White R M 1983 *Quantum Theory of Magnetism Springer Series in Solid-State Science* vol 32 2nd edn (Berlin: Springer)

[31] Hogreve H, Schrader R and Seiler R 1978 A conjecture on the spinor functional determinant *Nucl. Phys.* B **142** 525–34

[32] Avron J E and Seiler R 1979 Paramagnetism for nonrelativistic electrons and Euclidean Dirac particles *Phys. Rev. Lett.* **42** 931–43

[33] Avron J and Simon B 1979 A counterexample to the paramagnetic conjecture *Phys. Lett.* A **75** 41–2

[34] Tkachuk V M and Vakarchuk S I 1999 Broken supersymmetry for the electron in the magnetic field of straight current *J. Phys. Stud.* **3** 291–4

[35] Bjorken J D and Drell S D 1964 *Relativistic Quantum Mechanics* (New York: McGraw-Hill)

[36] Thaller B 1992 *The Dirac Equation* (Berlin: Springer)

[37] Laplace P-S 1799 *Traité de Mécanique Céleste. Tome I, Premiere Partie, Livre II* (Paris: Crapelet)

[38] Runge C 1919 *Vektoranalysis* vol 1 (Leipzig: Hirzel)

[39] Lenz W 1924 Über den Bewegungsverlauf und Quantenzustände der gestörten Keplerbewegung *Z. Phys.* **24** 197–207

[40] Pauli W 1926 Über das Wasserstoffspektrum vom Standpunkt der neuen Quantenmechanik *Z. Phys.* **36** 336–63

[41] Hirshfeld A 2012 *The Supersymmetric Dirac Equation* (London: Imperial College Press)

[42] Cardoso J L and Alvarez-Nodarse R 2003 Recurrence relations for radial wavefunctions for the Nth-dimensional oscillators and hydrogenlike atoms *J. Phys. A: Math Gen.* A **36** 2055–68

[43] Powell J L 1947 Recurrence formulas for Coulomb wave functions *Phys. Rev.* **72** 626

[44] Infeld L 1947 Recurrence formulas for Coulomb wave function *Phys. Rev.* **72** 1125

[45] Moshinsky M and Szczepaniak A 1989 The Dirac oscillator *J. Phys. A: Math. Gen.* **22** L817–9

[46] Ui H 1984 Supersymmetric quantum mechanics in three-dimensional space. I One-particle system with spin-orbit potential *Prog. Theor. Phys.* **72** 813–20

IOP Publishing

Supersymmetric Methods in Quantum, Statistical and Solid State Physics
Enlarged and revised edition
Georg Junker

Chapter 9

Supersymmetric Dirac Hamiltonians

Dirac [1, 2] derived the Pauli Hamiltonian from the Dirac Hamiltonian as a non-relativistic approximation. Since then it has been known that the quantity $\vec{\sigma} \cdot (\vec{p} - (e/c)\vec{A})$ is conserved in both Dirac's theory and Pauli's if the scalar field vanishes (see, e.g. problem 1 of lecture eleven in Feynman's lecture on quantum electrodynamics [3]). The idea that the conserved quantity plays the role of the supercharge for the supersymmetric Dirac Hamiltonian came much later. The connection of the Dirac Hamiltonian to the non-relativistic supersymmetric quantum mechanics was exploited for the first time by Jackiw [4] when studying a two-dimensional Dirac particle in an external field. See also the independent work by Ui [5]. This work triggered more extensive investigations of supersymmetry (SUSY) for the Dirac Hamiltonian in a variety of forms [6–8]. For example, the two-dimensional SUSY Dirac Hamiltonian plays a key role in analysing the electronic properties of topological superconductors [9] and graphene [10, 11]. The SUSY structure of the Dirac Hamiltonian can help to characterise heterojunctions in semiconductors. Namely, SUSY reveals the presence of localised states at a junction which are not spin-degenerate [12, 13]. For more references, see the books by Thaller [14] and Hirschfeld [15].

In this chapter we generalise the Dirac Hamiltonian in the SUSY framework by constructing the SUSY Dirac Hamiltonian which bears a close resemblance to the non-relativistic Schrödinger Hamiltonian, and discuss some applications in the presence of various external fields including a magnetic field, a Lorentz-scalar potential, an oscillator potential, a Coulomb potential, and others. To satisfy the interest in the formal aspect of the SUSY Hamiltonian, we touch upon its spectral representation and path integral formulation. We also explore the Dirac Hamiltonian with a rotational symmetry and a relativistic version of Witten's model.

doi:10.1088/2053-2563/aae6d5ch9

9.1 The Dirac Hamiltonian and SUSY

The Dirac Hamiltonian is the Hamiltonian for a particle of spin 1/2 and rest mass m in relativistic motion with a speed comparable to the speed of light c. In the absence of external forces, the Hamiltonian is called the free Dirac Hamiltonian, which is given by

$$H_0 = c\vec{\alpha} \cdot \vec{p} + \beta mc^2. \tag{9.1}$$

This is a linearised form of the energy–momentum relation $E^2 = c^2\vec{p}^2 + m^2c^4$ in relativistic mechanics. The condition $H_0^2 = E^2$ demands the entities $\vec{\alpha} = (\alpha_1, \alpha_2, \alpha_3)^T$ and β in equation (9.1) to obey the Dirac algebra,

$$\{\alpha_i, \alpha_j\} = 2\delta_{ij}, \qquad \{\alpha_i, \beta\} = 0, \qquad \beta^2 = 1, \qquad i, j = 1, 2, 3. \tag{9.2}$$

In the standard Dirac representation, they are given by 4×4 matrices

$$\vec{\alpha} = \begin{pmatrix} 0 & \vec{\sigma} \\ \vec{\sigma} & 0 \end{pmatrix}, \qquad \beta = \begin{pmatrix} 1 & 0 \\ 0 & -1 \end{pmatrix}, \tag{9.3}$$

where $\vec{\sigma} = (\sigma_1, \sigma_2, \sigma_3)^T$ is a vector whose components are the Pauli matrices. Note that 0 and 1 in equation (9.3) are the 2×2 null and unit matrices. Thus, the free Dirac Hamiltonian (9.1) acts on a four component vector, say $\psi = (\psi_1, \psi_2, \psi_3, \psi_4)^T$ with $\psi_i \in L^2(\mathbb{R}^3)$. In other words the free Dirac Hamiltonian then acts on the Hilbert space $\mathcal{H} = L^2(\mathbb{R}^3) \otimes \mathbb{C}^4$.

More generally, the Dirac Hamiltonian represents a spin one-half particle moving under external forces. However, by the Dirac Hamiltonian, we often mean the one for a fast moving spin-1/2 electron of charge e in an arbitrary electromagnetic field with vector potential $\vec{A}$ and scalar potential ϕ, which can be obtained from the free Dirac Hamiltonian H_0 with the help of the usual momentum replacement, $\vec{p} \to \vec{p} - (e/c)\vec{A}$, and the additional energy replacement, $H_0 \to H_0 - e\phi$. The Dirac Hamiltonian thus obtained with the minimal electromagnetic coupling is

$$H_D := c\vec{\alpha} \cdot \left(\vec{p} - \frac{e}{c}\vec{A}\right) + \beta mc^2 + e\phi \tag{9.4}$$

acting on the same Hilbert space $\mathcal{H} = L^2(\mathbb{R}^3) \otimes \mathbb{C}^4$.

Before entering the SUSY discussion, we wish to make some remarks on the relation between the Dirac Hamiltonian H_D and the Pauli Hamiltonian H_P. The Dirac Hamiltonian (9.4) can be squared in the form

$$(H_D - e\phi)^2 = c^2 \begin{pmatrix} \left[\vec{\sigma} \cdot \left(\vec{p} - \frac{e}{c}\vec{A}\right)\right]^2 & 0 \\ 0 & \left[\vec{\sigma} \cdot \left(\vec{p} - \frac{e}{c}\vec{A}\right)\right]^2 \end{pmatrix} + m^2c^4. \tag{9.5}$$

As has been mentioned in section 2.1.1, if the Pauli Hamiltonian of the form

$$H_P = \frac{1}{2m}\left[\vec{\sigma} \cdot \left(\vec{p} - \frac{e}{c}\vec{A}\right)\right]^2 + e\phi \tag{9.6}$$

and the Dirac Hamiltonian (9.4) are known, then it is trivial to obtain the relation between H_D and H_P,

$$(H_D - e\phi)^2 = 2mc^2\begin{pmatrix} H_P - e\phi & 0 \\ 0 & H_P - e\phi \end{pmatrix} + m^2c^4, \tag{9.7}$$

by substituting equation (9.6) into the right-hand side of equation (9.5). In fact, we can say that any two of the squared Dirac Hamiltonian (9.5), the Pauli Hamiltonian (9.6), and the Hamiltonian relation (9.7) give the third with no approximation. This statement is algebraically correct, but slightly misleading. Historically, what Pauli [16] originally proposed is the spin Hamiltonian $H_s = (e\hbar/2mc)\vec{B} \cdot \vec{\sigma}$ for the electron. The Schrödinger Hamiltonian $H_S = \frac{1}{2m}(\vec{p} - (e/c)\vec{A})^2$ plus the Pauli spin term, that is, $H_P = H_S + H_s$, is usually called the Pauli Hamiltonian. However, the so-called Pauli Hamiltonian is phenomenological in the sense that it contains the hand-picked value $g = 2$ for the g-factor. As we have seen in section 8.1, the Pauli Hamiltonian can be put into the form (9.6) only if $g = 2$. It was Dirac who deduced the value $g = 2$ from his Hamiltonian and verified the Pauli spin term [2]. Furthermore, the Hamiltonian relation (9.7) is empty by itself unless the implications of H_D and H_P are explicitly specified. Although it resembles the energy–momentum relation $(E - e\phi)^2 = c^2(\vec{p} - (e/c)\vec{A})^2 + m^2c^4$ in classical relativistic mechanics, equation (9.7) is by no means obtainable from the classical energy–momentum relation that cannot accommodate the non-classical notion of spin. In deriving the Pauli Hamiltonian, Dirac started with an expectation that Heisenberg's equation of motion for a slowly moving electron may be determined by a Hamiltonian of the form $H' + mc^2$, where H' is small in magnitude as compared with the rest energy of the electron mc^2. Use of this Hamiltonian for H_D and the non-relativistic approximation $(H' - e\phi)^2 \ll m^2c^4$ leads to

$$(H_D - e\phi)^2 = (H' - e\phi + mc^2)^2 \approx 2mc^2(H' - e\phi) + m^2c^4. \tag{9.8}$$

To Dirac this is the source of equation (9.7). Substituting equation (9.8) to the left-hand side of equation (9.5) yields equation (9.6) with $H' = H_P \otimes 1$. Naturally the resultant Hamiltonian of the form (9.6), identified as the Pauli Hamiltonian, must possess $g = 2$.

Now we look into the SUSY aspect of the Dirac Hamiltonian (9.4) in the absence of the scalar potential, i.e. $\phi = 0$. If we let

$$A := c\vec{\sigma} \cdot \left(\vec{p} - \frac{e}{c}\vec{A}\right), \qquad M := mc^2, \tag{9.9}$$

then the Dirac Hamiltonian may be expressed in the form

$$H_D = Q_1 + \beta M, \tag{9.10}$$

where

$$Q_1 := \begin{pmatrix} 0 & A \\ A^\dagger & 0 \end{pmatrix} \tag{9.11}$$

or, more explicitly,

$$H_D = \begin{pmatrix} M & A \\ A^\dagger & -M \end{pmatrix}. \tag{9.12}$$

Note that A given in equation (9.9) is self-adjoint, i.e. $A = A^\dagger$. Obviously H_D in either equation (9.10) or (9.12) acts on a Hilbert space $\mathcal{H} = \mathcal{H}^+ \oplus \mathcal{H}^-$ graded by β. With Q_1 of equation (9.11), we have $(H_D - \beta M)^2 = Q_1^2$. Hence the operator $(H_D - \beta M)^2$ can be viewed as a SUSY Hamiltonian for $N = 1$ with supercharge $Q_1/\sqrt{2} = (H_D - \beta M)/\sqrt{2}$. This observation that the Dirac Hamiltonian plays in essence a role of a supercharge [8] motivates the quest for a more general definition of the supersymmetric Dirac Hamiltonian.

9.1.1 Supersymmetric Dirac Hamiltonians

In the Dirac Hamiltonian (9.12), the matrix elements A and M are specified by equation (9.9). Here we define a generalised Dirac Hamiltonian by removing the conditions (9.9).

Definition 9.1. A generalised Dirac Hamiltonian is an operator acting on a graded Hilbert space $\mathcal{H} = \mathcal{H}^+ \oplus \mathcal{H}^-$ and possesses the form

$$H_D = \begin{pmatrix} M_+ & A \\ A^\dagger & -M_- \end{pmatrix}, \qquad M_\pm^\dagger = M_\pm \geqslant 0, \tag{9.13}$$

where A and $M_\pm$ are operators in $\mathcal{H}$. Note that $A^\dagger$ is the adjoint of A and that A is arbitrary and not necessarily self-adjoint. However, $M_\pm$ are required to be self-adjoint non-negative operators. The generalised Dirac Hamiltonian (9.13) is said to be *supersymmetric* if the following relations hold true:

$$AM_- = M_+A, \qquad A^\dagger M_+ = M_- A^\dagger. \tag{9.14}$$

Alternatively, introducing a grading operator W that separates $\mathcal{H}$ into $\mathcal{H}_+$ and $\mathcal{H}_-$, a self-adjoint supercharge operator Q_1 that anticommutes with W, and a non-negative operator $\mathcal{M}$ that commutes with W, we construct a generalised Dirac Hamiltonian by

$$H_D := Q_1 + W\mathcal{M}, \tag{9.15}$$

where

$$\{Q_1, W\} = 0, \qquad [M, W] = 0, \qquad W^2 = 1. \qquad (9.16)$$

Note that spec $W = \pm 1$ with $\mathcal{H}_\pm$ being the corresponding eigenspaces of the grading operator W. Now we define that the generalised Dirac Hamiltonian H_D thus constructed is supersymmetric if Q_1 and $\mathcal{M}$ commute, that is, if

$$[Q_1, \mathcal{M}] = 0. \qquad (9.17)$$

With the standard representation,

$$Q_1 = \begin{pmatrix} 0 & A \\ A^\dagger & 0 \end{pmatrix}, \qquad \mathcal{M} = \begin{pmatrix} M_+ & 0 \\ 0 & M_- \end{pmatrix}, \qquad W = \begin{pmatrix} 1 & 0 \\ 0 & -1 \end{pmatrix}, \qquad (9.18)$$

the generalised Dirac Hamiltonian (9.15) converts itself to the matrix form (9.13). The grading operator W is nothing other than the Witten parity operator introduced in section 2.2.1. A generalised Dirac Hamiltonian of the form (9.15) is called a Dirac operator, and a generalised Dirac Hamiltonian which is supersymmetric is a Dirac operator with supersymmetry in Thaller's book [19]. For more detailed descriptions on Dirac operators, see chapter 5 of the same book [19].

An obvious property of the supersymmetric Dirac Hamiltonian is that its square becomes block diagonal:

$$H_D^2 = Q_1^2 + \mathcal{M}^2 = \begin{pmatrix} AA^\dagger + M_+^2 & 0 \\ 0 & A^\dagger A + M_-^2 \end{pmatrix}. \qquad (9.19)$$

As has been shown in section 2.1, a SUSY Hamiltonian H_{SUSY} is usually composed of a supercharge Q_1 as $H_{SUSY} = 2Q_1^2$. For the Dirac operator (9.15), we are supposed to have $H_{SUSY} = 2(H_D^2 - \mathcal{M}^2)$. Here, however, in order to make H_{SUSY} similar in form to the Schrödinger Hamiltonian for a non-relativistic particle, we introduce a mass parameter $m > 0$, which is arbitrary in general but may be identified with the rest mass of the Dirac particle in most cases, and define the SUSY Hamiltonian for the Dirac operator by

$$H_{SUSY} := \frac{1}{2mc^2}\left(H_D^2 - \mathcal{M}^2\right) = \begin{pmatrix} H_+ & 0 \\ 0 & H_- \end{pmatrix}, \qquad (9.20)$$

with the partner Hamiltonians given by

$$H_+ := \frac{1}{2mc^2} AA^\dagger, \qquad H_- := \frac{1}{2mc^2} A^\dagger A. \qquad (9.21)$$

Together with the nilpotent complex supercharges,

$$Q := \frac{1}{\sqrt{2mc^2}} \begin{pmatrix} 0 & A \\ 0 & 0 \end{pmatrix}, \qquad Q^\dagger := \frac{1}{\sqrt{2mc^2}} \begin{pmatrix} 0 & 0 \\ A^\dagger & 0 \end{pmatrix} \qquad (9.22)$$

the partner Hamiltonians (9.21) form an $N = 2$ SUSY system with the Witten parity, satisfying the algebra,

$$H_{\text{SUSY}} = \{Q, Q^\dagger\}, \qquad \{Q, W\} = 0, \qquad Q^2 = 0 = (Q^\dagger)^2. \tag{9.23}$$

Note that $Q_1 = \sqrt{2mc^2}\,(Q + Q^\dagger)$.

9.1.2 Foldy–Wouthuysen and SUSY transformation

An important property of supersymmetric Dirac operators is that they can be diagonalised. In fact, it is possible to define a unitary matrix U such that

$$\tilde{H}_{\text{D}} := U H_{\text{D}} U^\dagger = \begin{pmatrix} \sqrt{AA^\dagger + M_+^2} & 0 \\ 0 & -\sqrt{A^\dagger A + M_-^2} \end{pmatrix} \tag{9.24}$$

and, therefore, positive- and negative-energy solutions are decoupled. That is, the two subspaces $\mathcal{H}^\pm$ belong to the eigenspaces of H_{D} with positive and negative eigenvalues, respectively. These kinds of transformations which diagonalise a Dirac Hamiltonian are usually called Foldy–Wouthuysen transformations [17]. In the case of a supersymmetric Dirac Hamiltonian such a unitary matrix U always exists and is explicitly given by [14, 18, 19]

$$U := a_+ + W\,\text{sgn}(Q_1)a_- \qquad \text{with} \qquad a_\pm := \sqrt{\frac{1}{2} \pm \frac{\mathcal{M}}{2|H_{\text{D}}|}}, \tag{9.25}$$

where $|H_{\text{D}}| := \sqrt{H_{\text{D}}^2}$ and $\text{sgn}(Q_1) := Q_1/|Q_1|$ on the orthogonal complement of ker Q_1. Note that $[|H_{\text{D}}|, \mathcal{M}] = 0$ and therefore the above definition of $a_\pm$ is not ambiguous. On ker Q_1 we have $\mathcal{M} = |H_{\text{D}}|$, that is, $a_+ = 1$ and $a_- = 0$ resulting in $U = 1$. Obviously, the operators $A^\dagger A$ and $AA^\dagger$ are essential iso-spectral. As a consequence, the positive and negative eigenvalues of H_{D} are closely related. In particular, for $M_+ = M_- = mc^2 > 0$ the spectrum of H_{D} is symmetric about zero with possible exceptions at $\pm mc^2$ and has a gap from $-mc^2$ to $+mc^2$. The value $+mc^2$ $(-mc^2)$ belongs to the spectrum of H_{D} if $AA^\dagger$ $(A^\dagger A)$ has zero eigenvalue(s). That is, when SUSY is unbroken.

Following the general discussion on $N = 2$ SUSY in chapter 2 we may also establish SUSY transformations between eigenstates of H_{D} for positive and negative eigenvalues. For this let us assume that the strictly positive eigenvalues $\varepsilon_n > 0$, and corresponding eigenstates $\phi_n^\pm$ of the partner Hamiltonians $H_\pm$ are known,

$$H_\pm \phi_n^\pm = \varepsilon_n \phi_n^\pm. \tag{9.26}$$

For simplicity we assume a purely discrete spectrum and the strictly positive eigenvalues are enumerated by $n = 1, 2, 3, \ldots$. Then the SUSY transformations, as discussed in chapter 2, read

$$\phi_n^+ = \frac{1}{\sqrt{2mc^2\varepsilon_n}} A\phi_n^-, \qquad \phi_n^- = \frac{1}{\sqrt{2mc^2\varepsilon_n}} A^\dagger \phi_n^+. \tag{9.27}$$

The spectral properties of $\tilde{H}_{\mathrm{D}}$ are given by

$$\tilde{H}_{\mathrm{D}}\tilde{\psi}_n^\pm = E_n^\pm \tilde{\psi}_n^\pm, \tag{9.28}$$

with

$$E_n^\pm = \pm\sqrt{2mc^2\varepsilon_n + M_\pm^2}, \qquad \tilde{\psi}_n^+ = \begin{pmatrix} \phi_n^+ \\ 0 \end{pmatrix}, \qquad \tilde{\psi}_n^- = \begin{pmatrix} 0 \\ \phi_n^- \end{pmatrix}. \tag{9.29}$$

In other words, the eigenvalue problem for a supersymmetric Dirac Hamiltonian can be reduced to that of the partner Hamiltonians $H_\pm$. Note that the eigenstates $\psi_n^\pm$ of H_{D} are related to those of $\tilde{H}_{\mathrm{D}}$ via the unitary transformation $\psi_n^\pm = U^\dagger \tilde{\psi}_n^\pm$. The above SUSY transformations induce similar transformations on eigenstates $\tilde{\psi}_n^\pm$ as follows:

$$\tilde{\psi}_n^- = \frac{1}{\sqrt{\varepsilon_n}} Q^\dagger \tilde{\psi}_n^+, \qquad \tilde{\psi}_n^+ = \frac{1}{\sqrt{\varepsilon_n}} Q\tilde{\psi}_n^-. \tag{9.30}$$

In case of unbroken SUSY we, in addition, have one or more SUSY ground states with eigenvalue $\varepsilon_0 = 0$ belonging to $\mathcal{H}^+$ and/or $\mathcal{H}^-$, that is,

$$E_0^+ = M_+ \quad \text{if} \quad \phi_0^+ \in \mathcal{H}^+ \quad \exists \quad \text{with} \quad A^\dagger \phi_0^+ = 0 \tag{9.31}$$

and/or

$$E_0^- = -M_- \quad \text{if} \quad \phi_0^- \in \mathcal{H}^- \quad \exists \quad \text{with} \quad A\phi_0^- = 0. \tag{9.32}$$

9.1.3 Resolvent and path integral representation

The resolvent of the Dirac operator (9.13) defined by

$$G(z) := \frac{1}{H_{\mathrm{D}} - z}, \qquad z \in \mathbb{C}\backslash\mathrm{spec}(H_{\mathrm{D}}), \tag{9.33}$$

is an analytical function of z in the complement of the spectrum of H_{D}. Let us write it in the form

$$G(z) = (H_{\mathrm{D}} + z)g(z^2), \tag{9.34}$$

where we have introduced the iterated resolvent

$$g(\zeta) := \frac{1}{H_{\mathrm{D}}^2 - \zeta}, \qquad \zeta \in \mathbb{C}\backslash\mathrm{spec}(H_{\mathrm{D}}^2), \tag{9.35}$$

which is the resolvent of the squared Dirac operator H_D^2, defined in the ζ-plane. As the square of a supersymmetric Dirac Hamiltonian is diagonal so is the iterated resolvent

$$g(\zeta) = \begin{pmatrix} g^+(\zeta) & 0 \\ 0 & g^-(\zeta) \end{pmatrix}, \tag{9.36}$$

with

$$g^+(\zeta) := \frac{1}{AA^\dagger + M_+^2 - \zeta}, \qquad g^-(\zeta) := \frac{1}{A^\dagger A + M_-^2 - \zeta}. \tag{9.37}$$

The resolvent $G(z)$ of the Dirac operator H_D may be obtained with the help of the diagonal elements of the iterated resolvent in the form

$$G(z) = (H_D + z)g(z^2) = \begin{pmatrix} (z + M_+)g^+(z^2) & Ag^-(z^2) \\ A^\dagger g^+(z^2) & (z - M_-)g^-(z^2) \end{pmatrix}. \tag{9.38}$$

Since H_D^2 and hence H_{SUSY} are positive semi-definite, their spectrum is on the non-negative real axis of the ζ-plane. Considering a contour C encircling counter-clockwise all points corresponding to $\text{spec}(H_D^2)$ and using Cauchy's integral formula, we can find

$$e^{-\text{i}tuH_D^2} = -\frac{1}{2\pi\text{i}} \oint_C d\zeta e^{-\text{i}tu\zeta} g(\zeta), \tag{9.39}$$

where t and u are arbitrary real constants. At this point, recall the Schrödinger-like operator H_{SUSY} is defined by equation (9.20), we replace H_D^2 on the left-hand side by $2mc^2 H_{\text{SUSY}} + \mathcal{M}^2$. Moreover, we let $u = (2mc^2\hbar)^{-1}$. Then the quantity on the left-hand side of equation (9.39) can be understood as the unitary evolution operator of the system with the Schrödinger Hamiltonian $H_{\text{SUSY}} + \mathcal{M}^2/2mc^2$ if t is identified with the time parameter. In the coordinate representation, equation (9.39) may be written as

$$K(\vec{r}'', \vec{r}'; t) = -\frac{1}{2\pi\text{i}} \oint_C d\zeta e^{-\text{i}t\zeta/2mc^2\hbar} g(\vec{r}'', \vec{r}'; \zeta), \tag{9.40}$$

which relates the resolvent kernel $g(\vec{r}'', \vec{r}'; \zeta) := \langle \vec{r}''|g(\zeta)|\vec{r}'\rangle$ to Feynman's kernel,

$$K(\vec{r}'', \vec{r}'; t) := \langle \vec{r}''|\exp\left\{-(\text{i}t/\hbar)(H_{\text{SUSY}} + \mathcal{M}^2/2mc^2)\right\}|\vec{r}'\rangle. \tag{9.41}$$

We also note that the iterated resolvent can be expressed as

$$g(\zeta) = \text{i}u \int_0^\infty dt \exp\left\{-\text{i}tu(H_D^2 - \zeta)\right\}. \tag{9.42}$$

The integral on the right-hand side converges for $\text{Im}(u\zeta) > 0$ and $\zeta \notin \text{spec}(H_D^2)$. With our choice $u = (2mc^2\hbar)^{-1}$ it converges on the upper half of the ζ plane. Hence, we consider the integral defined only for $\mathrm{Im}(\zeta) > 0$. In the coordinate representation, equation (9.42) takes the form

$$g(\vec{r}'', \vec{r}'; \zeta) = \frac{\mathrm{i}}{2mc^2\hbar} \int_0^\infty \mathrm{d}t \, P_\zeta(\vec{r}'', \vec{r}'; t), \qquad (9.43)$$

where

$$P_\zeta(\vec{r}'', \vec{r}'; t) = \langle \vec{r}''|\exp\left\{-(\mathrm{i}t/\hbar)\left(H_{\mathrm{SUSY}} + (\mathcal{M}^2 - \zeta)/2mc^2\right)\right\}|\vec{r}'\rangle, \qquad (9.44)$$

which we shall refer to as the promotor [20]. This promotor is the same in form as the propagator for the effective Hamiltonian $H_{\mathrm{eff}}(\zeta) := H_{\mathrm{SUSY}} + (\mathcal{M}^2 - \zeta)/2mc^2$. As the propagator is given in terms of Feynman's path integral, so is the promotor. While a Schrödinger Hamiltonian itself is self-adjoint, the effective Hamiltonian is not. However, the Hamiltonian has to be self-adjoint for the time evolution operator, but path integration of the promotor does not require self-adjointness of the effective Hamiltonian.

The corresponding diagonal elements of the iterated resolvent kernel are given by

$$g^\pm(\vec{r}'', \vec{r}'; \zeta) := \langle \vec{r}''|g^\pm(\zeta)|\vec{r}'\rangle = \frac{\mathrm{i}}{2 \, mc^2\hbar} \int_0^\infty \mathrm{d}t \, P_\zeta^\pm(\vec{r}'', \vec{r}'; t), \qquad (9.45)$$

where

$$P_\zeta^\pm(\vec{r}'', \vec{r}'; t) := \langle \vec{r}''|\exp\{-(\mathrm{i}t/\hbar)H_{\mathrm{eff}}^\pm(\zeta)\}|\vec{r}'\rangle \qquad (9.46)$$

and

$$H_{\mathrm{eff}}^+(\zeta) := \frac{1}{2mc^2}\left(AA^\dagger + M_+^2 - \zeta\right), \quad H_{\mathrm{eff}}^-(\zeta) := \frac{1}{2mc^2}\left(A^\dagger A + M_-^2 - \zeta\right). \quad (9.47)$$

One of the merits of utilising the iterated resolvent is that the kernel of its components can be represented by means of Feynman's path integral for effective non-relativistic systems. Let $\mathcal{L}_\zeta^\pm(\dot{\vec{r}}, \vec{r}, t)$ represent the classical Lagrangian associated with the effective Hamiltonian $H_{\mathrm{eff}}^\pm(\zeta)$. Then the promotors can be expressed as Feynman's path integral:

$$P_\zeta^\pm(\vec{r}'', \vec{r}'; t) = \int_{\vec{r}(0)=\vec{r}'}^{\vec{r}(t)=\vec{r}''} \mathcal{D}\vec{r} \, \exp\left\{\frac{\mathrm{i}}{\hbar} \int_0^t \mathrm{d}s \, \mathcal{L}_\zeta^\pm(\dot{\vec{r}}, \vec{r}, s)\right\}. \qquad (9.48)$$

If the system with the effective non-relativistic Lagrangian is path-integrable, the diagonal elements of the iterated resolvent kernel can be determined via equation (9.45). The resolvent kernel $G(\vec{r}'', \vec{r}'; z) = \langle \vec{r}''|G(z)|\vec{r}'\rangle$, or Green's function, of the Dirac operator is given as the coordinate representation of equation (9.38); namely

$$G(\vec{r}'', \vec{r}'; z) = \begin{pmatrix} (z + M_+)g^+(\vec{r}'', \vec{r}'; z^2) & A(\vec{r}'')g^-(\vec{r}'', \vec{r}'; z^2) \\ A^\dagger(\vec{r}'')g^+(\vec{r}'', \vec{r}'; z^2) & (z - M_-)g^-(\vec{r}'', \vec{r}'; z^2) \end{pmatrix}. \qquad (9.49)$$

Here $A(\vec{r}'')$ is the operator in the $\vec{r}''$-representation, i.e. $\langle \vec{r}'' | A | \vec{r}' \rangle =: A(\vec{r}'') \delta(\vec{r}'' - \vec{r}')$. If the two non-zero elements of the iterated resolvent kernel are found in closed form, then all the elements of the Green's function are obtained in closed form.

Let us conclude this section by mentioning that for obtaining the above results, the SUSY structure is a sufficient but not a necessary condition. The above approach is applicable even to the case where the Dirac operator is non-supersymmetric, i.e. not of the form (9.13), insofar as its squared Dirac operator is (block) diagonal [21].

9.2 The free Dirac Hamiltonian in three dimensions

The Hamiltonian of the free Dirac particle in the standard representation reads

$$H_0 = \begin{pmatrix} mc^2 & c\vec{\sigma} \cdot \vec{p} \\ c\vec{\sigma} \cdot \vec{p} & -mc^2 \end{pmatrix} \tag{9.50}$$

and represents a supersymmetric Dirac Hamiltonian with $A := c\vec{\sigma} \cdot \vec{p}$ and $M_\pm = mc^2$. The two associated SUSY partner Hamiltonians are that of the free Pauli particle with the same mass, $H_\pm = \vec{p}^2/2m$ acting on $\mathcal{H}^\pm = L^2(\mathbb{R}^3) \otimes \mathbb{C}^2$. The spectrum of the free particle is well-known and usually parameterised by a wave vector $\vec{k} \in \mathbb{R}^3$, $\varepsilon(\vec{k}) := \hbar^2\vec{k}^2/2m$. The eigenstates are plane waves and may be written as

$$\phi_{\vec{k},\lambda}^{\pm}(\vec{r}) = \left(\frac{1}{2\pi\hbar} \right)^{3/2} e^{i\vec{k}\cdot\vec{r}} \chi_\lambda(\vec{k}), \tag{9.51}$$

where $\lambda = \pm 1$ denotes an orthogonal basis in $\mathbb{C}^2$. These basis vectors can be chosen to be eigenstates of the helicity operator $\Lambda := \vec{\sigma} \cdot \vec{p}/|\vec{p}|$, that is, $\Lambda\chi_\lambda = \lambda\chi_\lambda$, and explicitly read [14] with $\vec{k} = (k_1, k_2, k_3)^T$ and $k = |\vec{k}|$

$$\chi_{+1}(\vec{k}) := \frac{1}{\sqrt{2k(k-k_3)}} \begin{pmatrix} k_1 - ik_2 \\ k - k_3 \end{pmatrix} \text{ for } k > k_3, \quad \chi_{+1}(k\vec{e}_3) := \begin{pmatrix} 1 \\ 0 \end{pmatrix},$$

$$\chi_{-1}(\vec{k}) := \frac{1}{\sqrt{2k(k-k_3)}} \begin{pmatrix} k_3 - k \\ k_1 + ik_2 \end{pmatrix} \text{ for } k > k_3, \quad \chi_{-1}(k\vec{e}_3) := \begin{pmatrix} 0 \\ 1 \end{pmatrix}. \tag{9.52}$$

The corresponding eigenvalues and eigenstates of the free Dirac Hamiltonian are then given by

$$E_{\vec{k},\lambda}^{\pm} = \pm\sqrt{\hbar^2 c^2 \vec{k}^2 + m^2 c^4}, \quad \tilde{\psi}_{\vec{k},\lambda}^{+} = \begin{pmatrix} \phi_{\vec{k},\lambda}^{+} \\ 0 \end{pmatrix}, \quad \tilde{\psi}_{\vec{k},\lambda}^{-} = \begin{pmatrix} 0 \\ \phi_{\vec{k},\lambda}^{-} \end{pmatrix}. \tag{9.53}$$

We leave it as an exercise to the reader to show that the Foldy–Wouthuysen transformation for the free particle is given by

$$U = a_+ + a_- \Lambda \begin{pmatrix} 0 & 1 \\ -1 & 0 \end{pmatrix} \text{ with } a_\pm = \left(\frac{1}{2} \pm \frac{mc^2}{2\sqrt{c^2\vec{p}^2 + mc^2}} \right)^{1/2} \tag{9.54}$$

and the SUSY transformations explicitly read

$$Q\tilde{\psi}_{\vec{k},\lambda}^{-} = \lambda\sqrt{\epsilon(\vec{k})}\tilde{\psi}_{\vec{k},\lambda}^{+}, \qquad Q^{\dagger}\tilde{\psi}_{\vec{k},\lambda}^{+} = \lambda\sqrt{\epsilon(\vec{k})}\tilde{\psi}_{\vec{k},\lambda}^{-}. \tag{9.55}$$

Let us note that in addition to the standard representations (9.3) of the Dirac matrices we may consider the so-called supersymmetric representation which reads

$$\vec{\alpha} = \begin{pmatrix} 0 & \vec{\sigma} \\ \vec{\sigma} & 0 \end{pmatrix}, \qquad \beta = \begin{pmatrix} 0 & -i \\ i & 0 \end{pmatrix}. \tag{9.56}$$

In this representation the free Dirac Hamiltonian takes the form

$$H_0 = \begin{pmatrix} 0 & c\vec{\sigma}\cdot\vec{p} - imc^2 \\ c\vec{\sigma}\cdot\vec{p} + imc^2 & 0 \end{pmatrix}. \tag{9.57}$$

Again this is a supersymmetric Dirac operator where now

$$A := c\vec{\sigma}\cdot\vec{p} - imc^2, \qquad M_{\pm} := 0 \tag{9.58}$$

and the partner Hamiltonians are those of the shifted free Pauli Hamiltonian

$$H_{\pm} = \frac{\vec{p}^2}{2m} + \frac{mc^2}{2} \geqslant \frac{mc^2}{2} > 0. \tag{9.59}$$

The corresponding eigenvalues are $\epsilon(\vec{k}) = \hbar^2\vec{k}^2/2m + mc^2/2$ and the associated eigenstates are given in equation (9.51) and lead to the same spectral properties (9.53). Note that here SUSY is broken as the partner potentials do not have a vanishing eigenvalue.

For the free Dirac Hamilton, in both representations discussed above, the effective Schrödinger-like Hamiltonians read

$$H_{\text{eff}}^{\pm}(\zeta) = \frac{\vec{p}^2}{2m} + \frac{mc^2}{2} - \frac{\zeta}{2mc^2} = \frac{\vec{p}^2}{2m} - \frac{\mu^2(\zeta)}{2m}, \tag{9.60}$$

with

$$\mu^2(\zeta) = \zeta/c^2 - m^2c^2. \tag{9.61}$$

Hence, the effective Hamiltonian is up to an additive constant simply that of the free non-relativistic Schrödinger Hamiltonian with effective classical Lagrangian

$$\mathcal{L}_{\zeta}^{\pm}(\dot{\vec{r}}, \vec{r}, t) = \frac{m}{2}\dot{\vec{r}}^2 + \frac{\mu^2(\zeta)}{2m}. \tag{9.62}$$

The path integral of the free particle characterised by equation (9.62) is well-known and can explicitly be calculated. Hence the promotors are found to be

$$P_{\zeta}^{\pm}(\vec{r}'', \vec{r}'; t) = \left(\frac{m}{2\pi i\hbar t}\right)^{3/2} \exp\left\{\frac{i}{\hbar}\left(\frac{m}{2t}(\vec{r}'' - \vec{r}')^2 + \frac{t}{2m}\mu^2(\zeta)\right)\right\} \tag{9.63}$$

and lead to the resolvent kernels for the squared Dirac Hamiltonian, $\vec{x} := \vec{r}'' - \vec{r}'$,

$$g^{\pm}(\vec{r}'', \vec{r}'; \zeta) = \frac{1}{4\pi|\vec{x}|(\hbar c)^2} \exp\{(i/\hbar)\mu(\zeta)|\vec{x}|\}. \tag{9.64}$$

With the help of equation (9.38) we can obtain the resolvent for the free Dirac operator in closed form

$$G(\vec{r}'', \vec{r}'; z) = \frac{e^{(i/\hbar)\mu(z^2)|\vec{x}|}}{4\pi|\vec{x}|(\hbar c)^2}\left(i\hbar c\frac{\vec{\alpha}\cdot\vec{x}}{|\vec{x}|^2} + c\mu(z^2)\frac{\vec{\alpha}\cdot\vec{x}}{|\vec{x}|} + \beta mc^2 + z\right). \tag{9.65}$$

9.3 The two-dimensional Dirac Hamiltonian with magnetic field

Restricting the configuration space to two dimensions the associated relativistic Dirac equation for an electron subjected to an external magnetic field is given by

$$H_{\mathrm{D}}^{(2)} := c\sigma_1\left(p_1 - \frac{e}{c}a_1\right) + c\sigma_2\left(p_2 - \frac{e}{c}a_2\right) + \sigma_3 mc^2. \tag{9.66}$$

Here the vector potential $\vec{A} = (a_1, a_2)^T$ is that defined in analogy to equation (8.10) and characterises a magnetic field $\vec{B} = B(x_1, x_2)\vec{e}_3$ being orthogonal to the x_1–x_2-plane. The field strength is given by $B(x_1, x_2) := \partial a_2/\partial x_1 - \partial a_1/\partial x_2$. Note that in two dimensions the Dirac matrices are replaced by Pauli matrices. Hence the Hilbert space is now $\mathcal{H} = L^2(\mathbb{R}^2) \otimes \mathbb{C}^2$.

Again this Dirac Hamiltonian is of supersymmetric form as in the standard representation of the Pauli matrices, cf. equation (1.3), it reads

$$H_{\mathrm{D}}^{(2)} := \begin{pmatrix} mc^2 & A \\ A^{\dagger} & -mc^2 \end{pmatrix} \quad \text{with} \quad A := (cp_1 - ea_1) - i(cp_2 - ea_2). \tag{9.67}$$

Note that the operator A is up to an overall factor $\sqrt{2mc^2}$ identical to the same operator (8.15) defined for the two-dimensional Pauli system. Hence the associated SUSY Hamiltonian (9.20) can be identified with the two-dimensional Pauli Hamiltonian (8.12), that is, $H_{\mathrm{SUSY}} = H_{\mathrm{P}}^{(2)}$ or

$$\left(H_{\mathrm{D}}^{(2)}\right)^2 = 2mc^2 H_{\mathrm{P}}^{(2)} + m^2 c^4. \tag{9.68}$$

This relation between the two-dimensional Pauli and Dirac Hamiltonians has first been observed by Feynman [3] and Jackiw [4]. As a consequence all our results of section 8.1 on the two-dimensional Pauli Hamiltonian can directly be carried over to the two-dimensional Dirac Hamiltonian. In particular, for a magnetic field strong enough, i.e. the net magnetic flux through the plane is larger than one flux quantum, SUSY is unbroken with ground states given by the Aharonov–Casher theorem (see problem 8.2 of chapter 8). The degeneracy d of this ground state energy level is the same as in the Pauli system (8.18).

In the special case of a constant magnetic field $B(x_1, x_2) = B$ the energy eigenvalues of the two-dimensional Pauli Hamiltonian are given by the well-known Landau levels [22, 23]

$$\varepsilon_n = \hbar\omega_c n, \qquad n \in \mathbb{N}_0, \tag{9.69}$$

with cyclotron frequency $\omega_c := |eB|/mc$. See also the previous chapter. The value $n = 0$ is only allowed for the spin-up subspace $\mathcal{H}^+$ if $eB > 0$, respectively, only for $\mathcal{H}^-$ when $eB < 0$. All Landau levels have the same degeneracy d. Knowing the eigenvalues of $H_{\mathrm{SUSY}} = H_{\mathrm{P}}^{(2)}$ we obtain those for $H_{\mathrm{D}}^{(2)}$ via equation (9.29). The so-called relativistic Landau levels are given by

$$E_n^{\pm} = \pm\sqrt{2c\hbar|eB|n + m^2c^4} = \pm mc^2\sqrt{1 + 2n\frac{\hbar\omega_c}{mc^2}}. \tag{9.70}$$

They have the same degeneracy d as the non-relativistic Landau levels.

The two-dimensional Dirac Hamiltonian, and in particular its massless variant the Dirac–Weyl Hamiltonian, currently plays a significant role in the characterisation of electronic transport properties of carbon-based nano-structures such as nano-tubes and graphene. See, for example, [10, 11]. In fact, SUSY has become a key concept in characterising condensed matter systems known as topological superconductors [9]. Here we will only briefly consider the SUSY aspects for the electronic properties of graphene. The low-energy band structure of a monolayer graphene is approximately described by so-called Dirac cones. These are at two different edges of the Brillouin zone usually denoted by K and K'. Here the intrinsic spin of the electrons is not considered but the so-called pseudospin which is locating the electron near the point K corresponding to a pseudospin up (+) or near K' related to a down-pseudospin (−). The corresponding positive- and negative-energy solutions characterise the electron and hole states. It turns out that the low-energy band structure can effectively be described by a two-dimensional supersymmetric Dirac Hamiltonian

$$H^{(\pm)} := \begin{pmatrix} m_{\mathrm{eff}}v_{\mathrm{F}}^2 & A_{\pm} \\ A_{\pm} & -m_{\mathrm{eff}}v_{\mathrm{F}}^2 \end{pmatrix}, \quad A_{\pm} := v_{\mathrm{F}}\big[(p_1 - ea_1/c) \mp \mathrm{i}(p_2 - ea_2/c)\big]. \tag{9.71}$$

Here v_{F} stands for the Fermi velocity which is of the order of 10^6 m s^{-1}. The effective mass m_{eff} is an effective description for the Coulomb interaction between the charge carriers and often is set to zero when this interaction is ignored, i.e. a non-interacting two-dimensional electron gas. The ± -sign stands for the pseudospin describing the low-energy band near the K and K' corner of the Brillouin zone, respectively. Note that here the operator A_+ is identical to A in equation (9.67), whereas the operator A_- is actually the alternative realisation of SUSY with supercharge $\tilde{Q}$ as defined in equation (8.20). Both Hamiltonians may be combined into a single one by introducing the Pauli matrix τ_3 acting on the pseudospin degree of freedom,

$$H_{\mathrm{G}} := v_{\mathrm{F}}\sigma_1(p_1 - ea_1/c) + \tau_3 v_{\mathrm{F}}\sigma_2(p_2 - ea_2/c) + \sigma_3 m_{\mathrm{eff}}v_{\mathrm{F}}^2. \tag{9.72}$$

which acts on 4-spinors realising the $SU(4)$-symmetry in graphene as well as its two-fold SUSY structure. In the case of a constant magnetic field, the spectrum is then given by the relativistic Landau levels (9.70)

$$E_n^\pm = \pm\sqrt{2m_{\text{eff}}v_F^2\varepsilon_n + m_{\text{eff}}^2 v_F^4} = \pm\hbar\omega_F\sqrt{n + \frac{m_{\text{eff}}^2 v_F^4}{\hbar^2\omega_F^2}}, \qquad (9.73)$$

where $\omega_F := v_F\sqrt{2|eB|/\hbar c}$. The SUSY structure of above graphene Hamiltonian H_G can, for example, be utilised to explain experimentally observed phenomena such as the unconventional quantum Hall effect in mono- and bi-layer graphene [24].

SUSY methods have also been utilised in studying magnetic confinement of the non-interacting, i.e. massless, charge carriers in graphene [25, 26]. In fact, due to the phenomenon called Klein tunnelling, it is not possible to use electrostatic barriers to confine charge carriers in case their dynamics is ruled by a Dirac Hamiltonian. In fact, graphene has been suggested to be an ideal environment where the Klein paradox may actually be tested [27]. This has experimentally been verified [28]. Furthermore, it can be shown that the graphene Hamiltonian with an external scalar potential has the free Dirac Hamiltonian as a SUSY partner [29]. Hence, it exhibits the same spectral properties, such as a vanishing reflection coefficient similar to what we have discussed in section 3.5.

A confinement of the charge carriers may be achieved with properly designed external magnetic fields. Here SUSY is the ideal tool to reach that goal. Let us consider an external magnetic field being translational invariant along the x_2-axis of the x_1–x_2-plane, that is, $B(x_1, x_2) = B(x_1)$. Such a magnetic field can be characterised by a vector potential with components $a_1 = 0$ and $a_2 = a(x_1)$, i.e. $B(x_1) = a'(x_1)$. With the magnetic field being constant along x_2 we make the ansatz $\psi(x_1, x_2) = e^{ik_2 x_2}\phi(x_1)$, which allows us to replace the operator p_2 by its eigenvalue $\hbar k_2$ with $k_2 \in \mathbb{R}$. The above Hamiltonian (9.71) then reduces to

$$H^{(\pm)} = \begin{pmatrix} m_{\text{eff}}v_F^2 & A_\pm \\ A_\pm & -m_{\text{eff}}v_F^2 \end{pmatrix}, \quad \text{with} \quad A_\pm := v_F\big[p_1 \mp i(\hbar k_2 - (e/c)a(x_1))\big]. \quad (9.74)$$

Note that here we do not restrict ourself to the massless case $m_{\text{eff}} = 0$. In analogy to equation (9.21) we can now define SUSY partner Hamiltonians by

$$H_+ := \frac{1}{2m_{\text{eff}}v_F^2} AA^\dagger, \qquad H_- := \frac{1}{2m_{\text{eff}}v_F^2} A^\dagger A. \qquad (9.75)$$

For simplicity, we consider here only the (+) pseudospin, that is, $A = A^{(+)}$. The case (−) then trivially follows by reversing the sign in front of k_2 and a. The above partner Hamiltonians explicitly reads

$$H_\pm := \frac{p_1^2}{2m_{\text{eff}}} + \frac{1}{2m_{\text{eff}}}\left[\frac{e}{c}a(x_1) - \hbar k_2\right]^2 \mp \frac{e\hbar}{2m_{\text{eff}}c}a'(x_1) \qquad (9.76)$$

and are standard SUSY partner Hamiltonians of the one-dimensional Witten model when identifying the SUSY potential via

$$\Phi(x_1) := \frac{1}{\sqrt{2m_{\text{eff}}}} \left[\frac{e}{c} a(x_1) - \hbar k_2 \right].$$ (9.77)

As a consequence one is able to design proper vector potentials for which the spectral properties of equation (9.76) are explicitly known. For example, one may select a shape-invariant SUSY potential or utilise one of the conditionally exactly solvable cases. This programme has been executed in [26]. To conclude, let ε_n denote an eigenvalue of equation (9.76) then the corresponding eigenvalues of equation (9.71) read

$$E_n^\pm = \pm \sqrt{2m_{\text{eff}} v_{\text{eff}}^2 \varepsilon_n + m_{\text{eff}}^2 v_{\text{eff}}^4}.$$ (9.78)

9.4 Dirac Hamiltonians with a Lorentz-scalar potential

As a generalisation of the free particle let us discuss the problem of the Dirac Hamiltonian with a Lorentz-scalar potential [14] $\beta\phi_s$, $\phi_s: \mathbb{R}^3 \mapsto \mathbb{R}$:

$$H_D := c\vec{\alpha} \cdot \vec{p} + \beta(mc^2 + \phi_s(\vec{r})).$$ (9.79)

In the supersymmetric representation (9.56) this is again a supersymmetric Dirac operator with

$$A := c\vec{p} \cdot \vec{\sigma} - \mathrm{i}(mc^2 + \phi_s(\vec{r})), \qquad M_\pm = 0.$$ (9.80)

The associated partner Hamiltonians can be put into the form

$$H_\pm = \frac{\vec{p}^2}{2m} + W^2(\vec{r}) \pm \frac{\hbar}{\sqrt{2m}} \vec{\sigma} \cdot (\vec{\nabla} W)(\vec{r}),$$ (9.81)

where we have defined the SUSY potential

$$W(\vec{r}) := \frac{1}{\sqrt{2mc^2}} (\phi_s(\vec{r}) + mc^2).$$ (9.82)

For further simplification we assume that W depends only on one coordinate, say x_3, $W = W(x_3)$. Such a model describes, for example, a x_3-dependent valence- and conduction-band edge of semiconductors near the Γ or L point in the Brillouin zone. The position dependent W characterises a position dependent band gap. Examples for such materials are $Pb_{1-x}Sn_xTe$, $Pb_{1-x}S_xSe$, and $Hg_{1-x}Cd_xTe$. In particular, all these semiconductors show band inversion, that is, conduction and valence band interchange [30]. A typical shape for the potential W, which characterises a $Pb_{1-x}Sn_xTe$ junction, is shown in figure 9.1.

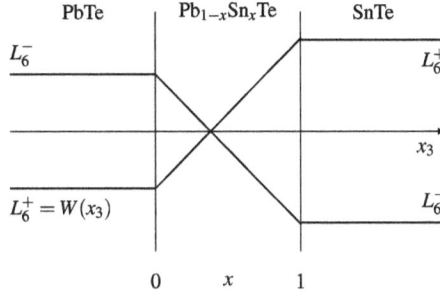

Figure 9.1. The structure of the $L_6^\pm$ band gap of a $Pb_{1-x}Sn_xTe$ junction as a function of the composition parameter x, which itself varies linearly with x_3 along the junction. For details see [30]. The scalar potential W may be identified with the L_6^+ band.

With this assumption the above partner Hamiltonians simplify to

$$H_\pm = \frac{p_1^2}{2m} + \frac{p_2^2}{2m} + H_\pm^{(1)}, \qquad (9.83)$$

where

$$H_\pm^{(1)} := \frac{p_3^2}{2m} + W^2(x_3) \pm \frac{\hbar}{\sqrt{2m}} W'(x_3)\sigma_3 \qquad (9.84)$$

is a pair of one-dimensional SUSY Hamiltonians acting on $L^2(\mathbb{R}) \otimes \mathbb{C}^2$. In fact, the Hamiltonians are of the type of Witten's model and differ only in the overall sign of the SUSY potential W. Our results on the Witten model can immediately be taken over. For the shape of W as given in figure 9.1, that is, having one zero, $H_\pm^{(1)}$ both have zero-energy eigenstates. They are given by

$$\phi_0^{(\pm)}(x_3) = C \exp\left\{\pm \frac{\sqrt{2m}}{\hbar} \int_0^{x_3} dz \; W(z)\right\}\chi_\pm, \qquad (9.85)$$

with spinors

$$\chi_+ := \begin{pmatrix} 1 \\ 0 \end{pmatrix}, \qquad \chi_- := \begin{pmatrix} 0 \\ 1 \end{pmatrix}. \qquad (9.86)$$

In other words, these semiconductors have unpaired spin-up and spin-down states in the conduction and valence band, respectively, which are localised at the junction [12, 13]. Finally, let us note that for $W(x) = mc^2 \tanh x$ the spectrum and eigenfunctions of H_D can explicitly be calculated due to shape invariance as we have shown in section 4.2. Of course, any other shape-invariant SUSY potential will also allow for an exact solution of the eigenvalue problem of H_D [7].

9.5 Dirac electrons in spherical symmetric potentials

In this section we will discuss the Dirac Hamiltonian with external potentials being spherically symmetric. That is, we consider

$$H_D = H_0 + V(r), \tag{9.87}$$

where

$$H_0 := c\vec{\alpha} \cdot \vec{p} + \beta mc^2 = -ic\vec{\alpha} \cdot \vec{\nabla} + \beta mc^2 \tag{9.88}$$

denotes the free Dirac Hamiltonian and

$$V(r) := \phi_e(r) + \beta\phi_s(r) - i\beta(\vec{\alpha} \cdot \vec{e}_r)\phi_t(r) \tag{9.89}$$

is a linear combination of rotationally invariant functions characterising an electric ϕ_e, a Lorentz-scalar ϕ_s, and a tensor potential ϕ_t, respectively [14]. Here $r := |\vec{r}|$, $\vec{e}_r := \vec{r}/r$ and for simplicity we will set $\hbar = 1$ throughout this section. Obviously, the spin angular momentum

$$\vec{S} := -\frac{i}{4}\vec{\alpha} \times \vec{\alpha} = \frac{1}{2}\begin{pmatrix} \vec{\sigma} & 0 \\ 0 & \vec{\sigma} \end{pmatrix}, \tag{9.90}$$

the orbital angular momentum $\vec{L} := \vec{r} \times \vec{p} = -i\vec{r} \times \vec{\nabla}$ and total angular momentum operator $\vec{J} := \vec{L} + \vec{S}$ are conserved quantities as they commute with H_D. In addition to this the relativistic spin–orbit operator

$$K := \beta\left(2\vec{S} \cdot \vec{L} + 1\right) = \beta K_{nr} \tag{9.91}$$

also commutes with H_D. In analogy to the non-relativistic relations (8.45) the relativistic spin–orbit operator obeys the relations

$$K = \beta\left(\vec{J}^2 - \vec{L}^2 + \frac{1}{4}\right), \quad \vec{L}^2 = K(K - \beta), \quad K^2 = \vec{J}^2 + \frac{1}{4}. \tag{9.92}$$

As in the Pauli case let us denote the eigenvalues of $\vec{L}^2$, $\vec{J}^2$, and K by $\ell(\ell + 1)$, $j(j + 1)$, and $-\kappa$. They also obey the relations (8.46) and (8.47) and we can introduce the same reduced quantum number $s = -\text{sgn}\,\kappa$. Note that each set of these quantum numbers is doubly degenerate. Hence, we are led to the decomposition

$$L^2(S^2) \otimes \mathbb{C}^4 = \bigoplus_{j=\frac{1}{2},\frac{3}{2},\dots}^{\infty} \bigoplus_{m_j=-j}^{j} \bigoplus_{s=\pm 1} \mathcal{H}_{jm_j}^{(s)}, \tag{9.93}$$

where $\mathcal{H}_{jm_j}^{(s)} \simeq \mathbb{C}^2$. A joint eigenstate of $\vec{J}^2$, J_3, and K is then given by $|j, m_j, s, \rangle_\pm$, where the additional subindex indicates the basis vectors within $\mathcal{H}_{jm_j}^{(s)}$. In the coordinate representation they may be expressed in terms of the Pauli spinors (8.50) and read [14, 31]

$$\Phi_{j,m_j}^{(s)+}(\theta, \varphi) := \langle\theta, \varphi|j, m_j, s\rangle_+ = \begin{pmatrix} i\varphi_{jm_j}^{(s)}(\theta, \varphi) \\ 0 \end{pmatrix},$$

$$\Phi_{j,m_j}^{(s)-}(\theta, \varphi) := \langle\theta, \varphi|j, m_j, -s\rangle_- = \begin{pmatrix} 0 \\ \varphi_{jm_j}^{(-s)}(\theta, \varphi) \end{pmatrix}. \tag{9.94}$$

These Dirac spinors obviously obey the relation

$$i(\vec{\alpha} \cdot \vec{e}_r)\Phi_{j,m_j}^{(s)\pm}(\theta, \varphi) = \mp\Phi_{j,m_j}^{(s)\mp}(\theta, \varphi) \qquad (9.95)$$

and, hence, the operator $i(\vec{\alpha} \cdot \vec{e}_r)$ leaves the subspace $\mathcal{H}_{jm_j}^{(s)}$ invariant. It transforms states between subindexes for fixed j, m_j, and s. As a consequence it can be represented within $\mathcal{H}_{jm_j}^{(s)}$ by a two-dimensional matrix:

$$i(\vec{\alpha} \cdot \vec{e}_r) = \begin{pmatrix} 0 & 1 \\ -1 & 0 \end{pmatrix}. \qquad (9.96)$$

Hence, it is an operator transforming between states with different indexes as is obvious from equation (9.95). In contrast to the relation (8.60), this operator transforms between positive- and negative-energy states. Note that the operator βmc^2 in this subspace is represented by a diagonal matrix, that is, $\text{diag}(mc^2, -mc^2)$. We also mention that this operator commutes with the spin–orbit operator. Or more general, for any vector operator $\vec{v}$ it is straightforward to show that

$$[K, (\vec{\alpha} \cdot \vec{v})] = 0. \qquad (9.97)$$

Finally let us express the free Dirac Hamiltonian in polar coordinates as follows [14]:

$$H_0 = -ic(\vec{\alpha} \cdot \vec{e}_r)\left[\partial_r + \frac{1}{r} - \frac{\beta K}{r}\right] + \beta mc^2. \qquad (9.98)$$

9.5.1 Supersymmetry of radial Dirac operators

For Dirac Hamiltonians of the form (9.87), that is, being invariant under rotation, it is convenient to decompose the Hilbert space into partial wave subspaces $\mathcal{H}_{jm_j}^{(s)}$ discussed above. In other words, we expand a general Dirac spinor in $\mathcal{H} = L^2(\mathbb{R}^3) \otimes \mathbb{C}^4$ as follows:

$$\Psi(\vec{r}) = \sum_{j,m_j,s} \frac{1}{r}\left[R_{jm_js}^{+}(r)\Phi_{j,m_j}^{(s)+}(\theta, \varphi) + R_{jm_js}^{-}(r)\Phi_{j,m_j}^{(s)-}(\theta, \varphi)\right], \qquad (9.99)$$

where the radial wave functions $R^{\pm}(r)$ are elements in $L^2(\mathbb{R}^+, dr)$. From now on we will drop the indices j, m_j, and s as they are kept constant. Note that here we have as Hilbert space $\mathcal{H}_{jm_j}^{(s)} = L^2(\mathbb{R}^+, dr) \otimes \mathbb{C}^2$ due to the extra $1/r$ factor in equation (9.99). This in effect results in replacing $\partial_r + 1/r$ on $L^2(\mathbb{R}^+, r^2dr)$ by ∂_r on $L^2(\mathbb{R}^+, dr)$. As a consequence the Dirac Hamiltonian restricted to this partial wave subspace explicitly reads [14]

$$H_r := \begin{pmatrix} mc^2 + \phi_s(r) + \phi_e(r) & -c\partial_r + c\kappa/r - \phi_t(r) \\ c\partial_r + c\kappa/r - \phi_t(r) & -mc^2 - \phi_s(r) + \phi_e(r) \end{pmatrix}. \qquad (9.100)$$

This radial Dirac operator may now be investigated by assuming it to be a supersymmetric Dirac operator. For this we identify according to definition 9.1

$$M_{\pm} := mc^2 + \phi_s(r) \pm \phi_e(r), \qquad A := -c\partial_r + c\kappa/r - \phi_t(r). \tag{9.101}$$

Remember that these operators are required to obey the conditions (9.14). These conditions turn out to be very restrictive and result in ϕ_e and ϕ_s both being constants. Without loss of generality we may set both to zero. Hence, SUSY restricts us to the case where $M_{\pm} = mc^2$ but the tensor potential remains arbitrary. The radial Dirac Hamiltonian which exhibits supersymmetry then reads

$$H_r = \begin{pmatrix} mc^2 & c(-\partial_r + W(r)) \\ c(\partial_r + W(r)) & -mc^2 \end{pmatrix}, \tag{9.102}$$

where we introduced the SUSY potential $W(r) := \kappa/r - \phi_t(r)/c$. The SUSY partner Hamiltonians associated with the above Dirac Hamiltonian turn out to be those of the Witten model on the half line

$$H_{\pm} = \frac{1}{2m}[-\partial_r^2 + W^2(r) \mp W'(r)]. \tag{9.103}$$

Or more explicitly

$$H_{\pm} = -\frac{1}{2m}\partial_r^2 + V_{\text{eff}}^{\pm}(r), \tag{9.104}$$

with

$$V_{\text{eff}}^{\pm}(r) := \frac{\kappa(\kappa \pm 1)}{2mr^2} + \frac{1}{2mc^2}\phi_t^2(r) - \frac{\kappa}{mcr}\phi_t(r) \pm \frac{1}{2mc}\phi'_t(r). \tag{9.105}$$

SUSY is unbroken if there exist states $R_0^{\pm}$ such that $A R_0^- = 0$ or $A^{\dagger} R_0^+ = 0$. This immediately results in solutions of the form

$$R_0^-(r) = Cr^{\kappa} \exp\{-U(r)\}, \qquad R_0^+(r) = Cr^{-\kappa} \exp\{U(r)\}, \tag{9.106}$$

where we have introduced the superpotential

$$U(r) := \frac{1}{c} \int^r dr' \, \phi_t(r'). \tag{9.107}$$

Note that in terms of U the effective potentials (9.105) read

$$V_{\text{eff}}^{\pm}(r) := \frac{\kappa(\kappa \pm 1)}{2mr^2} + \frac{1}{2m}U'^2(r) - \frac{\kappa}{mr}U'(r) \pm \frac{1}{2m}U''(r) \tag{9.108}$$

and are identical in form to the effective potentials $V_{\text{eff}}^{\mp}(r)$ of the corresponding Pauli system (8.82) upon replacing κ by ℓ.

Obviously for a superpotential $\lim_{r\to\infty} U(r) = +\infty$ fast enough SUSY is unbroken for $\kappa > 0$, that is, $\kappa = j + \frac{1}{2}$ and the ground state belongs to the negative-energy solutions $R_0^-(r) = Cr^{j+\frac{1}{2}}e^{-U(r)}$. In the other case $\lim_{r\to\infty} U(r) = -\infty$ we require

$\kappa = -j - \frac{1}{2} < 0$ and the SUSY ground state belongs to a positive energy solution $R_0^+(r) = Cr^{j+\frac{1}{2}}e^{U(r)}$.

Let us conclude this section by mentioning that for particular choices of the superpotential the spectral properties of equation (9.104) are explicitly known. Prominent examples of such superpotentials are

$$
\begin{aligned}
&\text{quad.pot.:} \quad U(r) = \frac{m}{2}\omega r^2, \quad V_{\text{eff}}^\pm = \frac{\kappa(\kappa \pm 1)}{2mr^2} + \frac{m}{2}\omega^2 r^2 - \omega\left(\kappa \mp \frac{1}{2}\right) \\
&\text{lin.pot.:} \quad U(r) = \gamma r, \quad V_{\text{eff}}^\pm = \frac{\kappa(\kappa \pm 1)}{2mr^2} + \frac{\gamma^2}{2m} - \frac{\gamma\kappa}{mr} \\
&\text{log.pot.:} \quad U(r) = \gamma \ln r, \quad V_{\text{eff}}^\pm = \frac{(\kappa - \gamma)(\kappa - \gamma \pm 1)}{2mr^2}.
\end{aligned}
\tag{9.109}
$$

The first two cases have been studied in some detail in [32] including the resolvent and the explicit path integration. Here we will limit our discussion in the next section to the first case which is related to the Dirac oscillator.

9.5.2 Spherical tensor potentials and the Dirac oscillator

In 1989 Moshinsky and Szczepaniak [33] introduced the so-called Dirac oscillator by replacing the momentum operator $\vec{p}$ in the free Dirac Hamiltonian by $\vec{p} + i\beta m\omega\vec{r}$. This in essence corresponds to introducing a complex vector potential $\vec{A} = (cm\omega/ie)\beta\vec{r}$ in equation (9.9). Hence, this leads to a Dirac Hamiltonian of the form

$$
H_D = \begin{pmatrix} mc^2 & c\vec{\sigma} \cdot (\vec{p} - im\omega\vec{r}) \\ c\vec{\sigma} \cdot (\vec{p} + im\omega\vec{r}) & -mc^2 \end{pmatrix},
\tag{9.110}
$$

which is of supersymmetric form upon the identification $A = c\vec{\sigma} \cdot (\vec{p} - im\omega\vec{r})$ and $M_\pm = mc^2$. A little calculation shows that the corresponding partner Hamiltonians are given by

$$
H_\pm = \frac{1}{2m}\vec{p}^2 + \frac{m}{2}\omega^2\vec{r}^2 \pm \hbar\omega\left(K_{nr} + \frac{1}{2}\right),
\tag{9.111}
$$

which is the standard non-relativistic harmonic oscillator Hamiltonian with an additional non-relativistic spin–orbit term. As the eigenvalues of both are known we immediately can write down the eigenvalues of $H_\pm$:

$$
\varepsilon_{n,j,s}^+ = \hbar\omega[2(n + 1) + j + sj], \quad \varepsilon_{n,j,s}^- = \hbar\omega[2n + j + 1 - s(j + 1)],
\tag{9.112}
$$

where $n \in \mathbb{N}_0$, $2j \in \mathbb{N}$ and $s \in \{-1, +1\}$. Obviously $\varepsilon_{0,j,1}^- = 0$ for all values of j. Hence, SUSY is unbroken with an infinite degenerate zero-energy eigenvalue belonging to H_-. We also observe the SUSY relation

$$
\varepsilon_{n,j,s}^- = \varepsilon_{n-1,j+1,-s}^+.
\tag{9.113}
$$

The corresponding SUSY transformations are discussed in [32]. Here we only mention that the eigenvalues of the Dirac oscillator Hamiltonian directly follow via equation (9.29) from the above eigenvalues (9.112):

$$E^+_{n, j, s} = mc^2 \left[1 + \frac{2\hbar\omega}{mc^2} [2n + 2 + j + sj] \right]^{1/2},$$

$$E^-_{n, j, s} = - mc^2 \left[1 + \frac{2\hbar\omega}{mc^2} [2n + j + 1 - s(j + 1)] \right]^{1/2}.$$

(9.114)

Following the previous section the Dirac oscillator actually corresponds to the quadratic potential $U(\vec{r}) = \frac{1}{2}m\omega\vec{r}^2$ and can be generalised by replacing the harmonic superpotential by a general $U(\vec{r})$. Hence, we consider now a supercharge defined by $A := c\vec{\sigma} \cdot (\vec{p} - i(\vec{\nabla}U))$, which leads us immediately to the partner Hamiltonians

$$H_\pm = \frac{\vec{p}^2}{2m} + \frac{1}{2m}[(\vec{\nabla}U)^2 \pm (\Delta U)] \pm \frac{1}{m}\vec{\sigma} \cdot (\vec{\nabla}U \times \vec{p}).$$

(9.115)

For the special case of a spherical symmetric superpotential $U(\vec{r}) = U(r)$ this simplifies to

$$H_\pm = \frac{\vec{p}^2}{2m} + \frac{1}{2m}[U'^2(r) \pm U''(r)] \pm \frac{U'(r)}{mr}K_{nr},$$

(9.116)

which are, up to an overall minus sign in U, identical to the Pauli Hamiltonians (8.81).

9.5.3 Dirac Hamiltonians in a spherical background

Let us consider a static spherical background space characterised by a metric of the form

$$ds^2 = w^2c^2dt^2 - \frac{w^2}{u^2}dr^2 - \frac{w^2}{v^2}r^2(d\theta^2 + \sin^2\theta d\varphi^2),$$

(9.117)

where u, v, and w are arbitrary dimensionless functions of the radial variable r. It can be shown that there exists a specific gauge where the free Dirac Hamiltonian in such a background field can be separated in these spherical coordinates just like in the flat space as discussed above. However, the only difference is the form of the radial Dirac Hamiltonian, which now reads [34]

$$H_r := \begin{pmatrix} mc^2w & - cu\partial_r + ck v/r \\ cu\partial_r + ck v/r & - mc^2w \end{pmatrix}$$

(9.118)

and acts on the Hilbert space $\mathcal{H}^{(s)}_{jm_j} = L^2(\mathbb{D}, dr/u) \otimes \mathbb{C}^2$ with $\mathbb{D} \subseteq \mathbb{R}^+$ being the domain of the radial variable r. The domain $\mathbb{D}$ depends on the functions u, v, and w and we have $(u\partial_r)^\dagger = -u\partial_r$ on $\mathcal{H}^{(s)}_{jm_j}$.

As it stands this radial Dirac Hamiltonian is not supersymmetric. However, it has been pointed out [34] that upon a r-dependent rotation of the basis in $\mathbb{C}^2$

$$R(\chi) := \begin{pmatrix} \cos\dfrac{\chi}{2} & -\sin\dfrac{\chi}{2} \\ \sin\dfrac{\chi}{2} & \cos\dfrac{\chi}{2} \end{pmatrix}, \qquad \chi = \chi(r), \tag{9.119}$$

the transformed Hamiltonian

$$\tilde{H}_r := R(\chi)H_r R^T(\chi) \tag{9.120}$$

acquires a supersymmetric form if there exist constants λ and μ such that the functions u, v, w, and χ are related by

$$\frac{1}{2mc}u\chi' = \mu, \quad w\cos\chi = \frac{\kappa v}{mcr}\sin\chi + \lambda. \tag{9.121}$$

Explicitly we have

$$\tilde{H}_r = \begin{pmatrix} M_+ & A \\ A^\dagger & -M_- \end{pmatrix} \tag{9.122}$$

with

$$M_\pm := mc^2(\lambda \pm \mu), \quad A := -cu\partial_r + \frac{c\kappa v}{r}\cos\chi + mc^2 w\sin\chi, \tag{9.123}$$

which represents a supersymmetric Dirac Hamiltonian. Note that here we do not require $M_\pm \geq 0$ as these are constant quantities and the condition $M_\pm \geq 0$ can always be met by a constant shift in Hamiltonian $\tilde{H}_r$.

As an illustrative example let us consider the case of an anti-de-Sitter metric which corresponds to a background space with a negative cosmological constant $\Lambda = -3\omega^2$, $\omega > 0$. Here we denote the radial variable by $\rho \in \mathbb{R}^+$ and the line element reads

$$ds^2 = (1 + \omega^2\rho^2)c^2dt^2 - (1 + \omega^2\rho^2)^{-1}d\rho^2 - \rho^2(d\theta^2 + \sin^2\theta d\varphi^2). \tag{9.124}$$

With a change of the radial variable $\tan\omega r := \omega\rho$, $r \in \mathbb{D} := [0, \pi/2\omega[$, the line element can be put into the form

$$ds^2 = \frac{c^2dt^2}{\cos^2\omega r} - \frac{dr^2}{\cos^2\omega r} - \frac{\tan^2\omega r}{\omega^2}(d\theta^2 + \sin^2\theta d\varphi^2). \tag{9.125}$$

Now we identify $u(r) = 1$, $v(r) = \omega r/\sin\omega r$, $w(r) = 1/\cos\omega r$, and set $\chi(r) = \omega r$. The conditions (9.121) are fulfilled with $\mu = \omega/2mc$ and $\lambda = 1 - \kappa\omega/mc$ and thus lead to a supersymmetric Dirac Hamiltonian where

$$M_\pm := mc^2\left(1 - \frac{\kappa\omega}{mc} \pm \frac{\omega}{2mc}\right), \quad A := -c\partial_r + cW(r), \tag{9.126}$$

with SUSY potential $W(r) := \omega\kappa \cot \omega r + mc \tan \omega r$. Note that this SUSY potential corresponds to a shape-invariant partner potential, the so-called Pöschl–Teller oscillator. Hence the eigenvalue problem for the free Dirac particle in an anti-de-Sitter background is related to the Schrödinger Hamiltonian for the exactly solvable Pöschl–Teller oscillator. In a similar way one can also solve the eigenvalue problem for the free particle in de Sitter space which corresponds to a positive cosmological constant $\Lambda = 3\omega^2$ and in essence follows from the above calculation by replacing ω by $i\omega$, that is, the trigonometric functions become hyperbolic functions. Here the rotation matrix $R \in SU(2)$ now becomes a non-unitary but irreducible two-dimensional representation of $SU(1,1)$. Explicit calculations and results for both cases can be found in [35].

9.6 Supersymmetry of the Dirac Coulomb Hamiltonian

The Dirac Coulomb Hamiltonian characterising an electron in the field of an atomic nucleus composed of neutrons and Z protons is given by

$$H_D := c\vec{\alpha} \cdot \vec{p} + \beta mc^2 - \gamma/r, \qquad \gamma = Ze^2. \tag{9.127}$$

This Hamiltonian does not represent a supersymmetric Dirac operator. However, similar to the Laplace–Runge–Lenz–Pauli vector in the non-relativistic case there exists a dimensionless Johnson–Lippmann–Biedenharn operator [36, 37] defined by

$$B := \vec{\sigma} \cdot \vec{e}_r - \frac{i}{\gamma mc} K\gamma_5(H_D - \beta mc^2), \qquad \gamma_5 := -i\alpha_1\alpha_2\alpha_3, \tag{9.128}$$

which commutes with H_D. It may be put into the form

$$B = \vec{\sigma} \cdot \left(\vec{e}_r + \frac{\beta}{2m\gamma}\left(\vec{L} \times \vec{p} - \vec{p} \times \vec{L}\right) \right) + \frac{i}{mcr}K\gamma_5 \tag{9.129}$$

and thus motivates the definition of a supercharge similar to that in the Pauli case (8.70)

$$Q_1 := \sqrt{\frac{m\gamma^2}{4K^2}} B. \tag{9.130}$$

The corresponding SUSY Hamiltonian then reads

$$H_{\text{SUSY}} = 2Q_1^2 = \frac{1}{2mc^2}\left(H_D^2 - m^2c^4\right) + \frac{m\gamma^2}{2K^2} \tag{9.131}$$

which is the relativistic version of equation (8.71).

The SUSY structure can also be utilised to explicitly solve the eigenvalue problem for H_D. This has first been discussed by Sukumar [38] followed by Jarvis and Stedman [39]. See also the books by Thaller [14] and Hirshfeld [15]. Here we conclude by mentioning that SUSY is unbroken, that is, there exists an eigen-subspace of K where $H_{\text{SUSY}} = 0$. This subspace belongs to $\mathcal{H}^+$. Replacing in this subspace K by its eigenvalue $-\kappa$ the relation (9.131) leads us to the ground state energy

$$E_{0,\kappa}^+ = mc^2\sqrt{1 - \frac{\gamma^2}{c^2\kappa^2}} = mc^2\left[1 + \frac{\gamma^2/c^2}{\kappa^2 - \gamma^2/c^2}\right]^{-1/2}. \qquad (9.132)$$

Note that the complete discrete positive part of the spectrum of the Dirac Coulomb Hamiltonian is given by

$$E_{n_r,\kappa}^+ = mc^2\left[1 + \frac{\gamma^2/c^2}{\left(n_r + \sqrt{\kappa^2 - \gamma^2/c^2}\right)^2}\right]^{-1/2}, \qquad n_r \in \mathbb{N}_0. \qquad (9.133)$$

9.7 The relativistic Witten model

Let us consider the following Dirac Hamiltonian in one dimension:

$$H_D := cp\sigma_1 + W(x)\sigma_2 + mc^2\sigma_3, \qquad (9.134)$$

which acts on the Hilbert space $\mathcal{H} := L^2(\mathbb{R}) \otimes \mathbb{C}^2$. Here p is the usual linear momentum operator and W is a differentiable real-valued function of the coordinate x. Writing this Hamiltonian in the standard representation of the Pauli matrices,

$$H_D = \begin{pmatrix} mc^2 & cp - iW(x) \\ cp + iW(x) & -mc^2 \end{pmatrix}, \qquad (9.135)$$

the function W may be viewed as the one-dimensional version of a tensor potential. This Hamiltonian obviously represents a supersymmetric Dirac operator with

$$A := cp - iW(x), \qquad M_\pm = mc^2 \qquad (9.136)$$

and the associated SUSY partner Hamiltonians are given by

$$H_\pm = \frac{p^2}{2m} + \frac{1}{2mc^2}W^2(x) \pm \frac{\hbar}{2mc}W'(x). \qquad (9.137)$$

With the identification of the SUSY potential $\Phi(x) := W(x)/\sqrt{2mc^2}$ we arrive at the one-dimensional Witten model extensively discussed in the previous chapters. Let us assume that the SUSY potential belongs to one of the shape-invariant ones or is one of the conditionally solvable ones. In that case we know the spectral properties of $H_\pm$ and can derive the spectral properties of the associated Dirac Hamiltonian by utilising the relations (9.26)–(9.30). For example, let $\varepsilon_n > 0$ be the common eigenvalues of $H_\pm$ then the associated eigenvalues of H_D are given by

$$E_n^\pm = \pm mc^2\sqrt{1 + \frac{2\varepsilon_n}{mc^2}}. \qquad (9.138)$$

In addition, in case of unbroken SUSY H_D has the eigenvalue $E_0^+ = mc^2$ or $E_0^- = -mc^2$ when $\varepsilon_0 = 0$ belongs to the spectrum of H_+ or H_-, respectively. Taking the non-relativistic limit it is obvious that

$$\lim_{c \to \infty} (E_n^+ - mc^2) = \varepsilon_n. \tag{9.139}$$

Hence it is justified to call the Hamiltonian (9.134) the relativistic version of Witten's model.

Finally let us note that whenever the Green's function of the non-relativistic Witten model is known one may be able to calculate also the Green's function of the corresponding relativistic model following our discussion in section 9.1.3. For the free particle the promotor (9.46) explicitly reads with $\mu^2(\zeta) := \zeta/c^2 - m^2c^2$

$$P_\zeta^\pm(x'', x'; t) = \sqrt{\frac{m}{2\pi i\hbar t}} \exp\left\{\frac{i}{\hbar}\left(\frac{m}{2t}|x'' - x'|^2 + \frac{t}{2m}\mu^2(\zeta)\right)\right\}. \tag{9.140}$$

The time-integration in equation (9.45) can be done and leads to the iterated resolvent

$$g^\pm(x'', x'; \zeta) = \frac{i}{2\mu(\zeta)c^2\hbar} \exp\{i\mu(\zeta)|x'' - x'|/\hbar\}, \tag{9.141}$$

which finally results in the Green's function of the relativistic free particle in one dimension:

$$G(x'', x', z) = \frac{i e^{i\mu(z^2)|x'' - x'|/\hbar}}{2\mu(z^2)c^2\hbar}[ic\mu(z^2)\mathrm{sgn}(x'' - x')\sigma_2 + mc^2\sigma_3 + z]. \tag{9.142}$$

Let us also note that the discussion on the quasi-classical SUSY approximation of the non-relativistic Witten model in chapter 6 can directly be applied to the relativistic model discussed here [40].

9.8 Problems

Problem 9.1. Supersymmetric representation of Dirac matrices

Show that the Dirac matrices in the supersymmetric representation

$$\vec{\alpha} = \begin{pmatrix} 0 & \vec{\sigma} \\ \vec{\sigma} & 0 \end{pmatrix}, \qquad \beta = \begin{pmatrix} 0 & -i \\ i & 0 \end{pmatrix}$$

obey the Dirac algebra (9.2).

Problem 9.2. The three-dimensional Dirac Hamiltonian with magnetic field

Consider a Dirac particle with rest mass m and charge e in an external magnetic field defined via a vector potential $\vec{A}: \mathbb{R}^3 \to \mathbb{R}^3$. Show that the corresponding Dirac Hamiltonian is supersymmetric with

$$A := c\left(\vec{p} - \frac{e}{c}\vec{A}\right) \cdot \vec{\sigma} = A^{\dagger}, \qquad M_{\pm} := mc^2.$$

Show that the corresponding partner Hamiltonians (9.21) are those of the non-relativistic Pauli Hamiltonian (8.5), that is,

$$H_{\pm} = \frac{1}{2m}\left[\left(\vec{p} - \frac{e}{c}\vec{A}\right) \cdot \vec{\sigma}\right]^2 = \frac{1}{2m}\left(\vec{p} - \frac{e}{c}\vec{A}\right)^2 - \frac{e\hbar}{2mc}\vec{B} \cdot \vec{\sigma}.$$

Problem 9.3. Proofs for the Foldy–Wouthuysen transformation

According to definition 9.1 a supersymmetric Dirac Hamiltonian is given by

$$H_D = Q_1 + \mathcal{M}W, \quad \text{with} \quad \{Q_1, W\} = 0, \quad [\mathcal{M}, W] = 0, \quad [Q_1, \mathcal{M}] = 0.$$

(a) Show that $[H_D^2, \mathcal{M}] = 0$ and hence $|H_D| := \sqrt{H_D^2}$ also commutes with $\mathcal{M}$, that is, $[|H_D|, \mathcal{M}] = 0$.

(b) Show that the operators $a_{\pm} := \sqrt{\frac{1}{2} \pm \frac{\mathcal{M}}{2|H_D|}}$ are self-adjoint, commute with each other as well as with H_D and Q_1, and obey the relations

$$a_+^2 + a_-^2 = 1, \quad a_+^2 - a_-^2 = \frac{\mathcal{M}}{|H_D|}, \quad 2a_+a_- = \frac{|Q_1|}{|H_D|}.$$

(c) Show that $\operatorname{sgn} Q_1 := Q_1|Q_1|^{-1} = |Q_1|^{-1}Q_1$ is an odd operator, that is,

$$\{\operatorname{sgn} Q_1, W\} = 0, \quad [\operatorname{sgn} Q_1, \mathcal{M}] = 0$$

and prove $\{H_D, W \operatorname{sgn} Q_1\} = 0$.

(d) Show that the operator $U := a_+ + W \operatorname{sgn} Q_1 a_-$ is unitary, i.e. $UU^{\dagger} = 1 = U^{\dagger}U$.

(e) Prove the Foldy–Wouthuysen transformation (9.24), that is, $UH_D U^{\dagger} = W|H_D|$.

Problem 9.4. The Helicity eigenstates of the free Dirac particle

Prove that the states $\chi_{\lambda}(\vec{k})$ defined in equation (9.52) are eigenstates of the helicity operator $\Lambda = \vec{\sigma} \cdot \vec{p}/|\vec{p}|$ with eigenvalue $\lambda = \pm 1$ and show that for fixed $\vec{k}$ they form an ortho-normal basis in $\mathbb{C}^2$. Show that for the free particle case the Foldy–Wouthuysen transformation operator U takes the form given in equation (9.54) and explicitly calculate the free particle solution for positive and negative energy.

Problem 9.5. Some relations for radial Dirac operators

(a) Show that for an arbitrary vector operator $\vec{v}$ the relativistic spin–orbit operator K obeys relation (9.97), i.e. $[K, (\vec{\alpha} \cdot \vec{v})] = 0$.

(b) Consider the operators $M_\pm$ and A as defined in equation (9.101) and show that SUSY requires that the electric potential and the Lorentz-scalar potential are constant.

(c) Show that for the first two examples in equation (9.109) SUSY is unbroken and calculate the corresponding ground states.

(d) Derive the eigenvalues (9.112) for the supersymmetric Dirac oscillator.

Problem 9.6. Dirac Hamiltonian in anti-de-Sitter space

Show that for the metric (9.125) the rotation angle $\chi(r) = \omega r$ obeys the conditions (9.121) with $\mu = \omega/2mc$ and $\lambda = 1 - \kappa\omega/mc$ and the free radial Dirac Hamiltonian in anti-de-Sitter space explicitly reads

$$H_r = c\omega \begin{pmatrix} k\sec\chi & -\partial_\chi + \kappa\csc\chi \\ \partial_\chi + \kappa\csc\chi & -k\sec\chi \end{pmatrix} \quad \text{with} \quad k := \frac{mc}{\omega}.$$

Calculate the transformed Hamiltonian $\tilde{H}_r := R(\chi)H_r R^T(\chi)$ and show that the result

$$\tilde{H}_r = c\omega \begin{pmatrix} k - \kappa + 1/2 & -\partial_\chi + \kappa\cot\chi + k\tan\chi \\ \partial_\chi + \kappa\cot\chi + k\tan\chi & -(k - \kappa - 1/2) \end{pmatrix}$$

represents a supersymmetric Dirac Hamiltonian according to equation (9.126) with SUSY potential $W(r) = \omega\kappa\cot\omega r + mc\tan\omega r$. Consider the limit for small ω and show that in first order $\tilde{H}_r$ represents the Dirac oscillator with frequency $\omega = \sqrt{-\Lambda/3}$.

References

[1] Dirac P A M 1928 The quantum theory of the electron *Proc. R. Soc. Lond.* A **117** 610–24
[2] Dirac P A M 1958 *The Principles of Quantum Mechanics* 4th edn (Oxford: University Press)
[3] Feynman R P 1961 *Quantum Electrodynamics* (New York: Benjamin)
[4] Jackiw R 1984 Fractional charge and zero modes for planar systems in a magnetic field *Phys. Rev.* D **29** 2375–7
[5] Ui H 1984 Supersymmetric quantum mechanics and fermion in a gauge field of (1+2) dimensions *Prog. Theor. Phys.* **72** 192–3
[6] Hughes R J, Kostelecky V A and Nieto M M 1986 Supersymmetric quantum mechanics in a first-order Dirac equation *Phys. Rev.* D **34** 1100–6
[7] Cooper F, Khare A, Musto R and Wipf A 1988 Supersymmetry and the Dirac equation *Ann. Phys.* **187** 1–28
[8] Beckers J and Debergh N 1990 Supersymmetry, Foldy–Wouthuysen transformations and relativistic oscillators *Phys. Rev.* D **42** 1255–9
[9] Grover T, Sheng D N and Vishwanath A 2014 Emergent space–time supersymmetry at the boundary of a topological phase *Science* **344** 280–3
[10] Nemec N 2007 Quantum transport in carbon-based nanostructures *PhD Thesis* Universität Regensburg
[11] Sarma S D, Adam S, Hwang E H and Rossi E 2011 Electronic transport in two-dimensional graphene *Rev. Mod. Phys.* **83** 407–70

[12] Pankratov O A 1987 Supersymmetric inhomogeneous semiconductor structures and the nature of a parity anomaly in (2 + 1) electrodynamics *Phys. Lett.* A **121** 360–6

[13] Pankratov O A, Pakhomov S V and Volkov B A 1987 Supersymmetry in heterojunctions: band-inverting contact on the basis of $Pb_{1-x}Sn_xTe$ and $Hg_{1-x}Cd_xTe$ *Solid State Commun.* **61** 93–6

[14] Thaller B 1992 *The Dirac Equation* (Berlin: Springer)

[15] Hirshfeld A 2012 *The Supersymmetric Dirac Equation* (London: Imperial College Press)

[16] Pauli W 1927 Zur Quantenmechanik des magnetischen Elektrons *Z. Phys.* **43** 601–23

[17] Foldy L L and Wouthuysen S A 1950 On the Dirac theory of spin 1/2 particles and its non-relativistic limit *Phys. Rev.* **78** 29–36

[18] Thaller B 1988 Normal forms of an abstract Dirac operator and applications to scattering theory *J. Math. Phys.* **29** 249–57

[19] Thaller B 1991 Dirac particles in magnetic fields *Recent Developments in Quantum Mechanics Mathematical Physics Studies Nr. 12* ed A Boutet de Monvel, P Dita, G Nenciu and R Purice (Dordrecht: Kluwer) pp 351–66

[20] Inomata A 1986 Remarks on the time transformation technique for path integration *Path Integrals from meV to MeV* ed M C Gutzwiller, A Inomata, J R Klauder and L Streit (Singapore: World Scientific) pp 433–48

[21] Kayed M A and Inomata A 1984 Exact path-integral solution of the Dirac–Coulomb problem *Phys. Rev. Lett.* **53** 107–10

[22] Fock V 1928 Bemerkung zur Quantelung des harmonischen Oszillators im Magnetfeld *Z. Phys.* **47** 446–8

[23] Landau L 1930 Diamagnetismus der Metalle *Z. Phys.* **64** 629–37

[24] Ezawa M 2008 Supersymmetry and unconventional quantum Hall effect in graphene *Phys. Lett.* A **372** 924–9

[25] De Martino A, Dell'Anna L and Egger R 2007 Magnetic confinement of massless Dirac fermions in graphene *Phys. Rev. Lett.* **98** 066802

[26] Midya B and Fernandez D J 2014 Dirac electron in graphene under supersymmetry generated magnetic fields *J. Phys. A: Math. Theor.* **47** 285302

[27] Katsnelson M I, Novoselov K S and Geim A K 2006 Chiral tunneling and the Klein paradox in graphene *Nat. Phys.* **2** 620–5

[28] Young A F and Kim P 2009 Quantum interference and Klein tunnelling in graphene heterojunctions *Nat. Phys.* **5** 222–6

[29] Jakubský V, Nieto L-M and Plyushchay M S 2001 Klein tunneling in carbon nanaostructures: a free-particle dynamics in disguise *Phys. Rev.* D **83** 047702

[30] Dornhaus R, Nimtz G and Schlicht B 1983 *Narrow-Gap Semiconductors, Springer Tracts in Modern Physics* vol 98 (Berlin: Springer)

[31] Bjorken J D and Drell S D 1964 *Relativistic Quantum Mechanics* (New York: McGraw-Hill)

[32] Junker G and Inomata A 2018 Path integral and spectral representations for supersymmetric Dirac–Hamiltonians *J. Math. Phys.* **59** 052301

[33] Moshinsky M and Szczepaniak A 1989 The Dirac oscillator *J. Phys. A: Math. Gen.* **22** L817–9

[34] Cotăescu I 1998 The Dirac particle on central backgrounds and the anti-de Sitter oscillator *Mod. Phys. Lett.* A **13** 2923–36

[35] Cotăescu I 2007 Dirac fermions in de Sitter and anti-de Sitter backgrounds *Rom. J. Phys.* **52** 895–940 http://www.nipne.ro/rjp/2007_52_9-10/0895_0940.pdf

[36] Johnson M H and Lippman B A 1950 Relativistic Kepler problem *Phys. Rev.* **78** 329

[37] Biedenharn L C 1962 Remarks on the relativistic Kepler problem *Phys. Rev.* **126** 845–51

[38] Sukumar C V 1985 Supersymmetry and the Dirac equation for a central Coulomb field *J. Phys. A: Math. Gen.* **18** L697–701

[39] Jarvis P D and Stedman G E 1986 Supersymmetry in second-order relativistic equations for the hydrogen atom *J. Phys. A: Math. Gen.* **19** 1373–85

[40] Haouat S and Chetouani L 2008 The (1 + 1)-dimensional Dirac equation with pseudoscalar potentials: quasi-classical approximation *Phys. Scr.* **78** 065005

IOP Publishing

Supersymmetric Methods in Quantum, Statistical and Solid
State Physics
Enlarged and revised edition
Georg Junker

Chapter 10

Concluding remarks and overview

During the last four decades supersymmetric quantum mechanics has attracted much attention in various branches of physics. The algebraic tools provided by supersymmetry can indeed be utilised in many areas of theoretical physics and have also become a valuable concept in mathematics. Starting with the previous version of this book [1] several authors have undertaken the challenge to write textbook dedicated exclusively to the topic of supersymmetric quantum mechanics and its applications [2–6]. All of these monographs have their own focus on particular aspects of this topic. Some concentrate in essence on the one-dimensional Witten model, others are more dedicated to supersymmetric aspects of particular systems. Some aim at an audience starting from an undergraduate level, others are more suitable for graduate students and professionals. The current book aims at a broader application of supersymmetric methods going beyond the quantum mechanical considerations, by covering topics from statistical and solid-state physics. In this respect the above mentioned monographs are more complementary to each other than competitors.

In the current book we start with a very general concept of supersymmetric quantum mechanics and discuss its implications for $N \geqslant 2$ SUSY. We introduce the Witten model, which is the prototype of a supersymmetric model in quantum mechanics. This model then reappears in many applications studied in the following chapters. The presented discussion on shape-invariance, its relation to Darboux's method, and the factorisation method of Schrödinger, Infeld, and Hull is limited to the so-called translational shape-invariance of the potential parameters. For a discussion on the scaling relations of these parameters see [3]. Here, we put some more focus on the modelling of exactly solvable problems. As an example, we construct a family of partner potentials related to the harmonic oscillator and the induced non-linear algebra obeyed by the corresponding creation and annihilation operators including the associated coherent states. This is followed by a discussion of a simple model of supersymmetric classical mechanics, which upon quantisation leads to Witten's model of supersymmetric quantum mechanics. Here we introduced the fermionic phase (5.19) characterising the classical solution of the Grassmannian degrees of freedom. It is this fermionic phase and its close relation to the Witten index which is the key ingredient of the resulting quasi-classical approximation (6.32) of the energy eigenvalues. The contribution of this additional phase is responsible for the fact that it respects the supersymmetric degeneracy of the eigenvalues. In combination with the shape-invariance condition this approximation even leads to exact eigenvalues. In chapter 7 we analyse the supersymmetric structure of classical systems obeying a stochastic dynamics ruled by the Langevin or Fokker–Planck equation. The implications of supersymmetry are discussed for both cases, good and broken SUSY. The idea of constructing conditionally exactly solvable quantum potentials is utilised to construct new drift potentials where the decay rates and decay modes can be explicitly calculated. The shape-invariance of drift potentials is utilised here to obtain identities between the transition probabilities of the associated stochastic processes. Chapter 8 is dedicated to the discussion of Pauli Hamiltonians. Here the problem of an electron moving in an external magnetic field is the prototype of a supersymmetric system in higher dimensions. In particular, the two-dimensional motion in an orthogonal magnetic field exhibits the $N = 2$ SUSY structure. This can be utilised to derive the paramagnetic susceptibility of a free electron gas in two and three space dimensions. What is less know is the fact that the Pauli electron moving in an external spherical symmetric potential also exhibits an $N = 2$ SUSY structure. This becomes visible when restricting the discussion onto the sub-Hilbert space with a fixed total angular momentum. Here the supercharges transform between spaces where orbital angular momentum and spin are parallel and anti-parallel, respectively. For the Coulomb problem this supercharge is in essence given by the projection of the spin onto the Laplace–Runge–Lenz–Pauli vector. Finally chapter 9 introduces so-called supersymmetric Dirac Hamiltonians. The square of a supersymmetric Dirac Hamiltonian can be related to a Pauli-like Hamiltonian and becomes block diagonal. This allows one to relate the supersymmetric relativistic Dirac Hamiltonian with a standard non-relativistic Schrödinger-like $N = 2$ SUSY Hamiltonian and enables us to provide a path-integral representation for the associated Green's function. Of particular interest here is the Dirac Hamiltonian in two space dimensions with an external magnetic field. This has found several applications in solid-state physics; examples are the transport properties of nanotubes and graphene. While high-energy experiments still lack any signature of supersymmetric elementary particles, the environment provided within topological superconductors is one where a space–time SUSY structure may emerge.

Obviously there are many more applications of supersymmetric methods in quantum mechanics and mathematical, statistical, solid-state, atomic, and nuclear physics. Only some of them are covered in the present monograph. For this reason we present an overview of topics, including many not covered here, in table 10.1 below. This list is by no means meant to be complete and the given references are certainly fragmentary. Clearly there are many more areas and references which

Table 10.1. Some applications of SUSY in theoretical and mathematical physics.

Quantum mechanics	References
Exactly solvable potentials, shape-invariance	section 4.2[a]
Duality	[8, 9]
Quasi-exactly solvable potentials	[10]
Conditionally exactly solvable potentials	section 4.3.1[a]
Singular potentials	[5, 11]
Complex potentials and PT symmetry	[12–17]
Quasi-classical approximation	[18, 19], chapter 6[a]
Tunnelling	[3, 20, 21] [a]
Variational approach	[3, 22]
δ-expansion method	[3, 23]
Large-N expansion	[3, 21, 24]
Inverse scattering method	[3, 25, 26]
Bounds on ground-state energy	[27]
Level ordering	[28–30] [a]
Pauli Hamiltonian	[6, 31] [a], chapter 8[a]
Dirac Hamiltonian	[6, 32] [a], chapter 9[a]
Klein–Gordon equation	[33]
Lamé equation	[34]
Coherent states	[35–38]
Supercoherent states, path integrals	[39–42]
N-fold SUSY	[43, 44]
Berry and other phases	[45–48]
Fermionic phase	section 5.4[a]
Integrability, many-body problems	[49]
SUSY breaking, instantons	[50–52]
Korteweg–de Vries equation, solitons	[32, 52–54] [a]
Fractional SUSY	[55–57]
Anyons	[58, 59]
Quantum chaos	[60]
Para- and orthosupersymmetry	[61, 62] [a]
Superconformal symmetry	[63–67]
q-deformation, deformed shape-invariance	[68–71]
SUSY and geometric motion	[72]
SUSY on Riemann surfaces	[73, 74]
Quantum optics	[75]

Mathematical physics	References
Pseudoclassical mechanics	chapter 5[a]
Exceptional orthogonal polynomials	[76, 77]
Morse theory	[78, 79]
Atiyah–Singer index theorem	[80, 81]
Localisation techniques	[82–84] [a]

(*Continued*)

Riemann zeta function	[85–88]
Sigma models	[89, 90]
Fermion algebras	[91]

Statistical physics	References
Fokker–Planck equation	section 7.2[a]
Langevin equation	section 7.3[a]
Nicolai map	[92, 93]
Coloured noise	[94]
Grassmann Brownian motion	[95, 96]
Random walks	[97, 98]
Pauli paramagnetism	[99], section 8.2
Supersymmetric Hubbard models	[100, 101]
Supersymmetric quantum spherical model	[102, 103]
Integrable quantum chains	[104, 105]
Fractional statistics	[106]
Random matrices	[107, 108]

Solid-state physics	References
Disordered systems	[109, 110]
Classical diffusion	[111] [a]
Electron localisation–delocalisation	[112, 113] [a]
Density of states	[114–116] [a]
Conductivity	[117] [a]
Correlation function	[118]
Jahn–Teller systems	[119]
Lattice models	[120, 121]
Band structure	[122–127], section 9.4
Nanophysics	[128–132]
Graphene	[133–135], section 9.3
Quantum Hall effect	[136–139]
Topological superconductors	[140, 141]

Atomic physics	References
Alkali-metal atoms	[142, 143]
Quantum-defect theory	[144, 145]
Stark effect	[146]
Penning trap	[147]
Rydberg atoms, superrevivals	[148–150] [a]

Nuclear physics	References
Dynamical supersymmetry	[151, 152] [a]
Classification of spectra	[153]

Nuclear–nuclear potential	[154]
Collective excitations	[155]
Superdeformed nuclei	[156]

Others	References
Cosmology	[157–161]
Black holes	[162]
QCD	[163, 164]
Gross–Neveu model	[165]
M-theory	[166, 167]
F-theory	[168]

[a] And references therein.

should have been mentioned in this table but are left out due to my ignorance. A broad collection of topics can also be found in [7].

This book does not cover any aspects of supersymmetric quantum field theories. There exists a vast number of excellent books on supersymmetric quantum field theories including gauge theories, string theory, and supergravity. As we have completely ignored this topic we provide here a, certainly incomplete, list:

- Buchbinder I L and Kuzenko S M 1998 *Ideas and Methods of Supersymmetry and Supergravity* revised edn (Boca Raton, FL: Taylor and Francis)
- Castellani L, D'Auria R and Fré P 1991 *Supergravity and Superstrings* (Singapore: World Scientific)
- Dine M 2015 *Supersymmetry and String Theory: Beyond the Standard Model* (Cambridge: Cambridge University Press)
- Freed D S 1999 *Five Lectures on Supersymmetry* (Providence, RI: American Mathematical Society)
- Freund P G O 1986 *Introduction to Supersymmetry* (London: Cambridge University Press)
- Gates S J, Grisaru M T, Roček M and Siegel W 1983 Superspace: one thousand and one lessons in supersymmetry *Front. Phys.* 58 1–54; see also 2009 (Gainesville: University Press of Florida) and arXiv:hep-th/0108200
- Gieres F 1988 *Geometry of Supersymmetric Gauge Theories* (*Lecture Notes in Physics* vol 302) (Berlin: Springer)
- Green M B, Schwarz J H and Witten E 2012 *Superstring Theory* vols 1 and 2, 25th edn (Cambridge: Cambridge University Press)
- Kaku M 1999 *Introduction to Superstrings and M-Theory* 2nd edn (New York: Springer)
- Lopusanski J 1991 *Introduction to Symmetry and Supersymmetry in Quantum Field Theory* (Singapore: World Scientific)
- Mohapatra R N 2003 *Unification and Supersymmetry* 3rd edn (New York: Springer)

- Müller M 1989 *Consistent Classical Supergravity Theories* (*Lecture Notes in Physics* vol 336) (Berlin: Springer)
- Müller-Kirsten H J W and Wiedemann A 1987 *Supersymmetry: An Introduction with Conceptual and Calculational Details* (Singapore: World Scientific)
- Olive D 2009 *Duality and Supersymmetric Theories* (Cambridge: Cambridge University Press)
- Piguet O and Sibold K 1986 *Renormalized Supersymmetry* (Boston, MA: Birkhäuser)
- van Nieuwenhuizen P 1989 *Anomalies in Quantum Field Theories: Cancellation of Anomalies in d = 10 Supergravity* (Leuven: Leuven University Press) Wess J and Bagger J 1992 *Supersymmetry and Supergravity* 2nd edn (Princeton, NJ: Princeton University Press)
- West P 1990 *Introduction to Supersymmetry and Supergravity* 2nd edn (Singapore: World Scientific)

References

[1] Junker G 1996 *Supersymmetric Methods in Quantum and Statistical Physics* 1st edn (Berlin: Springer)

[2] Kalka H and Soff G 1997 *Supersymmetrie* (Stuttgart: Teubner)

[3] Cooper F, Khare A and Sukhatme U 2001 *Supersymmetry in Quantum Mechanics* (Singapore: World Scientific)

[4] Bagchi B 2001 *Supersymmetry in Quantum and Classical Mechanics* (Boca Raton, FL: Chapman and Hall/CRC)

[5] Gangopadhyaya A, Mallow J V and Rasinariu C 2011 *Supersymmetric Quantum Mechanics: An Introduction* (Singapore: World Scientific)

[6] Hirshfeld A 2012 *The Supersymmetric Dirac Equation: The Application to Hydrogenic Atoms* (London: Imperial College Press)

[7] Duplij S, Siegel W and Bagger J 2004 *Concise Encyclopedia of Supersymmetry and Noncommutative Structures in Mathematics and Physics* (Dordrecht: Kluwer)

[8] Kostelecky V A, Nieto M M and Truax D R 1985 Supersymmetry and the relationship between the Coulomb and oscillator problems in arbitrary dimensions *Phys. Rev.* D **32** 2627–33

[9] Simon D S 2002 Duality and supersymmetric quantum mechanics *J. Phys. A: Math. Gen.* **35** 4143–50

[10] Jatkar D P, Kumar C N and Khare A 1989 A quasi-exactly solvable problem without $Sl(2)$ symmetry *Phys. Lett.* A **142** 200–2

[11] Znojil M 2011 Planarizable supersymmetric quantum toboggans *SIGMA* **7** 018

[12] Cannata F, Junker G and Trost J 1998 Schrödinger operators with complex potential but real spectrum *Phys. Lett.* A **246** 219–26

[13] Mostafazadeh A 2004 Statistical origin of pseudo-Hermitian supersymmetry and pseudo-Hermitian fermions *J. Phys. A: Math. Gen.* **37** 10193–207

[14] Znojil M 2004 PT-symmetric regularizations in supersymmetric quantum mechanics *J. Phys. A: Math. Gen.* **37** 10209–22

[15] Correa F and Plyushchay M S 2012 Self-isospectral tri-supersymmetry in PT-symmetric quantum systems with pure imaginary periodicity *Ann. Phys.* **327** 1761–83

[16] Rosas-Ortiz O, Castaos O and Schuch D 2015 New supersymmetry-generated complex potentials with real spectra *J. Phys. A: Math. Theor.* **8** 445302

[17] Ghosha P and Roy P 2016 An analysis of the zero energy states in graphene *Phys. Lett.* A **380** 567–9

[18] Robnik M and Salasnich L 1997 WKB expansion for the angular momentum and the Kepler problem: from the torus quantization to the exact one *J. Phys. A: Math Gen.* **30** 1719–29

[19] Salasnich L and Sattin F 1997 On the convergence of the WKB series for the angular momentum operator *J. Phys. A: Math. Gen.* **30** 7597–602

[20] Keung W-Y, Kovacs E and Sukhatme U 1988 Supersymmetry and double-well potentials *Phys. Rev. Lett.* **60** 41–4

[21] Roy B, Roy P and Roychoudhury R 1991 On solutions of quantum eigenvalue problems: a supersymmetric approach *Fortschr. Phys.* **89** 211–58

[22] Gozzi E, Reuter M and Thacker W D 1993 Variational methods via supersymmetric techniques *Phys. Lett.* A **183** 29–32

[23] Cooper F and Roy P 1990 δ expansion for the superpotential *Phys. Lett.* A **143** 202–6

[24] Imbo T and Sukhatme U 1985 Supersymmetric quantum mechanics and large-N expansion *Phys. Rev. Lett.* **54** 2184–7

[25] Nieto M M 1984 Relationship between supersymmetry and the inverse method in quantum mechanics *Phys. Lett.* B **145B** 208–10

[26] Sukumar C V 1985 Supersymmetric quantum mechanics and the inverse scattering method *J. Phys. A: Math. Gen.* **18** 2937–55

[27] Schmutz M 1985 The factorization method and ground state bounds *Phys. Lett.* A **108** 195–6

[28] Baumgartner B, Grosse H and Martin A 1984 The Laplacian of the potential and the order of energy levels *Phys. Lett.* B **146** 363–6

[29] Grosse H 1991 Supersymmetric quantum mechanics *Recent Developments in Quantum Mechanics Mathematical Physics Studies* vol 12 ed A Boutet de Monvel, P Dita, G Nenciu and R Purice (Dordrecht: Kluwer) pp 299–327

[30] Grosse H and Martin A 1997 *Particle Physics and the Schrödinger Equation* (Cambridge: Cambridge University Press)

[31] Gendenshteîn L É and Krive I V 1985 Supersymmetry in quantum mechanics *Sov. Phys. Usp.* **28** 645–66

[32] Thaller B 1992 *The Dirac Equation* (Berlin: Springer)

[33] Leung P T, van den Brink A M, Suen W M, Wong C W and Young K 2001 SUSY transformations for quasinormal modes of open systems *J. Math. Phys.* **42** 4802–20

[34] Correa F, Nieto L-M and Plyushchay M S 2007 Hidden nonlinear supersymmetry of finite-gap Lamé equation *Phys. Lett.* B **644** 94–8

[35] Fukui T and Aizawa N 1993 Shape-invariant potentials and an associated coherent state *Phys. Lett.* A **180** 308–13

[36] Bagrov V G and Samsonov B F 1996 Coherent states of anharmonic oscillators with a quasiequidistant spectrum *JETP* **82** 593–9

[37] Fernández D J, Hussin V and Rosas-Ortiz O 2007 Coherent states for Hamiltonians generated by supersymmetry *J. Phys. A: Math. Theor.* **40** 6491–511

[38] Chou C-C, Biamonte M T, Bodmann B G and Kouri D J 2012 New system-specific coherent states for bound state calculations *J. Phys. A: Math. Theor.* **45** 505302

[39] Balantekin A B, Schmitt H A and Barrett B R 1988 Coherent states for the harmonic oscillator representations of the orthosymplectic supergroup Osp(1/2N,R) *J. Math. Phys.* **29** 1634–9

[40] Balantekin A B, Schmitt H A and Halse P 1989 Coherent states for the noncompact supergroups Osp(2/2N,R) *J. Math. Phys.* **30** 274–9

[41] Fatyga B W, Kotelecký V A, Nieto M M and Truax D R 1991 Supercoherent states *Phys. Rev. D* **43** 1403–12

[42] Kochetov E A 1996 Path-integral formalism for the supercoherent states *Phys. Lett. A* **217** 65–72

[43] Aoyama H, Sato M and Tanaka T 2001 N-fold supersymmetry in quantum mechanics: general formalism *Nucl. Phys. B* **619** 105–27

[44] Sato M and Tanaka T 2002 N-fold supersymmetry in quantum mechanics—analyses of particular models *J. Math. Phys.* **43** 3484–510

[45] Iida S and Kuratsuji H 1987 Phase holonomy, zero-point energy cancellation and supersymmetric quantum mechanics *Phys. Lett. B* **198** 221–5

[46] Bhaumik D, Dutta-Roy B, Bagchi B K and Khare A 1994 Berry phase for supersymmetric shape-invariant potentials *Phys. Lett. A* **1931** 11–4

[47] Ilinski K N, Kalinin G V and Melezhik V V 1995 Berry phase and supersymmetric topological index *J. Math. Phys.* **36** 6611–24

[48] Pedder C, Sonner J and Tong D 2008 Geometric phase in supersymmetric quantum mechanics *Phys. Rev. D* **77** 025009

[49] Shastry B S and Sutherland B 1993 Super Lax pairs and infinite symmetries in the $1/r^2$ system *Phys. Rev. Lett.* **70** 4029–33

[50] Witten E 1981 Dynamical breaking of supersymmetry *Nucl. Phys. B* **188** 513–54

[51] Salomonson P and van Holten J W 1982 Fermionic coordinates and supersymmetry in quantum mechanics *Nucl. Phys. B* **196** 509–31

[52] Kulshreshtha D S, Liang J-Q and Müller-Kirsten H J W 1993 Fluctuation equations about classical field configurations and supersymmetric quantum mechanics *Ann. Phys.* **225** 191–211

[53] Arancibia A, Guilarte J M and Plyushchay M S 2013 Fermion in a multi-kink-antikink soliton background, and exotic supersymmetry *Phys. Rev. D* **88** 085034

[54] Arancibia A and Plyushchay M S 2015 Chiral asymmetry in propagation of soliton defects in crystalline backgrounds *Phys. Rev. D* **92** 105009

[55] Durand S 1993 Fractional supersymmetry and quantum mechanics *Phys. Lett. B* **312** 115–20

[56] Durand S 1993 Fractional superspace formulation of generalized mechanics *Mod. Phys. Lett. A* **8** 2323–34

[57] de Azcárraga J A and Macfarlane A J 1996 Group theoretical foundations of fractional supersymmetry *J. Math. Phys.* **37** 1115–27

[58] Sen D 1992 Some supersymmetric features in the spectrum of anyons in a harmonic potential *Phys. Rev. D* **46** 1846–57

[59] Roy P and Tarrach R 1992 Supersymmetric anyon quantum mechanics *Phys. Lett. B* **274** 59–64

[60] Guhr T 1996 Transition toward quantum chaos: With supersymmetry from Poisson to Gauss *Ann. Phys.* **250** 145–92

[61] Beckers J and Debergh N 1990 Parastatistics and supersymmetry in quantum mechanics *Nucl. Phys.* B **340** 767–76

[62] Cooper F, Khare A and Sukhatme U 1995 Supersymmetry and quantum mechanics *Phys. Rep.* **251** 267–385

[63] Akulov V P and Pashnev A I 1983 Quantum superconformal model in (1,2) space *Theor. Math. Phys.* **56** 862–66

[64] Fubini S and Rabinovici E 1984 Superconformal quantum mechanics *Nucl. Phys.* B **245** 17–44

[65] Anabalon A and Plyushchay M S 2003 Interaction via reduction and nonlinear superconformal symmetry *Phys. Lett.* B **572** 202–9

[66] Guilarte J M and Plyushchay M S 2017 Perfectly invisible PT-symmetric zero-gap systems, conformal field theoretical, kinks and exotic nonlinear supersymmetry *J. High Energ. Phys.* **12** 061

[67] Inzunza L and Plyushchay M S 2018 Hidden superconformal symmetry: where does it come from? *Phys. Rev.* D **97** 045002

[68] Ilinski K N and Uzdin V M 1993 Quantum superspace, q-extended supersymmetry and parasupersymmetric quantum mechanics *Mod. Phys. Lett.* A **8** 2657–70

[69] Debergh N 1993 On a q-deformation of the supersymmetric Witten model *J. Phys. A: Math. Gen.* **26** 7219–26

[70] Bonatsos D and Daskaloyannis C 1993 General deformation schemes and $N = 2$ supersymmetric quantum mechanics *Phys. Lett.* B **307** 100–5

[71] Bagchi B, Banerjee A, Quesne C and Tkachuk V M 2005 Deformed shape invariance and exactly solvable Hamiltonians with position-dependent effective mass *J. Phys. A: Math. Gen.* **38** 2929–45

[72] Boya L J, Wehrhahn R F and Rivero A 1993 Supersymmetry and geometric motion *J. Phys. A: Math. Gen.* **26** 5824–34

[73] Borisov N V and Ilinski K N 1994 $N = 2$ supersymmetric quantum mechanics on Riemann surfaces with meromorphic superpotentials *Commun. Math. Phys.* **161** 177–94

[74] Dolgallo A D and Ilinski K N 1994 Generalized supersymmetric quantum mechanics on Riemann surfaces with meromorphic superpotentials *J. Math. Phys.* **35** 2074–82

[75] Tomka M, Pletyukhov M and Gritsev V 2015 Supersymmetry in quantum optics and in spin-orbit coupled systems *Sci. Rep.* **5** 13097

[76] Quesne C 2009 Solvable rational potentials and exceptional orthogonal polynomials in supersymmetric quantum mechanics *SIGMA* **5** 084

[77] Marquette I and Quesne C 2013 New ladder operators for a rational extension of the harmonic oscillator and superintegrability of some two-dimensional systems *J. Math. Phys.* **54** 102102

[78] Witten E 1982 Supersymmetry and Morse theory *J. Diff Geom.* **17** 661–92

[79] Cycon H L, Froese R G, Kirsch W and Simon B 1987 *Schrödinger Operators with Application to Quantum Mechanics and Global Geometry* (Berlin: Springer)

[80] Alvarez-Gaumé L 1983 Supersymmetry and the Atiyah–Singer index theorem *Commun. Math. Phys.* **90** 161–73

[81] Friedan D and Windey P 1984 Supersymmetric derivation of the Atiyah–Singer index and the chiral anomaly *Nucl. Phys.* B **235** 395–416

[82] Blau M and Thompson G 1995 Localization and diagonalization: a review of functional integral techniques for low-dimensional gauge theories and topological field theories *J. Math. Phys.* **36** 2192–236

[83] Schwarz A and Zaboronsky O 1997 Supersymmetry and localization *Commun. Math. Phys.* **183** 463–76

[84] Behtash A 2018 More on homological supersymmetric quantum mechanics *Phys. Rev.* D **97** 065002

[85] Spector D 1990 Supersymmetry and the Möbius inversion function *Commun. Math. Phys.* **127** 239–52

[86] Castro C, Granik A and Mahecha J 2001 On SUSY-QM, fractal strings and steps towards a proof of the Riemann hypothesis, arXiv: hep-th/0107266

[87] Slater P B 2005 A numerical examination of the Castro–Mahecha supersymmetric model of the Riemann zeros, arXiv: math/0511188

[88] Castro C 2007 On strategies towards the Riemann hypothesis: fractal supersymmetric QM and a trace formula *Int. J. Geom. Methods Mod. Phys.* **04** 861–80

[89] Bagger J 1985 Supersymmetric sigma models *Proc. Bonn-NATO Advanced Study Institute on Supersymmetry* Dietz K, Flume R, von Gehlen G and Rittenberg V (New York: Plenum) 213–57

[90] Lindström U 2012 Supersymmetric sigma model geometry *Symmetry* **4** 474–506

[91] Narnhofer H and Thirring W 1999 Spontaneously broken symmetries *Ann. Inst. Henri Poincaré, Phys. Théor.* **70** 1–21

[92] Nicolai H 1980 Supersymmetry and functional integration measures *Nucl. Phys.* B **176** 419–28

[93] Ezawa H and Klauder J R 1985 Fermions without fermions: the Nicolai map revisited *Prog. Theor. Phys.* **74** 904–15

[94] Leiber T, Marchesoni F and Risken H 1988 Colored noise and bistable Fokker–Planck equations *Phys. Rev. Lett.* **59** 1381–4
Leiber T, Marchesoni F and Risken H 1988 Colored noise and bistable Fokker–Planck equations *Phys. Rev. Lett.* **60** 659 (erratum)

[95] Rogers A 1987 Fermionic path integration and Grassmann Brownian motion *Commun. Math. Phys.* **113** 353–68

[96] Corns R 1993 Path integrals in $N = 2$ supersymmetric quantum mechanics *J. Math. Phys.* **34** 2723–41

[97] Jauslin H R 1990 Supersymmetric partners for random walks *Phys. Rev.* A **41** 3407–10

[98] Rosu H C and Reyes M 1995 Supersymmetric time-continuous discrete random walks *Phys. Rev.* E **51** 5112–5

[99] Junker G 1995 Recent developments in supersymmetric quantum mechanics *Turk. J. Phys.* **19** 230–48

[100] Essler F H L, Korepin V E and Schoutens K 1990 New exactly solvable model of strongly correlated electrons motivated by high-T_c superconductivity *Phys. Rev. Lett.* **68** 2960–3

[101] Essler F H L, Korepin V E and Schoutens K 1993 Electronic model for superconductivity *Phys. Rev. Lett.* **70** 73–6

[102] Gomes P R S, Bienzobaz P F and Gomes M 2012 Supersymmetric extension of the quantum spherical model *Phys. Rev.* E **85** 061109

[103] Bienzobaz P F, Gomes P R S and Gomes M 2013 Stochastic quantization of the spherical model and supersymmetry *J. Stat. Mech.* **2013** P09018

[104] Baker T H and Jarvis P D 1994 Quantum superspin chains *Int. J. Mod. Phys.* B **8** 3623–35

[105] Bracken A J, Gould M D, Links J R and Zhang Y-Z 1995 A new supersymmetric and exactly solvable model of correlated electrons *Phys. Rev. Lett.* **74** 2768–71

[106] Girvin S M, MacDonald A H, Fisher M P A, Rey S-J and Sethna J P 1990 Exactly soluble models of fractional statistics *Phys. Rev. Lett.* **65** 1671–4

[107] Zuk J A 1994 Introduction to the supersymmetry method for the Gaussian random-matrix ensembles, arXiv: cond-mat/9412060

[108] Mirlin A D 2000 Statistics of energy levels and eigenfunctions in disordered systems *Phys. Rep.* **326** 259–382

[109] Efetov K B 1983 Supersymmetry and theory of disordered metals *Adv. Phys.* **32** 53–127

[110] Efetov K 1996 *Supersymmetry in Disorder and Chaos* (Cambridge: Cambridge University Press)

[111] Bouchaud J P, Comtet A, Georges A and Le Doussal P 1990 Classical diffusion of a particle in a one-dimensional random force field *Ann. Phys.* **201** 285–341

[112] Tosatti E, Zannetti M and Pietronero L 1988 Exponentiated random walks, supersymmetry and localization *Z. Phys.* B **73** 161–6

[113] Comtet A, Desbois J and Monthus C 1995 Localization properties in one dimensional disordered supersymmetric quantum mechanics *Ann. Phys.* **239** 312–50

[114] Wegner F 1983 Exact density of states for lowest Landau level in white noise potential superfield representation for interacting systems *Z. Phys.* B **51** 279–85

[115] Brézin E, Gross D J and Itzykson C 1984 Density of states in the presence of a strong magnetic field and a random impurity *Nucl. Phys.* B **235** 24–44

[116] Fischbeck H J and Hayn R 1990 On the density of states of a disordered Peierls–Fröhlich chain *Phys. Status Solidi* B **158** 565–72

[117] Hayn R and John W 1991 Instanton approach to the conductivity of a disordered solid *Nucl. Phys.* B **348** 766–86

[118] Tkachuk V M 1995 The supersymmetry representation for correlation functions of disordered systems *Condens. Matter Phys.* **6** 123–36

[119] Jarvis P D and Stedman G E 1984 Supersymmetry in Jahn–Teller systems *J. Phys. A: Math. Gen.* **17** 757–76

[120] Shapir Y 1985 Supersymmetric statistical models on the lattice *Physica* D **15** 129–37

[121] Berche B and Iglói F 1995 Realization of supersymmetric quantum mechanics in inhomogeneous Ising models *J. Phys. A: Math. Gen.* **28** 3579–90

[122] Pankratov O A, Pakhomov S V and Volkov B A 1987 Supersymmetry in heterojunctions: band-inverting contact on the basis of $Pb_{1-x}Sn_xTe$ and $Hg_{1-x}Cd_xTe$ *Solid State Commun.* **61** 93–6

[123] Tomić S, Milanović V and Ikonić Z 1997 Optimization of intersubband resonant second-order susceptibility in asymmetric graded $Al_xGa_{1-x}As$ quantum wells using supersymmetric quantum mechanics *Phys. Rev.* B **56** 1033–36

[124] Adagideli I, Goldbart P M, Shnirman A and Yazdani A 1999 Low-energy quasiparticle states near extended scatterers in d-wave superconductors and their connection with SUSY quantum mechanics *Phys. Rev. Lett.* **83** 5571–4

[125] Tomić S, Milanović V and Ikonić Z 2000 Optimization of gain in intersubband quantum well lasers by supersymmetry *Phys. Rev.* B **62** 16681–5

[126] Correa F, Jakubský V, Nieto L-M and Plyushchay M S 2008 Self-isospectrality, special supersymmetry, and their effect on the band structure *Phys. Rev. Lett.* **101** 030403

[127] Correa F, Jakubský V and Plyushchay M S 2008 Finite-gap systems, tri-supersymmetry and self-isospectrality *J. Phys. A: Math. Theor.* **41** 485303

[128] Huang B-L, Wu S-T and Mou C-Y 2004 Midgap states and generalized supersymmetry in semi-infinite nanowires *Phys. Rev.* B **70** 205408

[129] Nemec N 2007 Quantum transport in carbon-based nanostructures *PhD Thesis* Universität Regensburg

[130] Jakubský V, Nieto L-M and Plyushchay M S 2011 Klein tunneling in carbon nano-structures: a free particle dynamics in disguise *Phys. Rev.* D **83** 047702

[131] Jakubský V and Plyushchay M S 2012 Supersymmetric twisting of carbon nanotubes *Phys. Rev.* D **85** 045035

[132] Grover T, Sheng D N and Vishwanath A 2014 Emergent space–time supersymmetry at the boundary of a topological phase *Science* **344** 280–3

[133] Kailasvuori J 2009 Pedestrian index theorem à la Aharonov–Casher for bulk threshold modes in corrugated multilayer graphene *Europhys. Lett.* **87** 47008

[134] Sarma S D, Adam S, Hwang E H and Rossi E 2011 Electronic transport in two-dimensional graphene *Rev. Mod. Phys.* **83** 407–70

[135] Midya B and Fernández D J 2014 Dirac electron in graphene under supersymmetry generated magnetic fields *J. Phys. A: Math. Theor.* **47** 285302

[136] Ezawa M 2007 Supersymmetry and unconventional quantum Hall effect in monolayer, bilayer and trilayer graphene *Physica* E **40** 269–72

[137] Ezawa M 2008 Supersymmetry and unconventional quantum Hall effect in graphene *Phys. Lett.* A **372** 924–9

[138] Park K S and Yi K S 2007 Supersymmetric quantum mechanics in graphene *J. Korean Phys. Soc.* **50** 1678–82

[139] Sahoo S and Das S 2009 Supersymmetric structure of fractional quantum Hall effect in graphene *Indian J. Pure Appl. Phys.* **47** 186–91

[140] Grover T, Sheng D N and Vishwanath 2014 Emergent space–time supersymmetry at the boundary of a topological phase *Science* **344** 280–3

[141] Sato M and Ando Y 2017 Topological superconductors: a review *Rep. Prog. Phys.* **80** 076501

[142] Kostelecký V A and Nieto M M 1984 Evidence for a phenomenological supersymmetry in atomic physics *Phys. Rev. Lett.* **53** 2285–8

[143] Kostelecký V A and Nieto M M 1985 Evidence from alkali-metal-atom transition probability for a phenomenological atomic supersymmetry *Phys. Rev.* A **32** 1293–8

[144] Kostelecký V A and Nieto M M 1985 Analytic wave functions for atomic quantum-defect theory *Phys. Rev.* A **32** 1293–8

[145] Kostelecký V A, Nieto M M and Truax D R 1988 Fine structure and analytic quantum-defect wave functions *Phys. Rev.* A **38** 4413–8

[146] Bluhm R and Kostelecký V A 1993 Atomic supersymmetry and the Stark effect *Phys. Rev.* A **47** 794–808

[147] Kostelecký V A 1993 Atomic supersymmetry, oscillators and the Penning trap *Symmetries in Science VII: Dynamical Symmetries and Spectrum-Generating Algebras in Physics* ed B Gruber and T Otsuka (New York: Plenum)

[148] Bluhm R and Kostelecký V A 1994 Atomic supersymmetry, Rydberg wave packets, and radial squeezed states *Phys. Rev.* A **49** 4628–40

[149] Bluhm R and Kostelecký V A 1995 Long-term evolution and revival structure of Rydberg wave packets for hydrogen and alkali-metal atoms *Phys. Rev.* A **51** 4767–86

[150] Bluhm R, Kostelecký V A and Porter J A 1996 The evolution and revival structure of localized quantum wave packets *Am. J. Phys.* **64** 944–54

[151] Iachello F 1980 Dynamical supersymmetry in nuclei *Phys. Rev. Lett.* **44** 772–5

[152] Vervier J 1987 Boson–fermion symmetries and supersymmetries in nuclear physics *Riv. Nuovo Cim.* **10** 1–102

[153] Balantekin A B, Bars I and Iachello F 1981 $U(6/4)$ supersymmetry in nuclei *Nucl. Phys.* A **370** 284–316

[154] Urrutia L F and Hernádez E 1983 Long-range behavior of nuclear forces as a manifestation of supersymmetry in nature *Phys. Rev. Lett.* **51** 755–8

[155] Baake M, Reinicke P and Gelberg A 1986 A Hamiltonian with broken supersymmetry for the xenon isotopes *Phys. Lett.* B **166** 10–7

[156] Amado R D, Bijker R, Cannata F and Dedonder J P 1991 Supersymmetric quantum mechanics and superdeformed nuclei *Phys. Rev. Lett.* **67** 2777–9

[157] Socorro J and Medina E R 2000 Supersymmetric quantum mechanics for Bianchi class A models *Phys. Rev.* D **61** 087702

[158] García A, Guzmán W, Sabido m and Socorro J 2006 Iso-spectral potentials and inflationary quantum cosmology *Int. J. Theor. Phys.* **45** 2483–96

[159] Cotăescu I 2007 Dirac Fermions in de Sitter and anti-de Sitter backgrounds *Rom. J. Phys.* **52** 895–940

[160] Oikonomou V K 2012 de Sitter cosmic strings and supersymmetry *Gen. Relativ. Grav.* **44** 1285–97

[161] Rostami T, Jalalzadeh S and Moniz P V 2017 Quantum cosmology: the (supersymmetric) door, *Proceedings, 14th Marcel Grossmann Meeting on Recent Developments in Theoretical and Experimental General Relativity, Astrophysics, and Relativistic Field Theories (Rome, Italy, 12–18 July 2015)* pp 2791–800

[162] Claus P, Derix M, Kallosh R, Kumar J, Townsend P K and Van Proeyen A 1998 Black holes and superconformal mechanics *Phys. Rev. Lett.* **81** 4553–6

[163] Seeger M and Thies M 1998 QCD_{1+1} with static quarks as supersymmetric quantum mechanics *Phys. Rev.* D **58** 027701

[164] de Téramond G F, Dosch H G and Brodsky S J 2015 Baryon spectrum from superconformal quantum mechanics and its light-front holographic embedding *Phys. Rev.* D **91** 045040

[165] Feinberg J 2004 All about the static fermion bags in the Gross–Neveu model *Ann. Phys.* **309** 166–231

[166] de Wit B, Hoppe J and Nicolai H 1988 On the quantum mechanics of supermembrans *Nucl. Phys.* B **305** 545–81

[167] de Wit B 1997 Supersymmetric quantum mechanics, supermembranes and Dirichlet particles *Nucl. Phys. Proc. Suppl.* B **56** 76–87

[168] Oikonomou V K 2011 F-theory and the Witten index *Nucl. Phys.* B **850** 273–86

IOP Publishing

Supersymmetric Methods in Quantum, Statistical and Solid State Physics

Enlarged and revised edition
Georg Junker

Solutions to problems

Problems in chapter 2

Problem 2.1. The free particle on the real line

(a) For $N = 1$ the SUSY algebra (2.1) implies $H = 2Q_1^2$ which is obviously fulfilled. If we formally allow the state $|0\rangle$ to characterise the zero-momentum eigenstate, i.e. $\langle k|0\rangle = \delta(k)$ with $p|k\rangle = \hbar k|k\rangle$, SUSY is formally unbroken as $H|0\rangle = 0$.

(b) Obviously $Q_1|k\rangle = (\hbar k/\sqrt{4m})|k\rangle$ and $|Q_1||k\rangle = (\hbar|k|/\sqrt{4m})|k\rangle$. Hence, $\Lambda|k\rangle = \operatorname{sgn} k|k\rangle$ and thus the Hilbert space may be graded into subspaces of eigenstates with positive and negative wave number, i.e. right and left moving plane waves. That is, $\mathcal{H} = \mathcal{H}^+ \oplus \mathcal{H}^-$ with $\mathcal{H}^+ := \operatorname{span}\{|k\rangle: k > 0\}$ and $\mathcal{H}^- := \operatorname{span}\{|k\rangle: k \leqslant 0\}$ where we have added the SUSY ground state to $\mathcal{H}^-$ for convenience. We have the first two relations $[H, \Lambda] = 0$ and $\Lambda^2 = 1$ required for a Witten operator. However, the third $\{\Lambda, Q_1\} = 2\Lambda Q_1 = 2|Q_1| \neq 0$ is not fulfilled.

Note that the parity operator Π as defined in equation (3.31) may be identified as a grading operator and defines a second supercharge via $Q_2 := i\Pi Q_1$ obeying $H = 2Q_2^2$ and $\{Q_1, Q_2\} = 0$.

Problem 2.2. SUSY transformations for $N \geqslant 2$

Let $|q_1\rangle$ be an eigenstate of Q_1, that is, $Q_1|q_1\rangle = q_1|q_1\rangle$ with $q_1 \in \mathbb{R}$ as $Q_1 = Q_1^\dagger$ is self-adjoint. Obviously, $H|q\rangle = 2Q_1^2|q\rangle = 2q_1^2|q\rangle$. Defining $|\tilde{q}_1\rangle := CQ_2|q_1\rangle$ we have $\langle\tilde{q}_1|\tilde{q}_1\rangle = |C|^2 q_1^2$ and therefore the normalisation constant is given by $|C|^2 = 1/q_1^2$. It is straightforward to prove the following properties of $|\tilde{q}_1\rangle$:

eigenstate of Q_1: $Q_1|\tilde{q}_1\rangle = \dfrac{1}{|q_1|}Q_1Q_2|q_1\rangle = -\dfrac{1}{|q_1|}Q_2Q_1|q_1\rangle = -\dfrac{1}{|q_1|}Q_2q_1|q_1\rangle = -q_1|\tilde{q}_1\rangle.$

eigenstate of H: $H|\tilde{q}_1\rangle = 2Q_1^2|\tilde{q}_1\rangle = 2q_1^2|\tilde{q}_1\rangle$.

orthogonal to $|q_1\rangle$: $\langle q_1|\tilde{q}_1\rangle = \frac{1}{q_1}\langle q_1|Q_1|\tilde{q}_1\rangle = -\langle q_1|\tilde{q}_1\rangle \Rightarrow \langle q_1|\tilde{q}_1\rangle$.

Problem 2.3. The supersymmetric harmonic oscillator

(a) Using the definitions (2.15) and (2.20) the complex supercharge reads

$$Q = \frac{p}{\sqrt{2m}} \otimes \frac{\sigma_1 + i\sigma_2}{2} + \Phi(x) \otimes \frac{\sigma_2 - i\sigma_1}{2} = \begin{pmatrix} 0 & A \\ 0 & 0 \end{pmatrix} \quad \text{with} \quad A := \frac{p}{\sqrt{2m}} - i\Phi(x).$$

For the explicit case $\Phi(x) = \sqrt{m/2}\,\omega x$ this reads

$$A = -i\sqrt{\frac{\hbar\omega}{2}}\left(\sqrt{\frac{\hbar}{m\omega}}\,\partial_x + \sqrt{\frac{m\omega}{\hbar}}\,x\right).$$

Using units such that energy is given in units of $\hbar\omega$ and the length scale by units of $\sqrt{\hbar/m\omega}$ we may identify $A = -ia$ with a given in equation (2.23) up to an extra phase factor $-i = e^{-i\pi/2}$, i.e. $\alpha = \pi/2$ in equation (2.71).

(b) The condition (2.59) reads in coordinate representations $(\partial_x + \frac{m\omega}{\hbar}x)\phi_0^-(x) = 0$, which is easily integrated to $\phi_0^-(x) = \left(\frac{m\omega}{\pi\hbar}\right)^{1/4}\exp\{-m\omega x^2/2\hbar\}$, being the well-known harmonic oscillator ground state. Obviously condition (2.60) does not lead to a square integrable wave function. Hence, SUSY is unbroken with the zero-energy ground state belonging to $\mathcal{H}^-$.

(c) The partner Hamiltonians (2.18) for the harmonic oscillator explicitly read

$$H_\pm := \frac{p^2}{2m} + \frac{m}{2}\omega^2 x^2 \pm \frac{\hbar\omega}{2}$$

and the energy eigenvalues are obviously given by $E_n = \hbar\omega n$, $n = 0, 1, 2, 3, \ldots$, where $n = 0$ is only allowed for H_-. The corresponding eigenstates are given by $|\phi_{E_n}^-\rangle = |n\rangle$ and $|\phi_{E_n}^+\rangle = |n - 1\rangle$, where $\langle x|n\rangle = \left(\frac{m\omega}{\pi\hbar}\right)^{1/4}\exp\{-m\omega x^2/2\hbar\}H_n\left(\sqrt{\frac{m\omega}{\hbar}}\,x\right)/\sqrt{2^n n!}$ are the well-known harmonic oscillator eigenstates expressed in terms of the Hermite polynomials H_n. With this notation the relations (2.28) directly lead us to

$$A^\dagger|\phi_{E_n}^+\rangle = ia^\dagger|n - 1\rangle = i\sqrt{n}\,|n\rangle = i\sqrt{E_n}\,|\phi_{E_n}^-\rangle$$

$$A|\phi_{E_n}^-\rangle = -ia|n\rangle = -i\sqrt{n}\,|n - 1\rangle = -i\sqrt{E_n}\,|\phi_{E_n}^+\rangle,$$

which represents the SUSY transformation (2.57). Note again that the general SUSY transformations (2.57) are only defined up to an arbitrary phase factor.

Problem 2.4. The Witten operator W

(a) Recall the definition (2.32), i.e. $W := \frac{2}{H}QQ^\dagger - 1$. The proof of the relations (2.31) is as follows:
- Obviously $W = W^\dagger$ is self-adjoint as $[H, Q] = 0 = [H, Q^\dagger]$.
- $W^2 = \left(\frac{2}{H}QQ^\dagger - 1\right)^2 = \frac{4}{H^2}QQ^\dagger QQ^\dagger - \frac{4}{H}QQ^\dagger + 1 = \frac{4}{H^2}HQQ^\dagger - \frac{4}{H}QQ^\dagger + 1 = 1.$
- $\{W, Q\} = \left\{\frac{2}{H}QQ^\dagger - 1, Q\right\} = \frac{2}{H}QQ^\dagger Q - Q + \frac{2}{H}QQQ^\dagger - Q = 2Q - Q + 0 - Q = 0.$
- $\{W, Q^\dagger\} = \{W, Q\}^\dagger = 0.$

(b)
- $W = \frac{2}{H}QQ^\dagger - 1 = \frac{1}{H}QQ^\dagger - 1 + \frac{H - Q^\dagger Q}{H} = \frac{[Q, Q^\dagger]}{H} = \frac{[Q, Q^\dagger]}{\{Q, Q^\dagger\}}.$
- $[Q, Q^\dagger] = \frac{1}{2}(Q_1 + iQ_2)(Q_1 - iQ_2) - \frac{1}{2}(Q_1 - iQ_2)(Q_1 + iQ_2) = i[Q_2, Q_1]$
  $\Rightarrow W = \frac{1}{iH}[Q_1, Q_2].$

(c) The projection operators are defined by $P^\pm := \frac{1}{2}(1 \pm W) = (P^\pm)^\dagger$.
- $P^\pm P^\pm = \frac{1}{4}(1 \pm W)(1 \pm W) = \frac{1}{4}(1 \pm 2W + W^2) = \frac{1}{2}(1 \pm W) = P^\pm.$
- $P^\pm P^\mp = \frac{1}{4}(1 \pm W)(1 \mp W) = \frac{1}{4}(1 - W^2) = 0.$
- $P^+ + P^- = \frac{1}{2}(1 + W) + \frac{1}{2}(1 - W) = 1.$

Problem 2.5. Construction of supercharges from Witten operators

Using $Q_j := iW_j Q_1, \{W_i, W_j\} = 2\delta_{ij}$, and $\{Q_1, W_i\} = 0$ the proofs are straightforward.
- $Q_j^\dagger = -iQ_1^\dagger W_j^\dagger = -iQ_1 W_j = iW_j Q_1 = Q_j.$
- $\{Q_j, Q_k\} = -W_j Q_1 W_k Q_1 - W_k Q_1 W_j Q_1 = (W_j W_k + W_k W_j)Q_1^2 = 2Q_1^2 \delta_{jk} = H\delta_{jk}.$

Problem 2.6. Decomposition of the most general representation of supercharges

The proofs are via explicit calculations of all the commutators.

Problem 2.7. Generalised Fermion operators and matrix representation of $\mathcal{F}$

With the decomposition of an arbitrary state $|\psi\rangle = \sum_{E>0} (\alpha_E^+ |\psi_E^+\rangle + \alpha_E^- |\psi_E^-\rangle)$ on $\mathcal{H}\backslash \ker H$, the SUSY transformation (2.55), and the annihilation relations $Q|\psi_E^+\rangle = 0 = Q^\dagger|\psi_E^-\rangle$ the proofs are as follows:

(a) $Q^\dagger H^{-1/2}|\psi\rangle = \sum_{E>0}\alpha_E^- |\psi_E^-\rangle = H^{-1/2}Q^\dagger|\psi\rangle$, $QH^{-1/2}|\psi\rangle = \sum_{E>0}\alpha_E^+|\psi_E^+\rangle = H^{-1/2}Q|\psi\rangle$, $\{b, b^\dagger\} = \{H^{-1/2}Q^\dagger, H^{-1/2}Q\} = \frac{1}{H}\{Q^\dagger, Q\} = 1.$

(b) $b|\psi_E^+\rangle = \frac{1}{\sqrt{E}}Q^\dagger|\psi_E^+\rangle = |\psi_E^-\rangle$ and $b^\dagger|\psi_E^-\rangle = \frac{1}{\sqrt{E}}Q|\psi_E^-\rangle = |\psi_E^+\rangle.$

(c) Using the matrix representation

$$Q = \begin{pmatrix} 0 & A \\ 0 & 0 \end{pmatrix}, \qquad Q^\dagger = \begin{pmatrix} 0 & 0 \\ A^\dagger & 0 \end{pmatrix}, \qquad H = \begin{pmatrix} AA^\dagger & 0 \\ 0 & A^\dagger A \end{pmatrix}$$

one finds

$$\mathcal{F} = \frac{1}{H} QQ^\dagger = \begin{pmatrix} (AA^\dagger)^{-1} & 0 \\ 0 & (A^\dagger A)^{-1} \end{pmatrix} \begin{pmatrix} AA^\dagger & 0 \\ 0 & 0 \end{pmatrix} = \begin{pmatrix} 1 & 0 \\ 0 & 0 \end{pmatrix} = QQ^\dagger \frac{1}{H}$$

and

$$W = \frac{2}{H} QQ^\dagger - 1 = \frac{2}{H} \begin{pmatrix} AA^\dagger & 0 \\ 0 & 0 \end{pmatrix} - 1 = \begin{pmatrix} 2 & 0 \\ 0 & 0 \end{pmatrix} - 1 = \begin{pmatrix} 1 & 0 \\ 0 & -1 \end{pmatrix}.$$

Problem 2.8. On the internal energy relation of SUSY partners

Noting that $U_\pm(\beta)Z_\pm(\beta) = \text{Tr}\,(H_\pm e^{-\beta H_\pm})$ and $\text{Tr}\,(H_+ e^{-\beta H_+}) = \text{Tr}\,(AA^\dagger e^{-\beta AA^\dagger}) = \text{Tr}\,(A^\dagger e^{-\beta AA^\dagger}A) = \text{Tr}\,(A^\dagger A e^{-\beta A^\dagger A}) = \text{Tr}\,(H_- e^{-\beta H_-})$ immediately leads to the relation (2.68).

Problems in chapter 3

Problem 3.1. Real and complex supercharges of the Witten model

(a) Use the definitions (3.1) and (3.3) together with the standard representation of the Pauli matrices (1.3).

(b) In the matrix representation one finds

$$Q_1 Q_2 = \frac{1}{2} \begin{pmatrix} iAA^\dagger 0 & A \\ 0 & -iA^\dagger A \end{pmatrix}, \qquad Q_2 Q_1 = \frac{1}{2} \begin{pmatrix} -iAA^\dagger 0 & A \\ 0 & iA^\dagger A \end{pmatrix}$$

and therefore

$$[Q_1, Q_2] = i \begin{pmatrix} AA^\dagger & 0 \\ 0 & -A^\dagger A \end{pmatrix} = i \begin{pmatrix} AA^\dagger & 0 \\ 0 & A^\dagger A \end{pmatrix} \begin{pmatrix} 1 & 0 \\ 0 & -1 \end{pmatrix} = iH \otimes \sigma_3.$$

(c) Again use the definitions and explicit representation of the Pauli matrices.

Problem 3.2. SUSY transformation for continuum states revisited

(a) According to equation (3.9) the transformation between positive and negative parity states reads

$$\left(\frac{\hbar}{\sqrt{2m}} \frac{\partial}{\partial x} + \Phi(x) \right) \phi_E^-(x) = C\phi_E^+(x), \qquad |C|^2 = E.$$

Applying this to the asymptotic relation $\phi_E^\pm(x) \sim T^\pm(E)\exp\{-\kappa_R(E)x\}$ for $x \to \infty$ leads to the condition $T^+(E)C = T^-(E)(\Phi_+ - \hbar\kappa_R/\sqrt{2m})$. Now applying the SUSY transformation to the asymptotic form for $x \to -\infty$, cf. equation (3.38), and equating coefficients of right and left moving plane waves, results in the two relations, $C = \Phi_- + i\hbar k_L/\sqrt{2m}$ and $R^+(E)C = R^-(E)\,(\Phi_- + i\hbar k_L/\sqrt{2m})$. This then results in

$$T^+(E) = \frac{\Phi_+ - \dfrac{\hbar \kappa_R}{\sqrt{2m}}}{\Phi_- - i\dfrac{\hbar k_L}{\sqrt{2m}}} = \frac{\Phi_+ - \sqrt{\Phi_+^2 - E}}{\Phi_- + i\sqrt{E - \Phi_-^2}}T^-(E)$$

and the relation (3.41) for the reflection coefficients.

(b) Consider the ration $|T^+(E)|^2/|T^-(E)|^2 = \left[\Phi_+ - \sqrt{\Phi_+^2 - E}\right]^2 / E = f^2(\Phi_+/\sqrt{E})$

with $f(x) := x - \sqrt{x^2 - 1}$. Note that $f(1) = 1$ and $f'(x) < 0$ for all $x = \Phi_+/\sqrt{E} > 1$. Hence $f(x) < 1$ for all $x > 1$ and therefore $|T^+(E)|^2/|T^-(E)|^2 = f^2(\Phi_+/\sqrt{E}) < 1$ for $E < \Phi_+^2$.

Problem 3.3. Continuum states of the free particle and its SUSY partner

With the SUSY potential $\Phi(x) := \frac{\hbar}{\sqrt{2m}}\tanh(x)$ is it obvious that $V_+(x) = \hbar^2/2m$ and hence represents a free particle whose eigenstates are given by $\phi_E^+(x) = \frac{1}{\sqrt{2\pi}}e^{ikx}$ with eigenvalues $E = \hbar^2(k^2 + 1)/2m$ and wave number $k \in \mathbb{R}$. Applying the transformation formula (3.9),

$$\left(-\frac{\hbar}{\sqrt{2m}}\frac{\partial}{\partial x} + \Phi(x)\right)\phi_E^+(x) = C\phi_E^-(x),$$

one immediately finds the desired result $\phi_E^-(x) = N(\tanh x - ik)e^{ikx}$ with $|N|^2 = k^2/(k^2 + 1)$.

Problem 3.4. General power law potentials on the half line

As we assume $a, \lambda, \eta > 0$ the asymptotic behaviour of the SUSY potential $\Phi(x)$ at the boundaries of the configuration space $\mathbb{R}^+$ is given by

$$\Phi_- = \lim_{x \searrow 0} \Phi(x) = -\infty, \quad \Phi_+ = \lim_{x \to \infty} \Phi(x) = +\infty.$$

Hence, the Witten index (3.30) is obviously given by $\Delta = 1$ and SUSY is unbroken. The zero-energy ground state belongs to $\mathcal{H}^-$ and can explicitly be calculated via equation (3.11). The partner potentials directly follow from definition (3.6).

Problem 3.5. Spectral properties of the infinite square well and its SUSY partner

(a) The eigenvalues and eigenfunctions of the infinite square well described by Hamiltonian $H_- = (-\hbar^2/2m)(\partial_x^2 + 1)$ are given by

$$E_n = \frac{\hbar^2}{2m}(n^2 - 1), \quad \phi_{E_n}^-(x) = \sqrt{\frac{2}{\pi}}\sin(nx + n\pi/2)$$

as $\phi_{E_n}^-(\pm \pi/2) = 0$ for $n \in \mathbb{N}$.

(b) The ground state $\phi_{E_1}^-(x) = \phi_0^-(x) = \sqrt{2/\pi}\,\sin(x + \pi/2) = \sqrt{2/\pi}\,\cos x$ is annihilated by $A = (\hbar/\sqrt{2m})(\partial_x + \tan x)$, that is, $A\phi_0^-(x) = 0$.

(c) The eigenstates $\phi_{E_n}^+(x)$ directly follow from the SUSY transformation relation $A\phi_{E_n}^-(x) = \sqrt{E_n}\,\phi_{E_n}^+(x)$.

Problem 3.6. The square well with broken SUSY

The normalised ground states for $H_\pm$ are given by $\phi_0^\pm(x) = \cos(\pi/4 \pm x\pi/4)$ with energy eigenvalue $E_0 = \hbar^2\pi^2/8m > 0$. Hence SUSY is broken. Breaking of parity is also obvious as $\phi_0^\pm(-x) = \phi_0^\mp(x) \neq \phi_0^\pm(x)$ and is caused by the non-symmetric boundary conditions. Note that the SUSY transformations for the ground states explicitly read $\pm(\hbar/\sqrt{2m})\partial_x\phi_0^\mp(x) = (\hbar\pi/\sqrt{8m})\sin(\pi/4 \mp x\pi/4) = \sqrt{E_0}\,\phi_0^\pm(x)$.

Problem 3.7. Free motion on the unit circle $\mathbb{S}^1$ and its SUSY structure

(a) Verification of the eigenfunctions and eigenvalues are obvious.

(b) The parity operator as defined in equation (3.31) obeys the relation $\Pi\phi_n(x) = \phi_n(-x)$ for $n \neq 0$. Obviously, $\Pi\phi_n(x) = \phi_{-n}(x)$ and therefore $\Pi: \mathcal{H}^\pm \to \mathcal{H}^\mp$ on $\mathcal{H}\backslash\ker H$. Extension of the definition on $\mathcal{H}$ via $\Pi\phi_0(x) := 0$ will then allow for the definition of a proper supercharge $Q_1 := (i/\sqrt{4m})\Pi p = -(i/\sqrt{4m})p\Pi = Q_1^\dagger$ with $2Q_1^2 = H$ and $Q_1\phi_n(x) = i\,\mathrm{sgn}\,n\sqrt{E_n/2}\,\phi_{-n}(x)$, where the convention $\mathrm{sgn}\,0 = 0$ is used.

(c) This is obvious as with above property $\Pi\phi_{E_n}^\pm(x) = \pm\phi_{E_n}^\pm(x)$. This grading does not allow for a SUSY transformation in above sense.

Problems in chapter 4

Problem 4.1. Shape-invariance of the symmetric Pöschl–Teller potential

With SUSY potential $\Phi(a_0, x) := \frac{\hbar}{\sqrt{2m}}\,a_0\tan x$ we find the relation

$$V_+(a_0, x) = V_-(a_0 + 1, x) + \frac{\hbar^2}{2m}[(a_0 + 1)^2 - a_0^2]$$

and read off the functions $a_1 = F(a_0) = a_0 + 1$ and $R(a_1) = \frac{\hbar^2}{2m}\left[a_1^2 - a_0^2\right]$. The ground-state wave function is given by $\phi_0^-(a_0, x) = C\cos^{a_0} x$. Therefore, the energy eigenvalues and eigenfunctions are given by

$$E_n = \frac{\hbar^2}{2m}\sum_{s=1}^{n}\left(a_{s-1}^2 - a_s^2\right) = \frac{\hbar^2}{2m}\left[(a_0 + n)^2 - a_0^2\right], \qquad n \in \mathbb{N}_0,$$

$$\phi_n^-(a_0, x) = C_n(-\partial_x + a_0\tan x) \times \ldots \times (-\partial_x + (a_0 + n - 1)\tan x)\phi_0^-(a_n, x),$$

where $C_n := \prod_{s=0}^{n-1}[(a_0 + n)^2 - (a_0 + s)^2]^{-1/2}$ and $\phi_0^-(a_n, x) = C\cos^{n+a_0} x$.

Problem 4.2. Alternative approach to conditionally exactly solvable potentials

(a) Using the definitions $V_\pm(x) = \frac{1}{2}W^2(x) \pm \frac{1}{2}W'(x)$ and $W(x) = \Phi(x) + g(x)$, together with the condition

$$g^2(x) + 2\Phi(x)g(x) + g'(x) = b, \quad b = \text{const.}$$

directly leads to the desired results.

(b) With the ansatz given, one finds $g'(x) = \frac{u''(x)}{u(x)} - \left(\frac{u'(x)}{u(x)}\right)^2 - \Phi(x)$. Inserting this into above condition results in the Schrödinger-like equation

$$\left(-\frac{1}{2}\partial_x^2 + \frac{1}{2}\Phi^2(x) + \frac{1}{2}\Phi'(x)\right)u(x) = -\frac{b}{2}u(x).$$

Comparing this with equation (4.52) we may identify $b = -2\varepsilon$. Note that u should not have any zeros and therefore ε should be less than the ground-state energy E_0 of the unperturbed Hamiltonian $H_+^0 = -\frac{1}{2}\partial_x^2 + \frac{1}{2}\Phi^2(x) + \frac{1}{2}\Phi'(x)$. That is, $\varepsilon < E_0$ and $b > -2E_0$.

(c) For $\Phi(x) = x$ we have $V_+(x) = \frac{1}{2}(x^2 + b + 1)$ and $V_-(x) = \left(\frac{u'(x)}{u(x)}\right)^2 - \frac{1}{2}x^2 - \frac{b+1}{2}$. This is, up to the constant shift by $\frac{b+1}{2}$ in the energy, identical to the SUSY-partner family of the harmonic oscillator discussed in section 4.4. Here we have $E_0 = 1$ and therefore $b > -2$.

Problem 4.3. The harmonic oscillator SUSY partner for $\varepsilon = -5/2$ and $\beta = 0$

To be a bit more general let us consider the case $\varepsilon = -\frac{1}{2} - 2k$, $k \in \mathbb{N}_0$, and $\beta = 0$. With the help of the Kummer transformation $_1F_1(a, c, z) = e^z {}_1F_1(c - a, a, -z)$ the solution (4.60) can be put into the form $u(x) = e^{x^2/2} {}_1F_1(-k, \frac{1}{2}, -x^2) = e^{x^2/2}H_{2k}(ix)$, where H_n stands for the Hermite polynomial of order n. Noting that $H'_{2k}(z) = 4kH_{2k-1}(z)$ one finally arrives at

$$V_-(x) = \frac{x^2}{2} + 8ikx\frac{H_{2k-1}(ix)}{H_{2k}(ix)} - 16k^2\left(\frac{H_{2k-1}(ix)}{H_{2k}(ix)}\right)^2 - 4k - 1.$$

The special case $k = 1$ is easily derived with $H_1(ix) = 2ix$ and $H_2(ix) = -4x^2 - 2$.

Problem 4.4. A family of SUSY partners for the free particle

For the free particle case it is straightforward to show that $u(x) = e^{\kappa x} + \beta e^{-\kappa x}$ is a solution of the Schrödinger-like equation (4.52). As $\inf \text{spec} H_+ = 0$ we have the condition $\varepsilon < 0$, i.e. $\kappa > 0$. To assure a noteless $u(x)$ we also need to require in addition $\beta > 0$. The partner potential $V_-(x)$ is easily verified via relation (4.53). This class of potentials has exactly one bound state which is calculated via equation (4.54) to be

$$\phi_0^-(x) = \frac{C}{e^{\kappa x} + \beta e^{-\kappa x}}.$$

As this family of potentials has as SUSY partner the free particle on the real line they have a vanishing reflection coefficient and are thus all reflectionless. Note, the special case $\beta = 1$ corresponds to example 3.3 in section 3.6.1.

Problem 4.5. Generalised coherent states minimising the uncertainty relation

First let us calculate the commutator $[X, P] = \frac{\omega}{4i}[B + B^\dagger, B - B^\dagger] = \frac{\omega}{2i}[B^\dagger, B]$ and note that $\langle\mu|[B^\dagger, B]|\mu\rangle = |\mu|^2 - \langle\mu|BB^\dagger|\mu\rangle$. Now we evaluate the variances

$$\langle\Delta X\rangle_\mu = \frac{1}{4}\langle\mu|(B + B^\dagger)^2|\mu\rangle - \frac{1}{4}(\mu + \mu^*)^2 = \frac{1}{4}\left(\langle\mu|BB^\dagger|\mu\rangle - |\mu|^2\right)$$

$$\langle\Delta P\rangle_\mu = -\frac{\omega}{4}\langle\mu|(B - B^\dagger)^2|\mu\rangle - \frac{1}{4}(\mu - \mu^*)^2 = \frac{\omega^2}{4}\left(\langle\mu|BB^\dagger|\mu\rangle - |\mu|^2\right),$$

which result in

$$\langle\Delta X\rangle_\mu\langle\Delta P\rangle_\mu = \left(\frac{\omega}{4}\right)^2\left(\langle\mu|BB^\dagger|\mu\rangle - |\mu|^2\right)^2 = \frac{1}{4}|\langle\mu|[X, P]|\mu\rangle|^2.$$

Problem 4.6. Partition function and internal energy of the HO SUSY-partner family

The calculations are straightforward and here we only present the results:

$$Z_+(\beta) = \sum_{n=0}^\infty e^{-\beta(n+1/2-\varepsilon)} = \frac{\exp\{-\beta(1/2 - \varepsilon)\}}{1 - e^{-\beta}} = \frac{e^{\beta\varepsilon}}{2\sinh\frac{\beta}{2}}$$

$$U_+(\beta) = \frac{1}{2} - \varepsilon + \frac{e^{-\beta}}{1 - e^{-\beta}} = \frac{1}{2}\coth\frac{\beta}{2} - \varepsilon$$

$$Z_-(\beta) = 1 + Z_+(\beta) = \frac{2\sinh\frac{\beta}{2} + e^{\beta\varepsilon}}{2\sinh\frac{\beta}{2}}$$

$$U_-(\beta) = \frac{1}{2}\coth\frac{\beta}{2} - \frac{\cosh\frac{\beta}{2} + \varepsilon e^{\beta\varepsilon}}{2\sinh\frac{\beta}{2} + e^{\beta\varepsilon}} = e^{\beta\varepsilon}\frac{\frac{1}{2}\coth\frac{\beta}{2} - \varepsilon}{2\sinh\frac{\beta}{2} + e^{\beta\varepsilon}}$$

$$U_+(\beta)Z_+(\beta) = \left(\frac{1}{2}\coth\frac{\beta}{2} - \varepsilon\right)\frac{e^{\beta\varepsilon}}{2\sinh\frac{\beta}{2}} = U_-(\beta)Z_-(\beta).$$

Problem 4.7. The double quadratic field model via SUSY

Let us follow the programme as indicated at the end of section 4.5. First we calculate the zero mode $\psi_0(x) = \exp\{\int dx\, W(x)\} = e^{-|x|}$. In a second step we integrate this to get the static solution $\phi_s(x) = \int dx\,\psi_0(x) = (\text{sgn}\,x)(1 - e^{-|x|})$ with $\phi_\pm = \pm 1$. Now we utilise the relation (4.98) as follows:

$$U(\phi_s(x)) = \frac{1}{2}\psi_0^2(x) = \frac{1}{2}e^{-2|x|} = \frac{1}{2}(1 - |\phi_s|)^2.$$

Note that $\phi_s(x) > 0$ for $x > 0$ and $\phi_s(x) < 0$ for $x < 0$. Analytical continuation beyond the classical vacua $\phi_\pm = \pm 1$ leads to the desired double quadratic field model.

Problems in chapter 5

Problem 5.1. SUSY invariance of the supersymmetric classical model

Let us collect some of the results derived in the book by Kalka and Soff [1] using their notation. The effective Lagrangian derived in their equation (6.194) reads

$$L = \frac{1}{2}[\dot{q}^2 - W^2 + i(\bar\psi\dot\psi - \dot{\bar\psi}\psi) + W'(\bar\psi\psi - \psi\bar\psi)].$$

Noting that at this stage their auxiliary field h had been identified with the SUSY potential W, i.e. $h = W$, their SUSY transformations in equations (6.159) and (6.160) read

$$\delta q = \epsilon\psi + \bar\psi\bar\epsilon, \qquad \delta\psi = \bar\epsilon(W - i\dot{q}), \qquad \delta\bar\psi = \epsilon(W + i\dot{q})$$

leading to the variation in the Lagrangian given in their equation (6.218)

$$\delta L = \frac{i}{2}\frac{d}{dt}[(W - i\dot{q})\epsilon\psi - (W + i\dot{q})\bar\psi\bar\epsilon].$$

Upon the substitutions $q \to x$, $\epsilon \to \bar\epsilon$, $\bar\epsilon \to \epsilon$, and $W \to -\Phi$ it is easy to reproduce the stated results.

Problem 5.2. Supersymmetric classical equations of motion

With the Lagrangian (5.10) the Euler–Lagrange equations

$$\frac{d}{dt}\frac{\partial L}{\partial \dot{X}} = \frac{\partial L}{\partial X} \qquad \text{with} \qquad X \in \{x, \psi, \bar\psi\}$$

immediately lead to the equations of motion (5.13) and (5.14). Note that $\partial_\psi\bar\psi\psi = -\bar\psi$ and $\partial_{\bar\psi}\psi\bar\psi = -\psi$.

Problem 5.3. The $E = 0$ solution of the equations of motion

Obviously the ansatz $x(t) = x_k + q(t)\bar\psi_0\psi_0$ with $\Phi(x_k) = 0$ result in $x_{qc}(t) = x_k$, i.e. $\dot{x}_{qc} = 0$, and therefore equation (5.21) directly leads to $E = 0$. With $\Phi(x) = \Phi'(x_k)q\bar\psi_0\psi_0$ the conservation of energy (5.20) reads

$$F\bar{\psi}_0\psi_0 = \frac{1}{2}(\dot{q}\bar{\psi}_0\psi_0)^2 + \frac{1}{2}(\Phi'(x_k)q\bar{\psi}_0\psi_0)^2 + \Phi'(x_k)\bar{\psi}_0\psi_0 = \Phi'(x_k)\bar{\psi}_0\psi_0$$

and therefore $F = \Phi'(x_k)$. Let us also note that in this case the equation of motion (5.15) reduces to that of a forced harmonic oscillator

$$\ddot{q} + F^2 q = \Phi''(x_k).$$

Problem 5.4. Dirac brackets for Grassmann-valued degrees of freedom

Noting that $\deg\psi = \deg\pi = \deg\bar{\psi} = \deg\bar{\pi} = 1$ and $\deg x = \deg p = 0$ the result is obtained via explicit calculations.

Problems in chapter 6

Problem 6.1. WKB approximation for the radial harmonic oscillator

Consider the integral

$$I := \int_{r_L}^{r_R} dr\sqrt{2m(\tilde{E}-V(r))}, \quad V(r) := \frac{m}{2}\omega^2 r^2 + \frac{\hbar^2\lambda^2}{2mr^2}, \quad V(r_L) = \tilde{E} = V(r_R).$$

Substitution $x = r^2$ leads to

$$I = \frac{m\omega}{2}\int_{x_L}^{x_R} dx\frac{1}{x}\sqrt{bx - x^2 - c} \quad \text{with} \quad b := \frac{2\tilde{E}}{m\omega^2}, \quad c := \frac{\hbar^2\lambda^2}{m^2\omega^2},$$

$$x_{L/R} = \frac{b}{2} \mp \sqrt{\frac{b^2}{4} - c}.$$

Using the hint one explicitly arrives at the expression

$$I = I(\tilde{E}, \lambda) = \frac{m\omega\pi}{2}\left[\frac{\tilde{E}}{m\omega^2} - \frac{\hbar|\lambda|}{m\omega}\right],$$

where for the time being we assume that λ may also be negative.

(a) With $\lambda > 0$ and the quantisation condition $I(\tilde{E}, \lambda) = \hbar\pi(n + 1/2)$ one finds for the WKB spectrum $\tilde{E} = \hbar\omega(2n + \lambda + 1)$.

(b) With $\lambda = \sqrt{\ell(\ell + 1)} \to \ell + 1/2$ we obviously have $\tilde{E} \to E = \hbar\omega(2n + \ell + 3/2)$ which is the exact spectrum.

Problem 6.2. qc-SUSY approximation for the radial harmonic oscillator

Note that for the given SUSY potential its asymptotic values $\Phi_- = \lim_{r\searrow 0}\Phi(r)$ and $\Phi_+ = \lim_{r\to\infty}\Phi(r)$ lead to the Witten index (3.30) being of the form

$$\Delta = \frac{1}{2}[\text{sgn}(\Phi_+) - \text{sgn}(\Phi_-)] = \frac{1}{2} + \frac{1}{2}\,\text{sgn}\,\eta.$$

(a) Here we have $\eta = \ell + 1 > 0$ and therefore $\Delta = +1$ and the partner potentials follow by explicit calculation.

(b) Here we have $\eta = -\ell < 0$ and therefore $\Delta = 0$. Again the partner potentials follow by explicit calculation.

(c) Note that the integral in this case is of the same form as in the above problem 6.1. Hence the qc-SUSY approximation may be put into the form

$$I(E^\pm + \hbar\omega\eta,\, \eta) = \hbar\pi\left(n + \frac{1}{2} \pm \frac{\Delta}{2}\right).$$

With the result of problem 6.1, that is, $I(E^\pm + \hbar\omega\eta,\, \eta) = \frac{\pi\hbar}{2}\left[\frac{E^\pm + \hbar\omega\eta}{\hbar\omega} - |\eta|\right]$, this leads to

$$E^\pm = \hbar\omega(2n + 1 \pm \Delta - \eta + |\eta|).$$

As consequence we have in the case $\eta = \ell + 1 > 0$ and $\Delta = +1$ where SUSY is unbroken:

$$E_n^+ = \hbar\omega(2n + 2),\ E_n^- = \hbar\omega(2n).$$

The case of broken SUSY with $\eta = -\ell < 0$ and $\Delta = 0$ results in

$$E_n^\pm = \hbar\omega(2n + 2\ell + 1).$$

Hence for both cases the qc-SUSY approximation leads to the exact energy spectrum.

Problems in chapter 7

Problem 7.1. Derivation of the stationary distribution for unbroken SUSY

Assuming that a stationary distribution $P(x)$ exists it should of course also obey the Fokker–Planck equation (7.7),

$$0 = \frac{D}{2}\frac{\partial^2}{\partial x^2}P(x) + \frac{\partial}{\partial x}\,U'(x)P(x),$$

which upon integration yields

$$\frac{D}{2}\frac{\partial}{\partial x}P(x) + U'(x)P(x) = \text{const.},$$

where the const. actually has to be zero. This follows from the fact that $P(x)$ has to be normalisable and therefore for large $x \to \pm\infty$ it is reasonable to assume $P(x) \to 0$ and $P'(x) \to 0$ fast enough. Another integration then result in equation (7.9), which *a posteriori* is consistent with the assumptions made.

Problem 7.2. Derivation of the imaginary-time Schrödinger equation

With the ansatz (7.12) one finds

$$\partial_x m_t^\pm = e^{-[U_\pm(x)-U_\pm(x_0)]/D}(\partial_x K_t^\pm \pm \Phi(x)K_t^\pm/D),$$

$$\partial_x^2 m_t^\pm = e^{-[U_\pm(x)-U_\pm(x_0)]/D}\bigg(\partial_x^2 K_t^\pm \pm \frac{2}{D}\Phi(x)K_t^\pm$$

$$+ \frac{1}{D^2}\Phi^2(x)K_t^\pm \pm \frac{1}{D}\Phi'(x)K_t^\pm\bigg).$$

Inserting this into the Fokker–Planck equation directly leads to equation (7.13).

Problem 7.3. On the Cameron–Martin–Girsanov formula

Let $g(x) = g_{rs}(x)$ be the function as defined in equation (7.82) and $P = -i\partial_x$ be the momentum operator, that is, $gP = Pg + ig'$. Then we obviously have

$$\frac{1}{2}gP^2\frac{1}{g} = \frac{1}{2}P^2 + iP\frac{g'}{g} - \frac{g''}{2g}, \quad igPF_s\frac{1}{g} = iPF_s - F_s\frac{g'}{g}$$

and therefore

$$gM_s\frac{1}{g} = g\bigg(\frac{1}{2}P^2 + iPF_s\bigg)\frac{1}{g} = \frac{1}{2}P^2 + iP\bigg(F_s + \frac{g'}{g}\bigg) - \frac{g''}{2g} - F_s\frac{g'}{2g}.$$

Using the definition (7.82) we find $g' = (F_r - F_s)g$ and $g'' = (F'_r - F'_s)g + (F_r - F_s)^2 g$ which then results in equation (7.83). The proof of equation (7.84) then follows as shown below:

$$\langle x|e^{-t(M_s+V(Q))}|x_0\rangle = g^{-1}(x)\langle x|g(Q)\exp\{-t(M_s + V(Q))\}g^{-1}(Q)|x_0\rangle g(x_0)$$

$$= \frac{g(x_0)}{g(x)}\langle x|\exp\{-tg(Q)(M_s + V(Q))g^{-1}(Q)\}|x_0\rangle$$

$$= \frac{g(x_0)}{g(x)}\langle x|\exp\{-t(g(Q)M_sg^{-1}(Q) + V(Q))\}|x_0\rangle$$

$$= \frac{g(x_0)}{g(x)}\langle x|\exp\{-t(M_r + \tilde{V}(Q))\}|x_0\rangle,$$

with $\tilde{V}(x)$ as defined in equation (7.85).

Problem 7.4. An explicit example for a family of shape-invariant drifts

For the given family of drift potentials $\Phi(a_s, x) := a_s \tanh x$ with $a_s = a_0 - s$ we find

$$\exp\bigg\{\int_{x_0}^x dz[\Phi(a_{s+1}, x) + \Phi(a_s, x)]\bigg\} =$$

$$\exp\bigg\{\int_{x_0}^x dz(2a_0 - 2s - 1)\tanh z\bigg\} = \bigg(\frac{\cosh x}{\cosh x_0}\bigg)^{2a_0-2s-1}.$$

By noting that the residual term (7.92) reads $R(a_{s+1}) = a_0 - s - \frac{1}{2}$ we arrive at the desired result:

$$\langle x|\exp\left\{-t[P^2/2 + i(a_0 - s)P\tanh Q]\right\}|x_0\rangle = \left(\frac{\cosh x}{\cosh x_0}\right)^{2a_0 - 2s - 1} e^{-t(a_0 - s - 1/2)}$$

$$\times \langle x|\exp\left\{-t[P^2/2 - i(a_0 - s - 1)P\tanh Q]\right\}|x_0\rangle.$$

For $s = a_0$ this reduces to

$$\langle x|\exp\left\{-tP^2/2\right\}|x_0\rangle = \frac{\cosh x_0}{\cosh x}e^{t/2}\langle x|\exp\left\{-t[P^2/2 + iP\tanh Q]\right\}|x_0\rangle,$$

which is identical to equation (7.90) as $\langle x|\exp\left\{-tP^2/2\right\}|x_0\rangle = (2\pi t)^{-1/2}$ $\exp\{-(x - x_0)^2/2t\}$.

Problems in chapter 8

Problem 8.1. On the $N = 2$ SUSY of the two-dimensional Pauli operator

The proof is by explicit calculation as follows:

$$2mAA^\dagger = \left[\left(p_1 - \frac{e}{c}a_1\right) \mp i\left(p_2 - \frac{e}{c}a_2\right)\right]\left[\left(p_1 - \frac{e}{c}a_1\right) \pm i\left(p_2 - \frac{e}{c}a_2\right)\right]$$

$$= \left(p_1 - \frac{e}{c}a_1\right)^2 + \left(p_2 - \frac{e}{c}a_2\right)^2 \pm i\left[p_1 - \frac{e}{c}a_1, p_2 - \frac{e}{c}a_2\right]$$

$$= \left(p_1 - \frac{e}{c}a_1\right)^2 + \left(p_2 - \frac{e}{c}a_2\right)^2 \pm \frac{e\hbar}{c}(\partial_{x_2}a_1 - \partial_{x_1}a_2)$$

$$= \left(p_1 - \frac{e}{c}a_1\right)^2 + \left(p_2 - \frac{e}{c}a_2\right)^2 \mp \frac{e\hbar}{c}B$$

and similar for $2mA^\dagger A = \left(p_1 - \frac{e}{c}a_1\right)^2 + \left(p_2 - \frac{e}{c}a_2\right)^2 \pm \frac{e\hbar}{c}B$.

Problem 8.2. The Aharonov–Casher theorem

(a) $\Delta S(\vec{r}) = \int d^2\vec{r}' B(\vec{r}')\Delta\ln|\vec{r} - \vec{r}'| = \int d^2\vec{r}' B(\vec{r}')2\pi\delta(\vec{r} - \vec{r}') = 2\pi B(\vec{r})$, $B(\vec{r}) :=$ $\partial_{x_1}a_2(\vec{r}) - \partial_{x_2}a_1(\vec{r}) = \Delta S(\vec{r})/2\pi$.

(b) $\sqrt{2m}\,A = \hbar e^{S(\vec{r})/\Phi_0}[-\partial_{x_2} - i\partial_{x_1}]e^{-S(\vec{r})/\Phi_0} = \hbar[(\partial_{x_2}S)/\Phi_0 - \partial_{x_2} + i(\partial_{x_1}S)/\Phi_0 - i\partial_{x_1}]$
$= [-2\pi\hbar a_1/\Phi_0 - ip_2 + 2\pi\hbar a_2/\Phi_0 + p_1] = \left[\left(p_1 - \frac{e}{c}a_1\right) + i\left(p_2 - \frac{e}{c}a_2\right)\right]$.

(c) Unbroken SUSY implies $iA\psi_0^- = 0$ and/or $iA\psi_0^+ = 0$ with $\psi_0^\pm \in \mathcal{H}^\pm$. Hence $(\partial_{x_1} - i\partial_{x_2})e^{-S/\Phi_0}\psi_0^- = 0$ and/or $(\partial_{x_1} + i\partial_{x_2})e^{S/\Phi_0}\psi_0^+ = 0$. With $z = x_1 + ix_2$ this implies $\partial_z f_-(z^*) = 0$ and/or $\partial_z f_+(z) = 0$. That is f_- is analytic in z^* and f_+ in z.

(d) For large $\vec{r}$ with finite $\vec{r}'$ we have $\ln|\vec{r} - \vec{r}'| = \ln|\vec{r}|[1 + O(1/|\vec{r}|)]$. Therefore $S(\vec{r}) = F\ln|\vec{r}|[1 + O(1/|\vec{r}|)]$.

(e) With $eF > 0$ we have $\psi_0^-(\vec{r}) = e^{S/\Phi_0} f_-(z^*) = f_-(z^*)|z|^{F/\Phi_0} \, [1 + O(1/|z|)]$ $\in L^2(\mathbb{R}^2)$. Hence $f_-(z^*) \to 0$ for $|z| \to \infty$ in all directions. With f_- being analytic this implies $f_-(z^*) \equiv 0$. Therefore there exists no spin-down SUSY ground states as $\psi_{0,k}^-(\vec{r}) \equiv 0$.

(f) Following above reasoning we find $\psi_0^+(\vec{r}) = f_+(z)|z|^{-F/\Phi_0} \, [1 + O(1/|z|)]$ $\in L^2(\mathbb{R}^2)$. Hence f_+ has to be polynomially bound and analytic in z. Therefore f_+ is a polynomial in z with a degree $k < F/\Phi_0 - 1$. There exist exact $d := [[F/\Phi_0]]$ linearly independent polynomials z^k, $k = 0, 1, 2, 3, \ldots, d - 1$ and result in the spin-up SUSY ground states

$$\psi_{0,\,k}^+(\vec{r}) = C \exp\{-S(\vec{r})/\Phi_0\}(x_1 + ix_2)^k, \qquad k = 0, 1, 2, \ldots, d - 1.$$

Note that for $eF < 0$ the situation is reversed with $\psi_{0,k}^+(\vec{r}) \equiv 0$, i.e. no spin-up ground states but spin-down SUSY grounds states given by

$$\psi_{0,k}^-(\vec{r}) = C \exp\{-S(\vec{r})/\Phi_0\}(x_1 - ix_2)^k, \qquad k = 0, 1, 2, \ldots, d - 1.$$

Problem 8.3. Some relations of Pauli matrices

The proofs follow by explicit calculations:

(a) $(\vec{\sigma} \cdot \vec{A})(\vec{\sigma} \cdot \vec{B}) = \sigma_i \sigma_j A_i B_j = \delta_{ij} A_i B_j + i\varepsilon_{ijk}\sigma_k A_i B_j = \vec{A} \cdot \vec{B} + i\vec{\sigma} \cdot (\vec{A} \times \vec{B})$.

(b) These follow from (a) as special cases.

(c) $(\vec{\sigma} \cdot \vec{p}) = (\vec{\sigma} \cdot \vec{e}_r)^2 \vec{\sigma} \cdot \vec{p} = (\vec{\sigma} \cdot \vec{e}_r)[\vec{e}_r \cdot \vec{p} + i\vec{\sigma} \cdot (\vec{e}_r \times \vec{p})] = (\vec{\sigma} \cdot \vec{e}_r) \left[-i\partial_r + i\frac{\vec{\sigma}}{r} \cdot \vec{L} \right] = -i(\vec{\sigma} \cdot \vec{e}_r)\left[\partial_r - i\frac{K_{nr} - 1}{r} \right]$. Note that $\hbar = 1$.

(d) Let us prove here the hint of problem 8.5 below.

First we note that $\vec{r} \cdot \vec{p} f(r) = -i\hbar r f'(r) + f(r)\vec{r} \cdot \vec{p}$.

Now consider $(\vec{\sigma} \cdot \vec{p})(\vec{\sigma} \cdot \vec{r}) f(r) = [\vec{p} \cdot \vec{r} + i\vec{\sigma} \cdot (\vec{p} \times \vec{r})] f(r) = [\vec{p} \cdot \vec{r} - i\vec{\sigma} \cdot \vec{L}] f(r) = [-3i\hbar - i\vec{\sigma} \cdot \vec{L} + \vec{r} \cdot \vec{p}] f(r) = [-3i\hbar - i\vec{\sigma} \cdot \vec{L}] f(r) + f(r)\vec{r} \cdot \vec{p} - i\hbar r f'(r)$.

With $f(r)(\vec{\sigma} \cdot \vec{r})(\vec{\sigma} \cdot \vec{p}) = f(r)(\vec{r} \cdot \vec{p}) + if(r)\vec{\sigma} \cdot \vec{L}$ we obtain the commutation relation

$$[f(r)(\vec{\sigma} \cdot \vec{r}), (\vec{\sigma} \cdot \vec{p})] = 3i\hbar f(r) + 2if(r)\vec{\sigma} \cdot \vec{L} + i\hbar r f'(r). \qquad (11.1)$$

For $f(r) = 1/r$ and $\hbar = 1$ follows $[\vec{\sigma} \cdot \vec{e}_r, \vec{\sigma} \cdot \vec{p}] = \frac{2i}{r} K_{nr}$.

Problem 8.4. Properties of the non-relativistic spin–orbit operator

(a) $\vec{J}^2 = \vec{L}^2 + 2\vec{S} \cdot \vec{L} + \vec{S}^2 = \vec{L}^2 + \vec{\sigma} \cdot \vec{L} + \frac{3}{4} \Rightarrow \vec{\sigma} \cdot \vec{L} = \vec{J}^2 - \vec{L}^2 - \frac{3}{4} \Rightarrow K_{nr} = \vec{J}^2 - \vec{L}^2 + \frac{1}{4}$. $(\vec{\sigma} \cdot \vec{L})^2 = \vec{L}^2 + i\varepsilon_{ijk}\sigma_k L_i L_j = \vec{L}^2 + \frac{i}{2}\varepsilon_{ijk}\sigma_k[L_i, L_j] = \vec{L}^2 + \frac{i}{2}\sigma_k\varepsilon_{ijk}\varepsilon_{ijl}iL_l = \vec{L}^2 - \vec{\sigma} \cdot \vec{L}$, where we have made use of the relation $\varepsilon_{ijk}\varepsilon_{ijl} = 2\delta_{kl}$.

11-14

Hence $\vec{L}^2 = \vec{\sigma} \cdot \vec{L}(\vec{\sigma} \cdot \vec{L} + 1) = K_{nr}(K_{nr} - 1)$. Finally this and the previous result gives $K_{nr}^2 = \vec{L}^2 + K_{nr} = \vec{L}^2 + \vec{J}^2 - \vec{L}^2 + \frac{1}{4} = \vec{J}^2 + \frac{1}{4}$.

(b) $\{K_{nr}, (\vec{v} \cdot \vec{\sigma})\} = (\vec{\sigma} \cdot \vec{L})(\vec{v} \cdot \vec{\sigma}) + (\vec{v} \cdot \vec{\sigma})(\vec{\sigma} \cdot \vec{L}) + 2(\vec{v} \cdot \vec{\sigma}) = \mathrm{i}\varepsilon_{ijk}\sigma_k \left[L_i v_j + v_i L_j \right]$
$+ 2(\vec{v} \cdot \vec{\sigma}) = \mathrm{i}\varepsilon_{ijk}\sigma_k \left[v_j L_i + \mathrm{i}\varepsilon_{ijl}v_l + v_i L_j \right] + 2(\vec{v} \cdot \vec{\sigma}) = 2(\vec{v} \cdot \vec{\sigma}) - \varepsilon_{ijk}\varepsilon_{ijl}\sigma_k v_l = 0.$

Problem 8.5. SUSY Hamiltonians for spherically symmetric tensor potentials

Consider with $\hbar = 1$ and $f(r) = U'(r)/r$

$$
\begin{aligned}
2\,mAA^\dagger &= [\vec{\sigma} \cdot (\vec{p} + \mathrm{i}\vec{e}_r)U'][\vec{\sigma} \cdot (\vec{p} - \mathrm{i}\vec{e}_r)U'] \\
&= (\vec{\sigma} \cdot \vec{p})^2 + (\vec{\sigma} \cdot \vec{e}_r)^2 U'^2 + \mathrm{i}[f(r)(\vec{\sigma} \cdot \vec{r}), \vec{\sigma} \cdot \vec{p}] \\
&= \vec{p}^2 + U'^2 - 3f - 2f\vec{\sigma} \cdot \vec{L} - rf' \\
&= \vec{p}^2 + U'^2 - U'/r - 2(U'/r)K_{nr} - U'' + U'/r.
\end{aligned}
$$

Hence $H_+ = AA^\dagger = \frac{1}{2m}\vec{p}^2 + \frac{1}{2m}U'^2 - \frac{1}{2m}U'' - \frac{U'}{mr}K_{nr}$ and $H_- = A^\dagger A$ follows from H_+ by replacing U with $-U$.

Problems in chapter 9

Problem 9.1. Supersymmetric representation of Dirac matrices

$$
\begin{aligned}
\{\alpha_i, \alpha_j\} &= \begin{pmatrix} 0 & \sigma_i \\ \sigma_i & 0 \end{pmatrix}\begin{pmatrix} 0 & \sigma_j \\ \sigma_j & 0 \end{pmatrix} + \begin{pmatrix} 0 & \sigma_j \\ \sigma_j & 0 \end{pmatrix}\begin{pmatrix} 0 & \sigma_i \\ \sigma_i & 0 \end{pmatrix} \\
&= \begin{pmatrix} \{\sigma_i, \sigma_j\} & 0 \\ 0 & \{\sigma_i, \sigma_j\} \end{pmatrix} = 2\delta_{ij}, \\
\{\alpha_i, \beta\} &= \begin{pmatrix} 0 & \sigma_i \\ \sigma_i & 0 \end{pmatrix}\begin{pmatrix} 0 & -\mathrm{i} \\ \mathrm{i} & 0 \end{pmatrix} + \begin{pmatrix} 0 & -\mathrm{i} \\ \mathrm{i} & 0 \end{pmatrix}\begin{pmatrix} 0 & \sigma_i \\ \sigma_i & 0 \end{pmatrix} = 0.
\end{aligned}
$$

Problem 9.2. The three-dimensional Dirac Hamiltonian with magnetic field

In the standard representation (9.6) this Dirac Hamiltonian reads

$$
H_D = \begin{pmatrix} 0 & c\vec{\sigma} \cdot \left(\vec{p} - \dfrac{e}{c}\vec{A}\right) \\ c\vec{\sigma} \cdot \left(\vec{p} - \dfrac{e}{c}\vec{A}\right) & 0 \end{pmatrix} + \begin{pmatrix} mc^2 & 0 \\ 0 & -mc^2 \end{pmatrix} = \begin{pmatrix} mc^2 & A \\ A^\dagger & -mc^2 \end{pmatrix}
$$

and resembles the supersymmetric form (9.13) with A as defined in equation (9.9). The corresponding SUSY Hamiltonians (9.21) are given by

$$H_\pm = \frac{1}{2mc^2}A^2 = \frac{1}{2m}\left[\vec{\sigma}\cdot\left(\vec{p}-\frac{e}{c}\vec{A}\right)\right]^2$$

$$= \frac{1}{2m}\left[\left(\vec{p}-\frac{e}{c}\vec{A}\right)^2 + i\varepsilon_{ijk}\left(p_i-\frac{e}{c}A_i\right)\left(p_j-\frac{e}{c}A_j\right)\sigma_k\right]$$

$$= \frac{1}{2m}\left(\vec{p}-\frac{e}{c}\vec{A}\right)^2 - \frac{e\hbar}{2mc}\varepsilon_{ijk}\left(\partial_i A_j + A_i\partial_j\right)\sigma_k$$

$$= \frac{1}{2m}\left(\vec{p}-\frac{e}{c}\vec{A}\right)^2 - \frac{e\hbar}{2mc}(\vec{\nabla}\times\vec{A})\cdot\vec{\sigma}.$$

Problem 9.3. Proofs for the Foldy–Wouthuysen transformation

(a) $H_D^2 = (Q_1 + \mathcal{M}W)^2 = Q_1^2 + Q_1\mathcal{M}W + \mathcal{M}WQ_1 + \mathcal{M}W\mathcal{M}W = Q_1^2 + \mathcal{M}^2$.
With $[Q_1, \mathcal{M}] = 0$ also follows $[H_D^2, \mathcal{M}] = 0$. Hence $[|H_D|, \mathcal{M}] = 0]$. For the last step see section 8.3.4 in the book by Kalka and Soff [1].

(b) With above result it is obvious that $a_\pm = a_\pm^\dagger$ as $|H_D| = |H_D|^\dagger$ and $\mathcal{M} = \mathcal{M}^\dagger$. Furthermore, $[a_+, a_-] = 0$ as well as $[a_\pm, H_D] = 0 = [a_\pm, Q_1]$. With $a_\pm^2 = \frac{1}{2} \pm \frac{\mathcal{M}}{2|H_D|}$ also follows $a_+^2 + a_-^2 = 1$, $a_+^2 - a_-^2 = \frac{\mathcal{M}}{|H_D|}$ and

$$2a_+a_- = 2\sqrt{\frac{1}{4} - \frac{\mathcal{M}^2}{2H_D^2}} = \sqrt{1 - \frac{\mathcal{M}^2}{H_D^2}} = \frac{|Q_1|}{|H_D|}.$$

(c) As $[Q_1^2, W] = 0 \Rightarrow [|Q_1|, W] = 0$. Hence, $|Q_1|$ is an even operator whereas Q_1 is odd and therefore sgn Q_1 is also an odd operator, $\{sgn\ Q_1, W\} = 0$. See again section 8.3.4 in the book by Kalka and Soff [1]. The commutation relation $[sgn\ Q_1, \mathcal{M}] = 0$ directly follows from $[Q_1, \mathcal{M}] = 0$. With this we now calculate

$$H_D W\ sgn\ Q_1 = Q_1 W\ sgn\ Q_1 + \mathcal{M}W^2\ sgn\ Q_1$$
$$= -WQ_1\ sgn\ Q_1 - \mathcal{M}W\ sgn\ Q_1 W$$
$$= -W\ sgn\ Q_1\ Q_1 - W\ sgn\ Q_1\mathcal{M}W$$
$$= -W\ sgn\ Q_1\ H_D.$$

(d) Consider $UU^\dagger = (a_+ + W\ sgn\ Q_1\ a_-)(a_+ - W\ sgn\ Q_1\ a_-) = a_+^2 - W\ sgn\ Q_1\ a_-W\ sgn\ Q_1\ a_- = a_+^2 + a_-^2 = 1$. And similar $U^\dagger U = 1$.

(e) With above results we can now perform below calculation

$$UH_D U^\dagger = (a_+ + W \, \text{sgn} \, Q_1 \, a_-)H_D(a_+ - W \, \text{sgn} \, Q_1 \, a_-)$$
$$= (a_+ + W \, \text{sgn} \, Q_1 \, a_-)^2 H_D$$
$$= (a_+^2 + 2W \, \text{sgn} \, Q_1 \, a_+ \, a_- + W \, \text{sgn} \, Q_1 W \, \text{sgn} \, Q_1 \, a_-^2)H_D$$
$$= (a_+^2 + 2W \, \text{sgn} \, Q_1 \, a_+ a_- - a_-^2)H_D$$
$$= \left(\frac{\mathcal{M}}{|H_D|} + W \, \text{sgn} \, Q_1 \frac{|Q_1|}{|H_D|}\right)H_D$$
$$= W\left(\frac{W\mathcal{M} + Q_1}{|H_D|}\right)H_D = W\frac{H_D^2}{|H_D|} = W|H_D|.$$

Problem 9.4. The helicity eigenstates of the free Dirac particle

Consider the matrix

$$\vec{\sigma} \cdot \vec{k} = \begin{pmatrix} k_3 & k_1 - ik_2 \\ k_1 + ik_2 & -k_3 \end{pmatrix}.$$

Then we find

$$\vec{\sigma} \cdot \vec{k}\chi_{+1}(\vec{k}) = \frac{1}{\sqrt{2k(k-k_3)}}\begin{pmatrix} k_3 & k_1 - ik_2 \\ k_1 + ik_2 & -k_3 \end{pmatrix}\begin{pmatrix} k_1 - ik_2 \\ k - k_3 \end{pmatrix}$$

$$= \frac{k}{\sqrt{2k(k-k_3)}}\begin{pmatrix} k_1 - ik_2 \\ k - k_3 \end{pmatrix}$$

$$\vec{\sigma} \cdot \vec{k}\chi_{-1}(\vec{k}) = \frac{1}{\sqrt{2k(k-k_3)}}\begin{pmatrix} k_3 & k_1 - ik_2 \\ k_1 + ik_2 & -k_3 \end{pmatrix}\begin{pmatrix} k_3 - k \\ k_1 + ik_2 \end{pmatrix}$$

$$= \frac{-k}{\sqrt{2k(k-k_3)}}\begin{pmatrix} k_3 - k \\ k_1 + ik_2 \end{pmatrix}$$

and therefore $\vec{\sigma} \cdot \vec{k}\chi_\lambda = k\lambda\chi_\lambda$. Orthogonality $(\chi_{+1}^T)^* \chi_{-1} = 0$ and normalisation $|\chi_\lambda|^2 = (\chi_\lambda^T)^* \chi_\lambda = 1$ follow by explicit calculation.

For the Foldy–Wouthuysen transformation let us consider

$$Q_1 = \begin{pmatrix} 0 & c\vec{\sigma} \cdot \vec{p} \\ c\vec{\sigma} \cdot \vec{p} & 0 \end{pmatrix} = c|\vec{p}|\begin{pmatrix} 0 & \Lambda \\ \Lambda & 0 \end{pmatrix}, \quad Q_1^2 = \begin{pmatrix} c^2\vec{p}^2 & 0 \\ 0 & c^2\vec{p}^2 \end{pmatrix} = c^2\vec{p}^2\,1,$$

which leads us to

$$\text{sgn} \, Q_1 = \frac{Q_1}{\sqrt{Q_1^2}} = \Lambda\begin{pmatrix} 0 & 1 \\ 1 & 0 \end{pmatrix}, \quad W \, \text{sgn} \, Q_1 = \Lambda\begin{pmatrix} 1 & 0 \\ 0 & -1 \end{pmatrix}\begin{pmatrix} 0 & 1 \\ 1 & 0 \end{pmatrix} = \Lambda\begin{pmatrix} 0 & 1 \\ -1 & 0 \end{pmatrix}$$

resulting in equation (9.54). Finally with $\psi^{\pm}_{\vec{k},\lambda}(\vec{r}) = U^{\dagger}\tilde{\psi}^{\pm}_{\vec{k},\lambda}(\vec{r})$ the free electron (+) and positron (−) solutions explicitly read

$$\psi^{+}_{\vec{k},\lambda}(\vec{r}) = \frac{e^{i\vec{k}\cdot\vec{r}}}{(2\pi)^{3/2}}\begin{pmatrix} a_+(\vec{k})\chi_\lambda(\vec{k}) \\ \lambda a_-(\vec{k})\chi_\lambda(\vec{k}) \end{pmatrix}, \quad \psi^{-}_{\vec{k},\lambda}(\vec{r}) = \frac{e^{i\vec{k}\cdot\vec{r}}}{(2\pi)^{3/2}}\begin{pmatrix} -\lambda a_-(\vec{k})\chi_\lambda(\vec{k}) \\ a_+(\vec{k})\chi_\lambda(\vec{k}) \end{pmatrix},$$

with $a_\pm(\vec{k}) = \left(\frac{1}{2} \pm \frac{mc^2}{2\sqrt{c^2\hbar^2\vec{k}^2 + mc^2}}\right)^{1/2}$.

Problem 9.5. Some relations for radial Dirac operators

(a) With $K = \beta K_{nr} = \begin{pmatrix} K_{nr} & 0 \\ 0 & -K_{nr} \end{pmatrix}$ and $\vec{\alpha}\cdot\vec{v} = \begin{pmatrix} 0 & \vec{\sigma}\cdot\vec{v} \\ \vec{\sigma}\cdot\vec{v} & 0 \end{pmatrix}$ one finds

$$[K, (\vec{\alpha}\cdot\vec{v})] = \begin{pmatrix} \{K_{nr}, \vec{\sigma}\cdot\vec{v}\} & 0 \\ 0 & -\{K_{nr}, \vec{\sigma}\cdot\vec{v}\} \end{pmatrix} = 0,$$

where in the last step we made use of the result of problem 8.4(b) of chapter 8, i.e. $\{K_{nr}, \vec{\sigma}\cdot\vec{v}\} = 0$.

(b) The condition $AM_- = M_+A$ with the explicit form as given in equation (9.101) leads to the condition $\phi_e' = \phi_s'$. Whereas $A^{\dagger}M_+ = M_-A^{\dagger}$ gives the condition $\phi_e' = -\phi_s'$ and hence ϕ_s and ϕ_e are both constants. The constant ϕ_s can be absorbed in the mass parameter mc^2 and the constant ϕ_e is a simple additive constant to H_r and may be absorbed in a shift of the energy scale.

(c) By inspection of equation (9.106) it is obvious that for both cases $\lim_{r\to\infty}U(r) = +\infty$ and therefore $R_0^-(r)$ is the square integrable solution for $\kappa > 0$. Hence SUSY is unbroken and the ground state belongs to $\mathcal{H}^-$. The ground-state wave functions read $R_0^-(r) = Cr^{j+\frac{1}{2}}\exp\{-m\omega r^2/2\}$ and $R_0^-(r) = Cr^{j+\frac{1}{2}}\exp\{-\gamma r\}$, respectively.

(d) First let us note that the partner Hamiltonians $H_\pm$ given in (9.111) are basically the sum of standard harmonic oscillator H_{osc} and the non-relativistic spin–orbit operator, that is, $H_\pm = H_{osc} \pm \hbar\omega(K_{nr} + \frac{1}{2})$. The eigenvalues of H_{osc} are well-known and given by $\hbar\omega(2n + \ell + \frac{3}{2})$. The eigenvalues of K_{nr} are given by $s(j + \frac{1}{2})$ and let us remember the relation $\ell = j - \frac{s}{2}$. Then a simple calculation directly leads to equation (9.112).

Problem 9.6. Dirac Hamiltonian in anti-de-Sitter space

With $\chi = \omega r$ and $u = 1$ the first condition in equation (9.121) results in $\mu = \omega/2mc$. As in the present case $w\cos\chi = 1$ and $v\sin\chi = \omega r = \chi$, the second condition obviously reads $\lambda = 1 - \kappa\omega/mc$. The radial Hamiltonian (9.118) is given by

$$H_r := \begin{pmatrix} \dfrac{mc^2}{\cos \omega r} & -c\partial_r + \dfrac{c\kappa\omega}{\sin \omega r} \\ c\partial_r + \dfrac{c\kappa\omega}{\sin \omega r} & -\dfrac{mc^2}{\cos \omega r} \end{pmatrix} = \omega c \begin{pmatrix} k \sec \chi & -\partial_\chi + \kappa \csc\chi \\ \partial_\chi + \kappa \csc\chi & -k \sec \chi \end{pmatrix},$$

with $k = mc/\omega$, and the transformed Hamiltonian

$$\tilde{H}_r = R(\chi) H_r R^T(\chi) = c\omega \begin{pmatrix} k - \kappa + 1/2 & -\partial_\chi + \kappa \cot\chi + k \tan\chi \\ \partial_\chi + \kappa \cot\chi + k \tan\chi & -(k - \kappa - 1/2) \end{pmatrix}$$

follows from an explicit calculation. Upon the identification (9.126) this represents a supersymmetric Dirac Hamiltonian. Here the SUSY potential is given by $W(r) = \omega\kappa \cot \omega r + mc \tan \omega r$ which in first order in ω leads to $W(r) \simeq \kappa/r + mc\omega r$. Comparison with the general result (9.102) results in the identification of the tensor potential $\phi_t(r) = -mc^2\omega r$ and a superpotential (9.107) given by $U(r) = m\Omega r^2/2$ which is that of the Dirac oscillator, cf. equation (9.109) with $\Omega = -c\omega$. Note that here ω has the dimension of an inverse length. As $\Omega < 0$ the SUSY ground state belongs to $\mathcal{H}^+$.

Reference

[1] Kalka H and Soff G 1997 *Supersymmetrie* (Stuttgart: Teubner)

www.ingramcontent.com/pod-product-compliance
Lightning Source LLC
Chambersburg PA
CBHW080527220326
41599CB00032B/6233